# Photon, Electron, and Ion Probes of Polymer Structure and Properties

# Photon, Electron, and Ion Probes of Polymer Structure and Properties

**David W. Dwight,** EDITOR
*Virginia Polytechnic Institute and State University*

**Thomas J. Fabish,** EDITOR
*Ashland Chemical Company*

**H. Ronald Thomas,** EDITOR
*Xerox Corporation*

Based on a symposium
cosponsored by the Divisions of
Organic Coatings and Plastics
and Polymer Chemistry
at the 179th Meeting of the
American Chemical Society,
Houston, Texas,
March 25–27, 1980.

ACS SYMPOSIUM SERIES 162

AMERICAN CHEMICAL SOCIETY
WASHINGTON, D. C.      1981

Library of Congress CIP Data

Photon, electron, and ion probes of polymer structure
and properties.
   (ACS symposium series; 162, ISSN 0097-6156)
   "Based on a symposium co-sponsored by the Divisions of Organic Coatings and Plastics and Polymer
Chemistry at the 179th meeting of the American Chemical Society, Houston, Texas, March 25-27, 1980."

   Includes bibliography and index.

   1. Polymers and polymerization—Analysis—Con-
gresses. 2. Photoelectron spectroscopy—Congresses.
   I. Dwight, David W., 1941-    . II. Thomas, H.
Ronald, 1942-    . III. Fabish, Thomas J., 1938-    .
IV. American Chemical Society. Division of Organic
Coatings and Plastics Chemistry. V. American Chemi-
cal Society. Division of Polymer Chemistry. VI. Se-
ries: ACS symposium series; 162.

QD380.P53            547.7′046            81-10816
ISBN 0-8412-0639-2                        AACR2
                    ASCMC 8 162 1–442 1981

# ACS Symposium Series

**M. Joan Comstock,** *Series Editor*

# FOREWORD

The ACS Symposium Series was founded in 1974 to provide
a medium for publishing symposia quickly in book form. The
format of the Series parallels that of the continuing Advances
in Chemistry Series except that in order to save time the
papers are not typeset but are reproduced as they are submitted by the authors in camera-ready form. Papers are reviewed under the supervision of the Editors with the assistance
of the Series Advisory Board and are selected to maintain the
integrity of the symposia; however, verbatim reproductions of
previously published papers are not accepted. Both reviews
and reports of research are acceptable since symposia may
embrace both types of presentation.

# CONTENTS

# PREFACE

The last decade witnessed a dramatic growth in the use of energetic beam techniques to elucidate the electronic structures of atoms and molecules. Photon, electron, and ion spectroscopies applied to solids gave birth to a new level of surface sensitivity for studies of chemical structure and bonding. The time was right to provide a benchmark for the state of current knowledge and future possibilities in the field.

The first objective of the symposium upon which this book is based was to focus attention on the experimental and theoretical techniques currently being used to describe anion and cation states in large molecules and polymeric solids, with special emphasis being placed on the consequences to electronic structure incurred upon condensation from the gaseous into the solid state. It is, therefore, appropriate to begin with a review of the state of knowledge of anion states in large, isolated molecules as revealed by elastic and inelastic electron scattering, electron swarm, mass spectrographic, and ion-cyclotron resonance experiments. Insight into the effects of condensation on the nature of the lowest energy excitations that govern many of the chemical and physical properties of polymers is provided by results from high-energy electron energy loss spectroscopic studies on representative saturated polymers.

Energy and charge transport in saturated and conjugated polymeric solids represent limiting cases in the applicability of the precepts of band theoretical descriptions of the electronic structure of solids. Discussions of the nature of intrinsic localized electronic states and their consequences to treatments of transport phenomena in such materials comprise an important section of these proceedings.

Experimental studies of fundamental excitations in conjugated polymers are interpreted within the framework of current theoretical electronic structural calculations and physical structure characterizations. The instabilities peculiar to this class of materials that are responsible for their departure from metallic behavior are identified explicitly.

Considering the valence levels, the synergistic effect of combining spectroscopic measurements with theoretical calculations is illustrated by two pairs of chapters: (1) ultraviolet photoemission and optical absorption data compared to a spectroscopically parameterized CNDO/S3 model, and (2) x-ray photoemission compared to *ab initio* and intermediate approximation MO calculations.

The firm theoretical understanding of the spectra of model molecules has been a major impetus to apply the techniques to complex, unknown solid-state structures of technological importance. Characterization of atomic composition, structure, and bonding in the surface and subsurface of "practical" specimens is the first step. Most studies interpret changes in the core-level spectra in terms of surface–chemical mechanisms involved in, for example, processing conditions, exposure to inert or reactive gas plasmas, chemical reactions, or aggressive service environments.

Five chapters treat the basics as well as new developments in analytical methodology, emphasizing x-ray photoelectron spectroscopy for the identification of structure and bonding in polymer surfaces of increasing complexity. Each of the seven important levels of information available in the core-level data is illustrated in detail, with special emphasis upon: (1) Auger peak positions for chemical state identification; (2) angular and kinetic energy dependence of peak intensities for information on compositional variation with depth; and (3) enhanced resolution of functional groups with chemical derivatives.

Explicitly developed are models of several theoretical multiphase distributions, with corresponding depth-profile results on thin-film plasma polymers, phase-separated block copolymers, and chemical reactions on fiber surfaces. Ion impact is treated from three points of view: as an analytical fingerprint tool for polymer surface analysis via secondary ion mass spectroscopy, by forming unique thin films by introducing monomers into the plasma, and as a technique to modify polymer surface chemistry.

New experimental results on specific polymer material problems are presented in the last nine chapters. Several cases involve the study of polymers from commercial sources. The topics include: (1) surface chemistry as induced by (a) outdoor weathering, (b) chemical reactions, and (c) plasma exposure; (2) chemical bond formation at the polymer –metal interface; and (3) biomaterials characterization and relationship to blood compatibility.

In summary, these proceedings provide a survey of the fundamentals and applications of photon, electron, and ion probes to polymers up to 1980. The contributors include many pioneers—from the basic studies performed with specialized, custom apparatus on small, gas-phase molecules to those utilizing standard commercial spectrometers to characterize practical polymer systems. We hope this work will contribute to the attainment of the ultimate objective of a unified description of the electronic structures and properties of polymeric solids.

Acknowledgment is made to the donors of the Petroleum Research Fund (PRF #12290-SEO), administered by the American Chemical Society, for support in the way of travel expenses of the seven invited foreign speakers. Also Dr. Dwight gratefully acknowledges the support of the Polymer Program of the Materials Division, National Science Foundation (Grant DMR 78-05429) and the Army Research Office (Grant DAAG29-80-K-0093) for partial support during the organization of the meeting and the preparation of the manuscript. Drs. Fabish and Thomas gratefully acknowledge support from Ashland Chemical Company and Xerox Webster Research Center, respectively. The editors are indebted to Kathy Fuller for typing and organizing the manuscripts in this book.

DAVID W. DWIGHT
Virginia Polytechnic Institute
   and State University
Blacksburg, Virginia 24061

THOMAS J. FABISH
Ashland Chemical Company
Columbus, Ohio 43216

H. RONALD THOMAS
Xerox Webster Research Center
Rochester, New York 14644

March 27, 1981

# Resonant Electron Scattering and Anion States in Polyatomic Molecules

KENNETH D. JORDAN

Department of Chemistry, University of Pittsburgh, Pittsburgh, PA 15260

PAUL D. BURROW

Department of Physics, University of Nebraska, Lincoln, NE 68588

Photoelectron spectroscopy (PES) (1, 2) and more recently electron transmission spectroscopy (ETS) (3, 4) have provided much information on the cation and anion states, respectively, of many hydrocarbons. Within the context of the Koopmans' Theorem (KT) approximation, the cation states can often be associated with the filled orbitals and the anion states with the unfilled orbitals of a molecule. In this sense these two methods are complementary. However there are important distinctions between these two spectroscopic methods which arise in part from the very different lifetimes of the anions and cations.

In photoelectron spectroscopy one analyzes the energy of electrons photoejected from a molecule (or atom). The cation states associated with simple one-electron ionizations are usually stable or long lived. Hence the line widths in photoelectron spectra are generally determined by the experimental resolution, typically 0.02 eV for commercial spectrometers, or by the density of vibrational or rotational states in large molecules. The autoionization times associated with certain shake-up states may be much less than $10^{-12}$ sec., leading in some cases to spectral widths greater than the experimental resolution.

In electron scattering methods, such as ETS, one looks for the rapid variations in the scattering cross section associated with the temporary capture of an electron by a molecule. In these methods only anion states which lie above the ground state of the neutral species are accessible. For many prototypical molecules, (4) including benzene, butadiene, formaldehyde, naphthalene, and styrene, all anion states lie above the ground state of the neutral molecule, but for others such as anthracene, hexatriene, and glyoxal the ground state anions lie energetically below the ground state of the neutral molecule and cannot be studied by ETS. For TCNQ the ground and at least the first excited state of the anion are bound (5) and therefore inaccessible to study by electron scattering methods. In contrast, all cation states are in principle accessible in PES.

0097-6156/81/0162-0001$05.00/0

A second major difference between ETS and PES results from the unique properties of temporary anion states. The temporary anion states associated with the occupation of unfilled orbitals of most hydrocarbons have lifetimes of $10^{-14}$ - $10^{-15}$ s. Since the FWHM resolution attainable in ETS is 0.02 - 0.05 eV, this means that the line widths due to autodetachment are generally greater than the experimental resolution. It is precisely these short lifetimes which make the study of these anions in the gas phase exceedingly difficult by optical spectroscopy and which makes electron scattering techniques ideally suited.

Electron transmission spectroscopy can provide information which is not readily available or accessible by other techniques. Here we will give just a few illustrative examples. A more complete discussion of the applications of ETS can be found in a recent review (4).

## Use Of ETS To Study Substituent Effects

A frequently encountered problem in chemistry concerns the effect of a substituent on the properties of a parent molecule. Perhaps the most important, and certainly the most studied (6), class of substituents are alkyl groups. The highest-lying occupied $\pi$ orbitals of molecules such as ethylene, formaldehyde, and benzene are well known to be destabilized by alkyl substitution. The situation for the $\pi*$ orbitals is much less clear. From our ETS studies (4, 7, 8) we have found that alkyl substitution destabilizes the LUMO of ethylene and formaldehyde but slightly stabilizes that of benzene, toluene, and t-butylbenzene. This result is particularly significant because the conventional picture of resonance and inductive effects leads one to expect that alkyl substitution should also destabilize the benzene $\pi*$ orbitals (see Figure 1).

Another interesting observation is that alkyl substitution causes a substantial decrease in the lifetime of benzene anions (8). We believe that this is due to the presence of the $\ell = 1$ and $\ell = 2$ partial waves in the charge density of the LUMO of the alkylbenzenes. The benzene anion ($D_{6h}$ symmetry) itself has a leading partial wave of $\ell = 3$.

We have also employed ETS to study the effect of fluorine substitution on the $\pi*$ orbitals of benzene and ethylene (10). Here we briefly discuss the results for the fluoroethylenes. Fluorine substitution is known to cause only small shifts in $\pi$ ionization potentials (IP) of unsaturated hydrocarbons (11). For example, the vertical $\pi$ IP's of ethylene and perfluoroethylene agree to within 0.1 eV. The reason that has been most often forwarded to explain this is that the electron withdrawing inductive effect, which stabilizes the occupied orbitals, is nearly cancelled by the destabilizing resonance mixing of the fluorine $p_z$ orbitals with the $\pi$ orbitals of the ethylenic double bond.

ETS studies show that the situation is quite different for the $\pi^*$ orbital; fluorine substitution causes a <u>destabilization</u> of the $\pi^*$ orbital with the difference between the <u>EA</u>'s of ethylene and perfluoroethylene being 1.4 eV. This cannot be explained within the resonance-inductive model. Rather, we have shown that it can be qualitatively understood in terms of the variation in bondlengths in this series of compounds.

## Correlation Of Excitation Energies With IP's and EA's

There has been a long history of attempts to correlate electronic excitation energies with one-particle energies. In particular one can derive theoretical expressions (<u>12</u>, <u>13</u>) relating excitation energies with the quantitiy (IP-EA). In Figure 2 we present this correlation for the $\pi \rightarrow \pi^*$ singlet and triplet transitions of the fluorethylenes (<u>10</u>). For the triplet transitions the results fit a straight line to better than 0.05 eV. There is considerably more scatter in the results for the singlets. However, if fluoroethylene, 1,1-difluoroethylene, and trifluoroethylene are eliminated, the singlet transitions of the other four species can again be well represented by a straight line. The singlet transitions for the three asymmetrically substituted molecules fall well below this line. We believe that the "deviation" of the singlet transitions of these three molecules can be understood in terms of a simple configuration mixing model. Only for these three molecules can the $\pi \rightarrow \pi^*$ configuration mix with the $\pi^2 \rightarrow (\pi^*)^2$ configuration. This configuration mixing causes an energy lowering of the $\pi \rightarrow \pi^*$ configuration.

## Comparison Of Spectra Of Gas Phase And Condensed Phase Anions

In a solvent such as methyl tetrahydrofuran (MTHF), anion and cation states of an unsaturated hydrocarbon are typically stabilized by 1-1.5 eV. For example, the ground state anions of butadiene, benzene, naphthalene, and styrene, which are unstable by 1.1 eV or less in the gas phase, are all stable in MTHF glass at $77^{\circ}$K making possible their study by optical spectroscopy. On the other hand, the anions of ethylene and 1,4 cyclohexadiene which are unstable by about 1.8 eV in the gas phase have not been prepared in MTHF glass. Presumably, the use of a more polar solvent would allow the preparation of these anions in a glass. To illustrate the effect of solvation on temporary anions we will consider the naphthalene molecule. This molecule is particularly interesting because it is an alternant hydrocarbon (<u>14</u>), and for such molecules, the pairing theorem (<u>15</u>) predicts that the anion and cation spectra should be identical. This theorem is valid for both Huckel and PPP model Hamiltonians, but is not valid for <u>ab</u> initio or CNDO calculations. It has been found (<u>10</u>) to be true to a good approximation ( $\sim 0.1$ eV) in organic glasses (<u>16</u>). The ETS spectra allows an examination of the validity of this

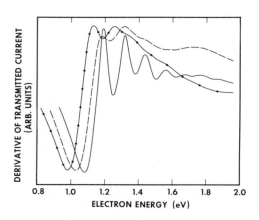

Journal of the American Chemical Society

Figure 1.   The derivative of transmitted current as a function of electron energy in benzene (———), toluene (– – –), and t-butylbenzene ( · · · ) (8)

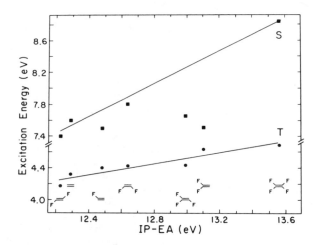

Chemical Physics Letters

Figure 2.   Vertical singlet and triplet $\pi \rightarrow \pi^*$ excitation energies of ethylene and the fluoroethylenes as a function of (IP-EA) (10)

theorem for gas phase species. Using differences between gas phase EA's, we construct (17) gas phase transition energies for comparison with the condensed phase values.

The ETS and PES spectra are presented in Figure 3 in such a way as to facilitate comparison of the spacings between the anion and cation energy levels. In Figure 4 we present the anion excitation energies obtained from ETS, from absorption studies in MTHF glass, semi-empirical PPP and Longuet-Higgins-Pople CI calculations, and finally from PES, assuming the validity of the pairing theorem.

Perhaps the most striking feature to be gleaned from this figure is that for the low-lying transitions the energies determined by all methods are in good agreement, while for the higher-lying transitions the gas phase results lie considerably above the others. We have interpreted this in terms of anion lifetimes, noting that, in general, the lifetime decreases as one goes to increasingly high-lying states of an anion. The lifetimes for the states near 5 eV in naphthalene are $\sim 10^{-15}$ s.

These times are only a factor of 2-3 greater than the time for an electron with 5 eV of kinetic energy to travel a distance equal to the "size" of the naphthalene molecule. Hence the wavefunctions associated with these very short-lived resonances have much less localization of charge in the molecular region than do those associated with long-lived resonances. As a result, the short-lived resonances undergo less electronic relaxation of the charge density associated with the neutral molecule.

In the condensed phase, where the lifetimes of the anions are increased and their wavefunctions should be considerably more localized, all the anion states should experience considerable reorganization. In other words, when compared to the condensed phase anions, the high-lying gas phase anions are "non-relaxed" and the low-lying anions essentially "fully relaxed". Clearly, any anion state which lives sufficiently long to display structure due to nuclear motion will be essentially fully relaxed electronically.

It has been noted by Duke and coworkers (18) that the various cation states of unsaturated hydrocarbons are stabilized to nearly the same extent going from the gas phase to the condensed phase. There is better pairing between the anion and cation states in MTHF glass than for the gas phase species. This is in agreement with the observation that the various anion states are stabilized by differing amounts going from the gas to the condensed phase.

## Conclusions and Discussion

In this paper we have tried to give the reader a general overview of the types of chemical information that can be obtained from electron transmission spectroscopy. ETS appears to be ideally suited for studying temporary anions formed by the cap-

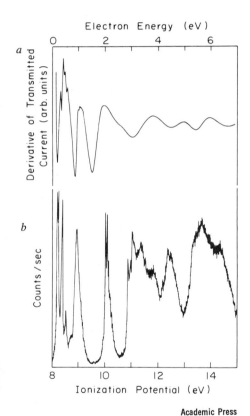

*Figure 3. Photoelectron and electron transmission spectra of naphthalene: (a) derivation of the transmitted current as a function of electron energy in naphthalene; (b) HeI spectra of naphthalene (21)*

**Academic Press**

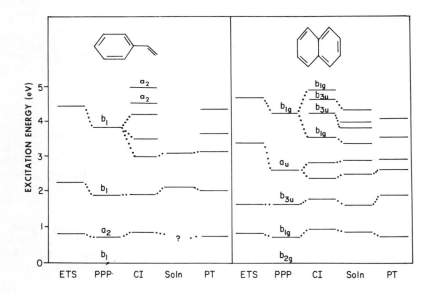

Chemical Physics

*Figure 4.    Anion excitation energies in eV for naphthalene. (ETS) Energies derived from electron transmission measurements in the gas phase; (Soln) optical absorption studies in anions in MTHF glass; (PPI, CI) theoretical energies; (PT) values derived from the cation spectrum by applications of the Pairing Theorem (17).*

ture of electrons into low-lying unfilled $\pi^*$ orbitals. Temporary anions associated with $\sigma^*$ orbitals apparently are too short-lived in general to be discernable in the total scattering cross-section. The only saturated hydrocarbon for which we have observed a shape resonance is cyclopropane (19). However, this probably should not be considered an exception since some of the sigma orbitals of this molecule are known to have appreciable pi character. For very short-lived anion states, the cross sections for vibrational excitation will provide a more sensitive means of detection. For example, Walker, et al. (20) have located a $\sigma^*$ resonance in ethylene between 3 and 6 eV.

## Acknowledgements

    This research has been supported by the Petroleum Research Fund, administered by the American Chemical Society and by Research Corporation.

## Abstract

    The temporary anions associated with low-lying unfilled orbitals generally have lifetimes in the range of $10^{-12} - 10^{-15}$ s in the gas phase. This paper discusses the use of electron transmission spectroscopy (ETS) to provide information on these short-lived anions. The sensitivity of the lifetimes to changes in symmetry is illustrated by comparing the electron transmission of benzene and various alkyl substituted benzenes.
    Knowledge of the gas phase electron affinities for a series of molecules, such as the fluoroethylenes, provides new insight into substituent effects. The correlation between the trends in the singlet and triplet excitation energies of the neutral molecules and the quantity (IP - EA) is also examined.
    The energy levels of the anion states of naphthalene and styrene as determined from the gas phase studies are compared to those obtained from optical absorption measurements on the anions in organic matrices as well as with theoretical predictions. It is demonstrated that the effect of solvation on the energy of an anion state depends on its lifetime, with larger shifts being observed for the shorter lived anions.

## Literature Cited

1.   Turner, D.W., Baker, C., Baker, A.D. and Brundle, C.R., "Molecular Photoelectron Spectroscopy" (John Wiley and Sons, New York, 1970).

2.   Rabalais, J.W., "Principles of Ultraviolet Photoelectron Spectroscopy", (John Wiley and Sons, New York, 1977).

3.   Sanche, L. and Schulz, G.J., J. Chem. Phys., 1973, 58, 479.

4.  Jordan, K.D. and Burrow, P.D., <u>Accts. of Chem. Res</u>., 1978, 11, 341.

5.  Klots, C.E., Compton, R.N., Raaen, V.F., <u>J. Chem. Phys</u>., 1974, 60, 1177.

6.  Libit, L. and Hoffman, R., <u>J. Amer. Chem. Soc</u>., 1974, 96, 1370; Mosclet, P., Grosjean, D., Mouvier, G. and Dubois, J., <u>J. Electron Spectry. and Related Phenom</u>., 1973, 2, 225.

7.  Chiu, N.S., Johnston, A.D., Burrow, P.D. and Jordan, K.D., to be submitted to <u>J. Electron Spectry. and Related Phenom</u>.

8.  Jordan, K.D., Burrow, P.D. and Michejda, J.A., <u>J. Am. Chem. Soc</u>., 1976, 98, 1245.

9.  Chiu, N.S., Burrow, P.D. and Jordan, K.D., to be submitted to <u>J. Chem. Phys</u>.

10. Chiu, N.S., Burrow, P.D. and Jordan, K.D., <u>Chem. Phys. Letters</u>, 1979, 68, 121.

11. Brundle, C.R., Robin, M.B., Kuebler, N.A. and Basch, H., <u>J. Am. Chem. Soc</u>., 1972, 94, 1451.

12. See, for example, Michl, J. and Becker, R.S., <u>J. Chem. Phys</u>., 1967, 46, 3889.

13. Jordan, K.D., Ph.D. Thesis (MIT, 1974), unpublished.

14. A system is defined as alternant if one can identify alternate positions of the skeleton (as with an asterick) and have no two adjacent positions both "starred" or "un-starred".

15. L. Salem, "The Molecular Orbital Theory of Conjugated Systems" (W.A. Benjamin, New York, 1966).

16. T. Shida and S. Iwata, <u>J. Am. Chem. Soc</u>., 1973, 95, 3473; Hoijtink, G.J., Velthorst, N.H. and Zandstra, P.J., <u>Mol. Phys</u>., 1960, 3, 533.

17. Jordan, K.D. and Burrow, P.D., <u>Chem. Phys</u>., 1980, 45, 171.

18. Duke, C.B., Salaneck, W.R., Fabish, T.J., Ritsko, J.J., Thomas, H.R. and Paton, A., <u>Phys. Rev. B</u>., 1978, 18, 5717.

19. Burrow, P.D. and Jordan, K.D., unpublished results.

20. Walker, I.C., A. Stamatovic, A. and Wong, S.F., <u>J. Chem. Phys</u>., 1978, 69, 5532.

21.  Heilbronner, E. and Maier, J.P., "Electron Spectroscopy:
     Theory, Techniques and Applications, Vol. I", ed. C.R.
     Brundle and A.D. Baker (Academic Press, Inc., New York,
     1977) p. 228.

RECEIVED December 22, 1980.

# Negative Ion States of Polyatomic Molecules[1]

L. G. CHRISTOPHOROU

Atomic, Molecular and High Voltage Physics Group, Health and Safety Research Division, Oak Ridge National Laboratory, Oak Ridge, TN 37830 and Department of Physics, University of Tennessee, Knoxville, TN 37916

In this paper I shall elaborate on recent knowledge concerning (i) the number and energies, (ii) the cross sections and decompositions, and (iii) the lifetimes of negative ion states (NISs) of polyatomic organic molecules. These are formed by capture of an electron in either the field of the ground or the field of an excited electronic state through a number of mechanisms (1) and are metastable with lifetimes ranging from $\sim 10^{-15}$ to $>10^{-2}$ sec. I shall restrict myself only to two cases, namely, those NISs which are formed in the field of the ground electronic state via a shape resonance and via a nuclear-excited Feshbach resonance mechanism. These two mechanisms are illustrated in Fig. 1. In the former type the electron is trapped in the attractive region of the potential, which results from the combined effect of the long-range polarization potential and the short-range repulsion due to the centrifugal motion of the two particles, and the NIS lies above the corresponding parent neutral state [i.e., its electron affinity, EA, is negative (<0 eV)]; it decays by electron emison and by dissociative attachment (if energetically possible) usually in times $<10^{-12}$ sec. The latter type involves coupling of the kinetic energy of the captured electron to molecular vibration and in this case the NIS lies energetically below [i.e., its EA is positive (>0 eV)] the corresponding parent neutral state; it thus can be stabilized, leading to stable parent negative ions. In the very initial step of their formation (i.e, prior to stabilization) these NISs can also decay by autoionization and autodecomposition (if energetically possible). The lifetimes of these NISs toward autodetachment (and at times toward autodissociation) are long, often $> 10^{-6}$ sec. Their cross

[1]Research sponsored by the Office of Health and Environmental Research of the US Department of Energy under contract W-7405-eng-26 with Union Carbide Corporation.

sections are large (often $> 10^{-14}$ cm$^2$) and as a rule reach their maximum value at thermal energies ($\underline{2}$).

We usually learn of the properties of NISs through their decay channels. Thus knowledge on (i) is normally obtained either from the energy dependence of the total electron scattering cross section, or the cross section for elastic scattering, or the cross section for inelastic scattering involving vibrational and/or electronic excitation--at an angle or in the forward direction-- via the decay of the NIS by electron emission; these cross sections are measured conveniently using electron scattering and electron transmission techniques. Knowledge on (i) can be acquired also from studies of the energy dependence of the cross sections for specific dissociative attachment fragment ions, total dissociative attachment cross sections or cross sections for parent negative ion formation; the latter cross sections are usually determined (especially at low energies) by the swarm and the swarm-beam techniques. Knowledge on (ii) is obtained from electron beam, mass spectrometric and swarm techniques, the last method being most appropriate for low energies and for nuclear-excited Feshbach resonances. Our knowledge on (iii) comes from electron scattering methods when EA $< 0$ eV and $\tau_a \gtrsim 10^{-13}$ sec, from high pressure electron swarm studies when EA $> 0$ eV and $\tau_a$ in the range $10^{-7}$-$10^{-12}$ sec, from time-of-flight mass spectrometric studies when EA $> 0$ eV and $\tau_a \gtrsim 10^{-6}$ sec, and from ion-cyclotron resonance techniques when EA$^a > 0$ eV and $\tau_a \gtrsim 10^{-3}$ sec (see Ref. 2).

Molecular negative ion states are abundant; their energies, cross sections, and lifetimes are strongly affected by the details of the molecular structure as well as the medium which surrounds the metastable anions. The cross sections and lifetimes of the NISs are functions of the electron energy $\varepsilon$. Often the NISs are described--their energies approximated and their numbers rationalized--in terms of the unoccupied molecular orbitals of the neutral molecule. These aspects of NISs will be apparent from the selected data presented in this paper on the negative ion shape and nuclear-excited Feshbach resonances of three basic groups of polyatomic organic molecules, namely, nonaromatic double-bonded systems, benzene and benzene derivatives, and selected halocarbons. The new knowledge on NISs provides for effective probing of molecular structure and reactions, and for the testing of theoretical treatments of molecular electronic structure; it is also directly relevant to a multiplicity of applied fields and technologies ($\underline{3}$, $\underline{4}$).

## Nonaromatic Double-Bonded Structures

_Effect of Double Bonds and Their Mutual Interaction on the Energy of NISs Formed via a Shape Resonance Mechanism._ The lowest NIS of $C_2H_4$ is located (adiabatic; $0 \to 0$ transition) at 1.55 eV above the ground state and is due to electron capture into the $\pi*$

($b_{2g}$) orbital. Since to our knowledge this is the lowest NIS of the isolated molecule, the EA of $C_2H_4$ is $-1.55$ eV. Comparing the positions of the reported lowest NISs OF $C_2H_4$ and $C_2H_6$ (Table 1) it is seen that the ethylenic double bond stabilizes the NIS (makes EA less negative). The presence of an additional double bond in the molecule (Table 1) lowers the NIS energy even more. This observation appears to be valid, also, for cyclic compounds where again two double bonds lower the negative ion resonance energy much more than does a single double bond. Furthermore, this lowering is a function of the mutual interaction of the two double bonds, which, in turn, depends on their relative positions. Localized double bonds can interact directly through space or indirectly through other bonds ($\underline{7}$). The changes in the energies of the NISs apparently are sensitive probes of such interactions.

The position of the NISs in Table 1 are from electron scattering studies. On the basis of these data, the EA of the listed molecules is negative and the NISs short lived. No knowledge of the cross section for formation of these NISs is available.

Effect of $CH_3$ Group. Substitution of $CH_3$ for H lowers the molecular symmetry, and the data in Table 2 show that it raises the position of the NIS (makes the EA smaller). The same effects are observed when one of the $CH_2$ groups of the ethylene molecule is replaced by atomic oxygen, which is the united atom equivalent of $CH_2$. Similar effects are observed for aromatics (e.g., the position of the lowest NIS of aniline, methyl aniline and dimethyl aniline, were reported to be respectively, $-1.13$ ($\underline{10}$), $-1.19$ ($\underline{11}$) and $-1.24$ ($\underline{11}$) eV). These data were again from electron scattering experiments; the NISs are very short-lived ($<10^{-12}$ sec) and the EA $< 0$ eV.

Effects of Electron Withdrawing Substituents. The successive substitution of $CH_2$ by O, the united atom equivalent, or substitution of the H atom by electronegative groups or atoms lowers the position of the negative ion resonance (increases the EA of the molecule). Actually for the last two molecules in Table 3 the EA is positive and large. At thermal energies the parent negative ions $C_2(CN)_4^{-*}$ and $C_2Cl_4^{-*}$ are formed and are long-lived (the lifetimes, $\tau_a$, toward autodetachment are $> 10^{-6}$ sec) ($\underline{2}$, $\underline{16}$). The formation of these ions is attributed to the large electron withdrawing ability of CN and Cl, which increases the effective positive charge at the carbon atoms by withdrawing electrons from the double bond and hence increasing the EA of the molecule.

The yield of the $C_2(CN)_4^{-*}$ ion and its $\tau_a$ are shown in Fig. 2 as a function of $\varepsilon$. Both the long $\tau_a$ and its variation with $\varepsilon$ have been discussed earlier ($\underline{2}$, $\underline{17}$). In Fig. 3 the yield of $C_2Cl_4^{-*}$ is compared with that of $SF_6^{-*}$. Both peak at thermal energies. They are very narrow, and their shapes in Fig. 3 are

Table 1:    Effect of double bonds and their mutual interaction on the position (in eV)[*] of the NISs of nonaromatic double-bonded polyatomic molecules formed via a shape resonance mechanism.

| Molecule | Formula | Reported Position of First NIS[**] | Reported Position of Second NIS[**] |
|----------|---------|-----------------------------------|-------------------------------------|
| Ethane | $H_3C-CH_3$ | ~-2.3[a] | |
| Ethylene | $H_2C=CH_2$ | -1.78 (-1.55)[b] | |
| 1,3-Butadiene | $H_2C=CH-CH=CH_2$ | -0.62 (0.62)[b] | -2.80[b] |
| Cyclohexene | | -2.07[b] | |
| 1,3-Cyclohexadiene | | -0.80 (-0.80)[b] | -3.43[b] |
| 1,4-Cyclohexadiene | | -1.75 (-1.75)[b] | -2.67[b] |
| 1,5-Cyclooctadiene | | -1.83[b] | -2.33[b] |

[*]    The uncertainty usually is $\pm$ 0.1 eV.

[**]   Values listed are vertical except those in parentheses which are adiabatic.  The minus (-) sign is used to indicate that the EA is negative, equal to the number indicated.

a     Ref. 5.

b     Ref. 6.

Table 2: Effect of $CH_3$ substitution on the position of the first NIS of polyatomic molecules.

| Molecule | Formula | Reported position of First NIS* |
|----------|---------|-------------------------------|
| Propene | $CH_3HC=CH_2$ | $-1.99^a$ |
| cis-Butene | $CH_3HC=CHCH_3$ | $-2.22^a$ |
| Formaldehyde | 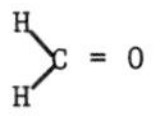 | $-0.7^{b**}$   $(-0.65)^c$ |
| Acetaldehyde | 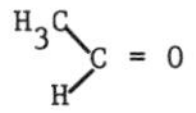 | $-1.3^{b**}$ |
| Acetone | 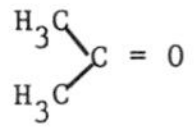 | $-1.5^{b**}$ |

\*   See footnote ** in Table 1.

\*\*  Maximum of shape resonance.

a   Ref. 6.

b   Ref. 8.

c   Ref. 9.

Table 3.  Effects of electron withdrawing groups or atoms on the position of the first NIS of some ethylenic-type structures.

| Molecule | Formula | Reported position of first NIS (eV) | Comments |
|----------|---------|-------------------------------------|----------|
| Ethylene | $H_2C=CH_2$ | $(-1.55)^a$ | Isoelectronic sequence; replace $CH_2$ by united atom equivalent, 0 |
| Formaldehyde | $H_2C=0$ | $(-0.65)^b$ | |
| Oxygen | $0_2$ | $(+0.44)^c$ | |
| Tetracyanoethylene | $(CN)_2C=C(CN)_2$ | $(+2.88)^c$ | Nuclear-excited Feshbach |
| Tetrachloroethylene | $Cl_2C=CCl_2$ | $(+2.12?)^c$ | resonance |

a   See Table 1

b   See Table 2

c   Equated to the EA value reported and thus to the position of the lowest NIS [$0_2$ (Ref. 13), $C_2(CN)_4$ (Ref. 14), $C_2Cl_4$ (Ref. 15)]

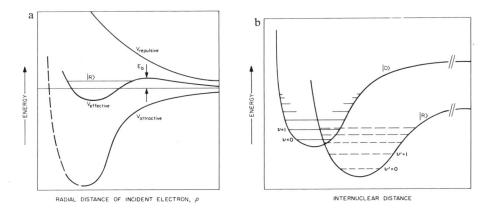

*Figure 1.   Schematic illustration of (a) shape and (b) nuclear-excited Feshbach resonances.   The symbols |0> and |R> designate, respectively, the electronic ground state of the neutral molecule and the NIS (1).*

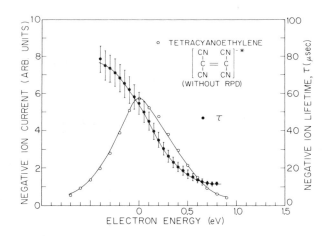

*Figure 2.   Negative ion autodetachment lifetime (●) and negative ion current (measured without the retarding potential difference method) (○) for $C_2(CN)_4^{-*}$ as a function of electron energy (17)*

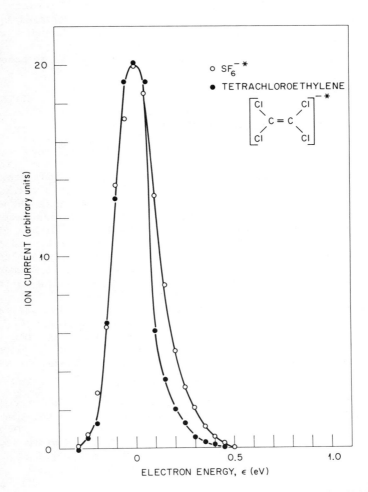

*Figure 3.    Comparison of the energy dependence of the $C_2Cl_4^-$* ion with that of $SF_6^-$*. The yields of the two ions were normalized at the peak (16).*

instrumental. At $\sim 0.0$ eV the lifetime of $C_2Cl_4^{-*}$ toward autodetachment is (16) $14{+}3$ $\mu$ sec. Very interestingly the transient
$C_2Cl_4^{-*}$ ion can multiply fragment yielding large quantities of
$Cl^-$ and also small amounts of $Cl_2^-$. The work of Johnson et al.
(16) has shown that the transient anions, $M^-$, of chloroethylenes
and chloroethanes fragment (even when these are formed by the
capture of electrons with virtually zero energy) through a multiplicity of channels. Depending on $M^-$, all or part of the
decomposition channels shown below were found (16) to be possible.

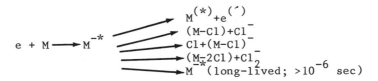

The cross section for $Cl^-$ production (Fig. 4) indicates the
existence of five maxima below $\sim 2.0$ eV for the chloroethylenes
(and also for the chloroethanes) (16). It seems that these are
not due to vibrational excitation, but rather due to separate NISs
associated (16) with orbitals dominated by the p-orbitals of the
Cl atom. Their positions are relatively insensitive, and their
cross sections very sensitive to the details of the molecular
structure. A similar behavior is exhibited by a number of
perfluorocarbon structures (see Refs. 18 and 19 and the following
sections).

Although for the molecules in Fig. 4 dissociative attachment
processes are expected to be fast ($<<10^{-6}$ sec), for $C_2Cl_4^{-*}$ at
$\sim 0.0$ eV (and for a number of other autodissociating long-lived
parent negative ions) both the autodissociation and the autodetachment processes are slow, as a result of vibrational redistribution of internal energy (2).

Effect of Perfluorination. As discussed in the preceeding
section, the effect of substitution of an H atom by a halogen atom
is a general lowering of the energy of the NIS. This is
dramatized in the case of total replacement of the H atoms in the
molecule by fluorines which, for the two perfluorocarbons in
Table 4 and for other similar structures with four or more carbon
atoms, (3) lowers the lowest negative ion state such that the EA
attains a positive value. This, in turn, has a profound effect on
the magnitude of the attachment cross section and the autodetachment lifetime of the parent anions. Thus, although the nonfluorinated forms do not attach low-energy electrons, the perfluorinated molecules have very large electron attachment cross
sections (close to the maximum s-wave capture cross section; see
Fig. 5), and the lifetime of the parent ions are $>10^{-6}$ sec. i.e.,
orders of magnitude longer than those ($<10^{-12}$ sec) of the NISs of
the nonfluorinated analogues.

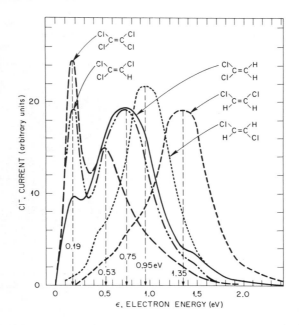

Journal of Chemical Physics

*Figure 4.    Comparison of the energy dependence of the Cl⁻ yield for the chloro-ethylenes (16)*

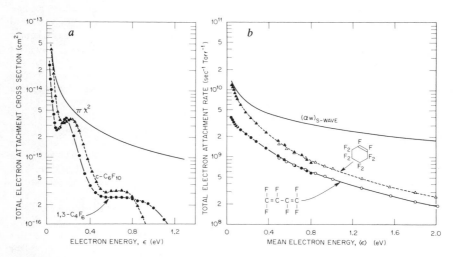

*Figure 5.    (a) Total electron attachment cross section as a function of electron energy for $1,3\text{-}C_4F_6$ and $c\text{-}C_6F_{10}$. These were unfolded by McCorkle et al. (22) from the $\alpha w$ vs. $E/P$ data of Ref. 18 on $1,3\text{-}C_4F_6$ and of Ref. 19 on $c\text{-}C_6F_{10}$. (b) Plot of $\alpha w$ vs. $\langle\epsilon\rangle$ from Refs. 18 and 19. The $\sigma_{s\text{-}wave}$ and $(\alpha w)_{s\text{-}wave}$ are, respectively, the maximum s-wave capture cross section and capture rate.*

Table 4.            Effect of perfluorination on the energies and lifetimes of NISs.

| Molecule | Formula | Position of lowest observed NIS (eV) | Ev (eV) | Autodetachment lifetime (μsec) |
|---|---|---|---|---|
| 1,3-Butadiene | $H_2C=CH-CH=CH_2$ | $-0.62$[a] | - | $\ll 1$ |
| 1,3-Perfluoro-butadiene | $F_2C=CF-CF=CF_2$ | b | $>0$[c] | $7$[d] |
| Cyclohexene | $c-C_6H_{10}$ | $-2.07$[a] | - | $\ll 1$ |
| Perfluoro-cyclohexene | $c-C_6F_{10}$ | b | $>1.4\pm0.3$[e] | $106;113$[d] |

a     From Table 1

b     The $\sigma_a(\varepsilon)$ shows at least three resonances below $\sim 1$ eV (see Fig. 5).

c     EA $(2-C_4F_6) = 0.7 - 1.45$ eV (20)

d     Ref. 2

e     Ref. 21

It is interesting to note the structure in the total electron attachment cross section $\sigma_a$ ($\varepsilon$) below $\sim 1$ eV in Fig. 5, which indicates the existence of three NISs in this energy range. Since at thermal and epithermal energies the ions formed are long-lived and the measured attachment rates showed (18, 19) no pressure dependence, the $\sigma_a$ ($\varepsilon$) in Fig. 5 are taken to give a true measure of the rate of formation of the lowest NIS of these molecules. Studies of these and other perfluorocarbons (18, 19) indicated that the number and energy positions of the NISs of these molecules do not vary considerably, contrary to the large variations in the respective $\sigma_a$ ($\varepsilon$). It would, then, seem that the magnitude of $\sigma_a$ ($\varepsilon$) for the formation of a NIS is a rather sensitive probe of the details of molecular structure.

For a number of perfluorocarbons, at thermal energies the largest contribution to the magnitude of $\sigma_a(\varepsilon)$ comes form the formation of parent ions (3, 18, 19). At higher energy, $\sigma_a(\varepsilon)$ contains varied contributions from dissociative attachment fragments which depend on the details of molecular structure (23). For example, the polyatomic molecule $1,3-C_4F_6$--which is non-planar with two ethylene groups lying in different planes--is more easily decomposed (23) by dissociative attachment than its isomers $2-C_4F_6$ and $c-C_4F_6$. This is clearly shown in Fig. 6 where the fragment and parent ions are shown for $1,3-C_4F_6$ and $2-C_4F_6$. The lifetimes of $1,3-C_4F_6^{-*}$ and $2-C_4F_6^{-*}$ at thermal energies are (23) 7 and 6 μsec, respectively.

For $C_2F_4$ the effect of perfluorination seems not to be sufficient to lower the lowest NIS of the isolated molecule below that of the neutral. $\{C_2F_4^-$ (or $C_2F_4^{-*}$) was not observed in electron impact studies. The anion was observed in solid solution, however, and its EPR spectrum was consistent with electron capture into a $\sigma$ orbital (24). Recently, it was reported [N.S. Chiu, P.D. Burrow and K.D. Jordan, Chem. Phys. Lett. 68, 121

Table 5.   Effect of structure on the positions (in eV)[†] of the negative-ion resonances of benzene and some of its derivatives [§][#]

| Compound | Formula | First $\pi$-NIS | Second $\pi$-NIS | Third $\pi$-NIS | $\pi^*_{CO}$ |
|---|---|---|---|---|---|
| Benzene | | -1.35 (-1.13)[a] | -1.35 (-1.13)[a] | -4.80[a] | |
| **Substituted benzenes with electron donating substituents** | | | | | |
| Phenol | OH | -0.61[b] (-1.01)[c] | -1.67[b] (-1.73)[c] | -4.92[c] | |
| Aniline | NH$_2$ | -0.55[b] (-1.13)[c] | -1.88[b] (-1.72)[c] | -5.07[c] | |
| **Substituted benzenes with electron accepting substituents introducing additional NIR states** | | | | | |
| Benzaldehyde | CHO | -0.71[b] | -1.12[b] | -4.61[d] | -2.22[b] |
| Benzoic Acid | COOH | -0.63[b] | -1.33[b] | $\approx$-4.4** | -2.64[b] |
| **N-heterocyclic derivatives** | | | | | |
| Pyridine | N | -0.62[e]; -0.84[f] | -1.20[e]; -1.30[f] | -4.58[e] | |

---

† The uncertainty is usually $\pm 0.1$ eV; values listed are vertical except those in parentheses which are adiabatic. As in earlier tables the minus (-) sign is used to indicate that EA for the NIS listed is negative, equal to the number indicated.

§ The first three NIRs are associated with the three unoccupied $\pi$ orbitals. The $\pi^*_{CO}$ is associated with an additional orbital resulting from an interaction of the carbonyl $\pi^*$ orbital with one (symmetric) of the two lowest degenerate $\pi$ orbitals of benzene.

# The position of the lowest negative ion state (traditionally referred to as the electron affinity, EA) for some of the molecules in this table may be positive ($>0$ eV), i.e., the lowest negative ion state may lie energetically below the first $\pi$-NIR listed in the table. The EA of $C_6H_5CHO$, for example, has been reported to be $\sim$+0.42 eV (25).

** On the basis of the data in Table II of Ref. 26.

a-Ref. 26, b-Ref. 27, c-Ref. 10, d-Ref. 12, e-Ref. 28, f-Ref. 29.

(1979); see also preceeding paper ] that for $C_2H_4$ fluorination destabilizes the $\pi^*$ anions with respect to that of ethylene.} Apparently it does appear that four or more carbon atoms are necessary for a perfluorocarbon molecule to have a positive EA and thus to be capable of forming long-lived NISs at low energies (3).

## Benzene and Derivatives

**Benzene.** The observed NISs of benzene have been attributed to capture of an electron in the degenerate $\pi_4$, $\pi_5$ ($e_{2u,1}$ $e_{2u,2}$) and the $\pi_6(b_{2g})$ orbitals. On the basis of the data in Table 5 on the energy of the first $\pi$-NIS of benzene, it may be concluded that the $\pi$-electron affinity of the <u>isolated</u> benzene molecule is -1.13 eV.

**Monosubstituted Benzenes.** The NISs of monosubstituted benzenes observed in electron scattering experiments were also understood on the basis of the simple $\pi$-molecular orbital picture

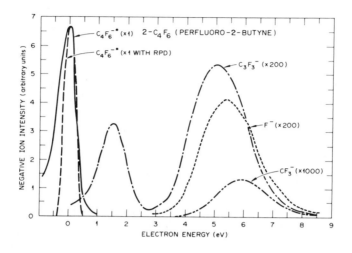

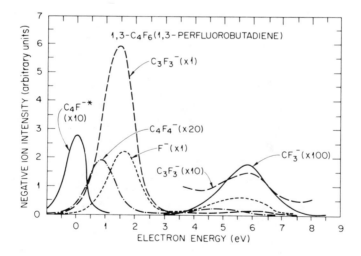

*Figure 6. Parent and fragment negative ion intensity as a function of electron energy on $1,3$-$C_4F_6$ and $2$-$C_4F_6$ (note the multiplication factors for the various ions). The asterisk means that the ion was found to be metastable with respect to auto-detachment (23).*

and the effect of substitution on the $\pi$-orbitals of benzene. For electron donating substituents, three NISs were observed due to electron capture in the $\pi_4$, $\pi_5$ and $\pi_6$ orbitals, the degeneracy of the $\pi_4$, $\pi_5$ orbitals in benzene being removed by substitution (Table 5).[5] Electron accepting substituents introduce additional NISs due to further splitting ([1]). The energies of the NISs associated with the $\pi_4$, $\pi_5$ orbitals depend rather strongly on the electron accepting/donating properties of the substituent, but the energy of the NIS associated with the $\pi_6$ orbital is relatively insensitive to the nature of the substituent ([26]).

The energies of the NISs in Table 5 were determined from electron scattering experiments and refer to isolated molecules and to anionic states reached in the very initial step of the electron-molecule interaction. Although these show that the $\pi$-electron affinity of the isolated molecules of these compounds are negative, they do not preclude the possibility of positive EA values in relaxed states or in non-isolated molecule environments. For strongly electron withdrawing substituents (e.g., $NO_2$) the parent molecule (e.g. $C_6H_5NO_2$) has a positive ($>0.5\%$ eV) ([2]) EA and forms long-lived ($\tau_a > 20 \times 10^{-6}$ sec) ([2]) parent negative ions at thermal energies. The lifetimes and cross sections for formation of the NISs in Table 5 are not known.

Heterocyclics. The degeneracy of the $\pi_4$, $\pi_5$ benzene orbitals can also be removed by replacement of a C-atom in the benzene skeleton by a heteroatom such as N. Energies of NISs of N-heterocyclics were obtained by electron scattering methods, and an example is shown in Table 5.

Fluorobenzenes. Partial or total replacement of H atoms on the benzene periphery by F atoms affects considerably the energies of the two lowest NISs but has only a minor effect on the energy of the third NISs, as is shown by the electron transmission data in Table 6. All three NISs observed in electron transmission have been associated with $\pi$-orbitals. On the basis of these data, all fluorobenzenes have negative $\pi$-electron affinities. It is, however, known that at least the EA of $C_6F_6$ is positive ($>1.8\pm0.3$ eV) ([21]) and that $C_6F_6^-$ is formed with a very large cross section at near thermal energies (Fig. 7); the autodetachment lifetime, $\tau_a$, of $C_6F_6^-$ is long ($\sim 12$ $\mu$sec)([2]). This, as was discussed earlier, ([26]) suggests that the lowest NIS of $C_6F_6$ is not connected with a $\pi-$ but most probably with a $\sigma$-orbital.

The effect of perfluorination on EA, $\tau_a$, and $\sigma_a$ is perhaps dramatized by a comparison of the magnitudes of these quantities for $C_6H_6$ and $C_6H_5CH_3$ and their perfluorinated analogues (Table 7). The EA of the isolated molecules are probably negative for the former, but they are large and positive for the latter; the $\tau_a$ for the former are most likely $<10^{-12}$ sec, but they are $>10^{-5}$ sec for the latter; the thermal electron attachment rates, $(\alpha w)_{th}$, for the former are orders of magnitude smaller compared to those for the latter.

Table 6.    π-negative ion states of fluorobenzenes[*][a]

| | | | | |
|---|---|---|---|---|
| Fluorobenzene | | -0.91 (-0.82) | -1.40 | -4.66 |
| p-Difluoro-benzene | | -0.62 (-0.53) | -1.41 | -4.51 |
| 1,3,5-Trifluo-robenzene | | -0.77** | -0.77** | -4.48 |
| 2,3,5,6-Tetra-fluorobenzene | | -0.50 (-0.34?) | -1.29 | -4.51 |
| Pentafluo-robenzene | | -0.36 | -1.19 | -4.53 |
| Hexafluo-robenzene | | -0.42** | -0.42** | -4.50 |

*    The uncertainty is usually ±0.1 eV; values listed are vertical except those in parentheses which are adiabatic.  As in earlier tables, the minus (-) sign is used to indicate that EA for the NIS listed is negative, equal to the number indicated.

**   Note the degeneracy of these NISs due to the symmetry of the molecule.

a    Ref. 26.

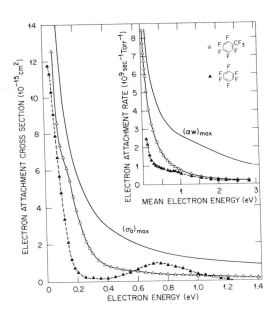

*Figure 7.  Total electron attachment rates as a function of mean electron energy, and swarm-unfolded electron attachment cross sections as a function of electron energy for $C_6F_6$ (30) and $C_6F_5CF_3$ (19)*

Table 7.     Comparison of the energy of lowest $\pi$-NIS, EA, $\tau_a$ and $(\alpha w)_{th}$ for $C_6H_6$ and $C_6F_6$, and $C_6H_5CH_3$ and $C_6F_5CF_3$.

| Molecule | Lowest $\pi$-NIS (eV) | EA (eV) | $\tau_a$ (sec) | $(\alpha w)_{th}$ (sec$^{-1}$ torr$^{-1}$) |
|---|---|---|---|---|
| $C_6H_6$ | $-1.35$[a*] | $< 0^+$ (?) | $<10^{-12}$[b] | $<10^4$[b**] |
| $C_6F_6$ | $-0.42$[a*] | $\geq 1.8 \pm 0.3$[c] | $1.3 \times 10^{-5}$[d] | $3.3 \times 10^9$[e] |
| $C_6H_5CH_3$ | $-0.4$(?)[f]; $-1.11$[g] | $< 0$(?) | $<10^{-12}$ | very small |
| $C_6F_5CF_3$ | $0.0$(?)[h] | $\geq 1.7 \pm 0.3$[c] | $>1.2 \times 10^{-5}$[d] | $9.2 \times 10^9$[h] |

*     Position of maximum of NIS in transmission; see footnote + in Table 5.

+     See discussion in Ref. 31.

**    For $C_6H_6$ in $N_2$ at total pressures $\leq$ 9500 torr.

a-Ref. 26, b-Ref. 31, c-Ref. 21, d-Ref. 2, e-Ref. 30, f-Ref. 27, g-Ref. 32, h-Ref. 19.


Table 8.     Negative ion states of pyrrole, furan and thiophene[*a]

| Compound | Formula | First $\pi$-NIS (eV)** | Second $\pi$-NIS (eV)** |
|---|---|---|---|
| Pyrrole | N-H | $-2.38$ | $-3.44$ |
| Furan | O | $-1.76$ | $-3.14$ |
| Thiophene | S | $-1.17$ | $-2.67$ |

*     The electron affinities of N, O and S are, respectively, <0, 1.465 and 2.07 eV (see Ref. 34).,

**    Position of the center of the NIR in an electron transmission study.[a] The − sign is used to indicate that the EA is negative.

a     From Ref. 35.

The effect of perfluorination is certainly drastic for aromatic as it is for aliphatics. It is also noted that--as for the aliphatic hydrocarbons--replacement of F by $CF_3$ on the benzene periphery (Fig. 7) increases $\sigma_a(\varepsilon)$ considerably.

Other Conjugated Systems. There are other basic molecular structures such as styrene (link between the ethylenic and benzene-type molecules), naphthalene (link to higher aromatics), biphenyl (typical of nonfused-ring aromatics) that could help elucidate further our discussion of the effects of structure on the properties of the NISs of organic molecules. The first two types of molecules have been discussed in the preceeding paper, and the available information (33) on the last one is incomplete. In this section, therefore, I shall refer only to the reported values on the energies of the lowest two NISs of a related group of heteroatom-containing conjugated organics, namely pyrrole, furan and thiophene (Table 8). These molecules, as benzene, have six $\pi$-electrons, but these are now spread over five centers rather than over six centers as in benzene. The observed lowest two NISs have been associated (35) with shape resonances in which the extra electron is captured in the lowest two $\pi$-unoccupied molecular orbitals. On the basis of this data, the $\pi$-electron affinities of these systems are negative. The data in Table 8 also show that the energy of each of the lowest two NISs of these heterocyclics is lowered, i.e., the anionic state becomes more stable, with increasing electron affinity of the heteroatom.

## Selected Halocarbons

In this section I shall refer to some of our recent work on the attachment of slow electrons to chlorofluoroethanes and chlorofluoromethanes as it pertains to the NISs of these molecules. In Fig. 8 are presented the attachment rates as functions of $<\varepsilon>$ for $1,1,1-C_2F_3Cl_3$, $1,1,2-C_2F_3Cl_3$, $1,1,1-C_2Cl_3H_3$, and $1,1,2-C_2Cl_3H_3$. The data in Fig. 8 and Table 9 show the effect of F substitution as well as the effect of the number and relative positions of the Cl atoms in the molecule on the energies and corresponding cross sections of the NISs of these compounds. For a given number of Cl atoms in the molecule, replacement of H by F lowers the positions of the resonance maxima and increases the magnitude of $\sigma_a(\varepsilon)$; the magnitude of $\sigma_a(\varepsilon)$ increases with increasing number of Cl atoms in the molecule. The $1,1,1-$ or $1,1-$isomers capture electrons much more efficiently than the $1,1,2-$ or $1,2-$isomers, and the positions of the resonance maxima lie at lower energies for the former than for the latter. It is noted that the dipole moments of the $1,1,1-$ or $1,1-$ compounds are as a rule larger than those of the $1,1,2-$ or $1,2-$ isomers.

Although the attachment cross sections reported here are total (i.e., for all negative ions produced), earlier work on

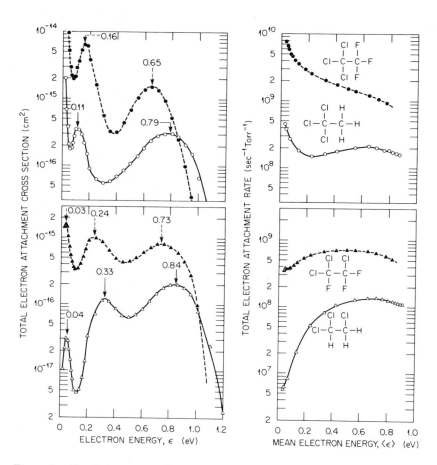

*Figure 8.   Total electron attachment rates as a function of mean electron energy,
and swarm-unfolded electron attachment cross sections as a function of electron
energy for chlorofluoroethanes (36)*

Table 9.   Positions of electron attachment cross section maxima, values of $\sigma_a(\epsilon)$ at these maxima, energy integrated cross sections, and thermal electron attachment rates for haloethanes[a]

| Haloethane compound | $\epsilon_1$ (eV) | $\epsilon_2$ (eV) | $\epsilon_3$ (eV) | $\sigma_a(\epsilon_2)$ ($10^{-16}$ cm$^2$) | $\sigma_a(\epsilon_3)$ ($10^{-16}$ cm$^2$) | $\int_{0.04}^{1.4\ eV}\sigma_a(\epsilon)dE$ ($10^{-16}$ eV cm$^2$) | Thermal attachment rate (sec$^{-1}$ torr$^{-1}$) |
|---|---|---|---|---|---|---|---|
| $1,1,1\text{-}C_2F_3Cl_3$ | <0.05 | 0.16 | 0.65 | 64 | 14.6 | 13.5 | $8.93 \times 10^9$ |
| $1,1,1\text{-}C_2H_3Cl_3$ | <0.05 | 0.10 | 0.79 | 3.6 | 3.1 | 1.74 | $4.72 \times 10^8$ |
| $1,1,2\text{-}C_2F_3Cl_3$ | <0.05 | 0.25 | 0.73 | 10 | 8.1 | 5.94 | $3.59 \times 10^8$ |
| $1,1,2\text{-}C_2H_3Cl_3$ | 0.05 | 0.33 | 0.85 | 1.1 | 2 | 1.02 | $5.8 \times 10^6$ |
| $1,1\text{-}C_2F_4Cl_2$ | <0.05 | 0.25 | 0.80 | 3.0 | 3.8 | 3.22 | $1.57 \times 10^8$ |
| $1,2\text{-}C_2H_4Cl_2$ | 0.2[b] | 0.75[b] | 1.08[b] | | | | |
| $1,2\text{-}C_2F_4Cl_2$ | ∿0.05 | 0.33 | 0.95 | 2.1 | 2.1 | 1.85 | $2.28 \times 10^7$ |
| $1,2\text{-}C_2H_4Cl_2$ | 0.2[b] | 0.53[b] | 1.06[b] | | | | |

a   From Ref. 36 unless otherwise indicated.

b   Ref. 16.

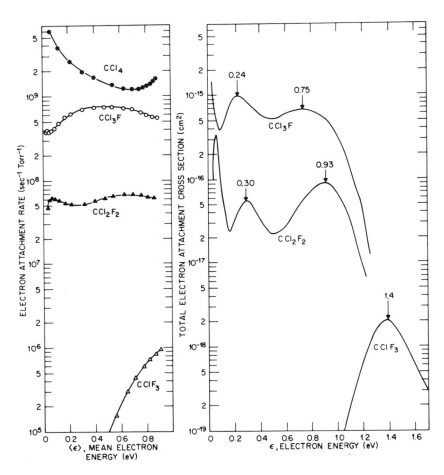

*Figure 9.    Total electron attachment rates as a function of mean electron energy, and swarm-unfolded electron attachment cross sections as a function of electron energy for chlorofluoromethanes (22)*

similar compounds ($\underline{16}$) indicated that the predominant ion is $Cl^-$. Other ions [e.g., $\overline{Cl_2}$ or (parent molecule less one chlorine atom)$^-$ ] are produced but with much lower yields. The present results, as well as earlier studies on aliphatic halocarbons, ($\underline{16}$, $\underline{23}$) show that the cross sections for electron attachment to these compounds are large and that the electron attachment process proceeds through a multiplicity of negative ion states.

In Fig. 9 the electron attachment rates as a function of $<\varepsilon>$ for $CCl_4$, $CCl_3F$, $CCl_2F_2$ and $CClF_3$ and the swarm-unfolded total attachment cross sections as a function of electron energy for $CCl_3F$, $CCl_2F_2$ and $CClF_3$ are presented. These, and other findings (see, for example, Refs. $\underline{3}$, $\underline{16}$), show that the magnitude of $\sigma_a(\varepsilon)$ increases with increasing number of Cl atoms in the molecule, becoming very large for multiply chlorinated methanes, and that the attachment of slow electrons proceeds through a multiplicity of low-lying NISs which--on the basis of the positions of the maxima in $\sigma_a(\varepsilon)$ (Fig. 9)--becomes stabilized (their positions shift to lower energies) with increasing Cl substitution. Finally, the extensive fragmentation of these compounds upon electron impact at these low energies demonstrates their extreme fragility towards the presence of low energy electrons, including those of thermal energy.

## Effect of Medium on the Negative Ion States of Molecules

Negative ion states, as other molecular properties and processes involving charges, are dramatically influenced by the nature and the density of the medium in which they are embedded. These environmental influences are still not well understood, and our knowledge on the transition from gaseous to condensed-phase behavior is still incomplete, although progress is being made (see, for example, Refs. 37-39).

## Abstract

Recent work on the negative ion states of organic molecules formed in the field of the ground electronic state via a shape or a nuclear-excited Feshbach resonance mechanism is discussed. Electron scattering and electron attachment data on three basic groups of polyatomic organic molecules (nonaromatic double-bonded structures, benzene and benzene derivatives, selected halocarbons) are used to indicate the rather delicate and strong effects of molecular structure on the energy, cross section and lifetime of the negative ion states of organic molecules. The fragility of anionic species toward (multiple) fragmentation is shown using recent data on perfluorinated and other halocarbons.

## Literature Cited

1.  See, for example, Christophorou, L.G., Grant, M.W., and

McCorkle, D.L., "Advances in Chemical Physics," Wiley-Inter-science, New York, Vol. 36, 1977, p. 413; Schulz, G.J. Rev. Mod. Phys., 1973, 45, 378, 423.

2.  Christophorou, L.G., "Advances in Electronics and Electron Physics," ed. by L. Marton, Academic Press, N.Y., Vol. 46, 1978, p. 55.

3.  Christophorou, L.G., "Negative Ions Of Polyatomic Mole-cules", Environmental Health Perspectives (in press).

4.  Christophorou, L.G., "Proceedings of the XIIIth Internation-al Conference on Phenomena in Ionized Gases", Berlin, G.D.R., 1977, Invited Lectures, pp. 51-73.

5.  Pisanias, M.N., Christophorou, L.G. and Carter, J.G., Oak Ridge National Laboratory Report ORNL/TM-3904 (1972).

6.  Jordan, K.D., Michejda, J.A. and Burrow, P.D., Chem. Phys. Lett., 1976, 42, 227.

7.  Hoffmann, R., Acc. Chem. Res., 1971, 4, 1.

8.  van Veen, E.H., van Dijk, W.L. and Brongersma, H.H., Chem. Phys., 1976, 16, 337.

9.  Jordan, K.D. and Michejda, J.A., Chem. Phys. Lett., 1976, 42, 223.

10. Jordan, K.D., Michejda, J.A. and Burrow, P.D., J. Am. Chem. Soc., 1976, 98, 7189.

11. Nenner, I., quoted in Reference 12.

12. Jordan, K.D. and Burrow, P.D., Acc. Chem. Res., 1978, 11, 341.

13. Celotta, R.J., Bennett, R.A., Hall, J.L., Siegel, M.W. and Levine, J., Phys. Rev., 1972, A6, 631.

14. Farragher, A.L. and Page, F.M., Trans. Faraday Soc., 1967, 63, 2369.

15. Gaines, A.F., Kay, J. and Page, F.M., Trans. Faraday Soc., 1966, 62, 874.

16. Johnson, J.P., Christophorou, L.G. and Carter, J.G., J. Chem. Phys., 1977, 67, 2196.

17. Christophorou, L.G., Hadjiantoniou, A. and Carter, J.G., J.C.S. Faraday Trans. II, 1973, 69, 1713.

18. Christodoulides, A.A., Christophorou, L.G., Pai, R.Y.   and Tung, C.M., J. Chem Phys., 1979, 70, 1156.

19. Pai, R.Y., Christophorou, L.G. and Christodoulides, A.A., J. Chem Phys., 1979, 70, 1169.

20. Hammond, P.R., J. Chem. Phys., 1971, 55, 3468.

21. Lifshitz, C., Tiernan, T.O. and Hughes, B.M., J. Chem. Phys., 1973, 59, 3182.

22. McCorkle, D.L., Christodoulides, A.A., Christophorou, L.G. and Szamrej, I., J. Chem. Phys., 1980, 72, 4049.

23. Sauers, I., Christophorou, L.G. and Carter, J.G., J. Chem. Phys., 1979, 71, 3016.

24. McNeil, R.I., Shiotani, M., Williams, F. and Yim, M.B., Chem. Phys. Lett., 1977, 51, 433.

25. Wentworth, W.E., Kao, L.W. and Becker, R.S., J. Phys. Chem., 1975, 79, 1161.

26. Frazier, J.R., Christophorou, L.G., Carter, J.G. and Schweinler, H.C., J. Chem. Phys., 1978, 69, 3807.

27. Christophorou, L.G., McCorkle, D.L. and Carter, J.G., J. Chem. Phys., 1974, 60, 3779.

28. Nenner, I. and Schulz, G.J., J. Chem. Phys., 1975, 62, 1747.

29. Pisanias, M.N., Christophorou, L.G., Carter, J.G. and McCorkle, D.L., J. Chem. Phys., 1973, 58, 2110.

30. Gant, K.S. and Christophorou, L.G., J. Chem. Phys., 1976, 65, 2977.

31. Christophorou, L.G. and Goans, R.E., J. Chem. Phys., 1974, 60, 4244.

32. Jordan, K.D., Michejda, J.A. and Burrow, P.D., J. Am. Chem. Soc., 1976, 98, 1295.

33. Frazier, J.R., Ph.D. dissertation, University of Tennessee (1978).

34. Christophorou, L.G., "Atomic and Molecular Radiation Physics", Wiley-Interscience, N.Y., 1971, Chap. 7.

35.  van Veen, E.H., Chem. Phys. Lett., 1976, 41, 535.

36.  McCorkle, D.L., Szamrej, I. and Christophorou, L.G., (to be published).

37.  Christophorou, L.G., Chem. Revs., 1976, 76, 409.

38.  Christophorou, L.G., Int. J. Rad. Phys. Chem., 1975, 7, 205.

39.  Christophorou, L.G. and McCorkle, D.L., Can. J. Chem., 1977, 55, 1876.

RECEIVED February 28, 1981.

# Electronic States and Excitations in Polymers

JOHN J. RITSKO

Xerox Webster Research Center, Webster, NY 14580

The most common probe of the electronic structure of matter is light. However, for many polymeric materials the fundamental electronic excitation spectrum is beyond the range of common optical spectrometers. For these materials electron energy loss spectroscopy is a particularly attractive technique for measuring the spectrum of electronic excitations, thus providing key information for the basic understanding of the nature of the electronic states. The experiments reported here measured energy loss spectra of an 80 keV electron beam transmitted through thin samples ($\sim$100 nm thick) which were prepared by standard solvent casting techniques or polymerized in thin film form. The special purpose high resolution energy loss spectrometer used in these studies was operated with a resolution of 0.1 eV. Spectra were recorded as a function of scattering angle (or momentum transfer) over the energy range 0.2 to 400 eV, thus encompassing both valence and core electronic excitations.

Radiation induced changes in the electronic structure of all samples were evident as changes in energy loss spectrum with increased exposure to the electron beam. The spectrum of radiation induced chromophores could thus be studied (1). Spectra recorded at the earliest exposure times compared favorably with optical results and are believed to contain primarily intrinsic electronic excitations (1, 2, 3, 4). An analysis of these intrinsic spectra is the subject of the bulk of this paper. The spectra of radiation damaged polymers are described briefly after the intrinsic excitations are discussed.

The energy loss probability for fast electrons, or the differential scattering cross-section per unit energy loss, E, per unit solid angle, $\Omega$, can be written:

$$\frac{d^2\sigma}{dEd\Omega} \sim \frac{1}{q^4} \; | \; <\psi_f \; | \; e^{i\vec{q}\cdot\vec{r}} \; | \; \psi_o > \; |^2 \qquad (1)$$

where q is the momentum transferred to the sample in the inelastic

0097-6156/81/0162-0035$05.00/0

collision and $\psi_o$ and $\psi_f$ are the ground and excited states of the
sample respectively.  This expression can also be written as:

$$\frac{d^2\sigma}{d\overline{E}d\Omega} \sim \frac{1}{q^2} \quad \text{Im} \quad \frac{-1}{\varepsilon(\vec{q},E)} \tag{2}$$

where $\varepsilon$ is the momentum dependent dielectric function.  q depends
on the scattering angle and can be varied from 0 to $1\text{Å}^{-1}$ - the
momentum associated with typical Brillouin zone boundaries in
solids.  From the energy loss probability the real and imaginary
parts of the dielectric function can be calculated by a Kramers-
Kronig analysis (1, 2, 3, 4).  For small $\vec{q}$, $\varepsilon(\vec{q},E)$ is equivalent
to the optical dielectric function and observables such as the
optical reflectivity and absorption coefficient can be calculated.  The momentum dependence of the matrix element in Equation
1 is the source of the unique information which energy loss
spectroscopy can provide.  For small scattering angles (small q),
the energy loss spectrum is dominated by electric dipole excitations as in optical measurements.  However, at large q dipole
forbidden excitations are measured (5).  This is of particular
importance for the study of energy bands in solids.  Optically
only vertical interband transitions are measured due to the small
momentum of photons.  But, in electron scattering the spectra of
non-vertical excitations of arbitrary q can be obtained.  This can
provide essential information about the width and shape of energy
bands in solids and about the local or extended nature of electronic excitations (1, 6).

In many polymers, functional groups which are repeated along
the polymer chain are far enough apart that energy bands formed
from degenerate electronic states on different groups are very
narrow and the electronic structure of the polymer is essentially
that of the individual monomeric unit repeated many times.  Examples are pendant group polymers (7) and polymers such as Lexan
(bis-phenol-A-(polycarbonate)).  Energy bands due to the overlap
of pendant groups should be only about 0.1 eV wide (7).  Some
energy loss spectra are given in Figure 1 where the similarity
between polystyrene, poly(2-vinylpyridine), and Lexan is quite
apparent.  The reason for this is that in this spectral region the
strongest excitations are $\pi \rightarrow \pi^*$ bonding to anti-bonding transitions associated with individual phenyl rings.  The C=O excitations are relatively weak (2) and excitations of the $\sigma$ bonds
generally occur at higher energy.  The valence excitation spectrum can be calculated with semiempirical molecular orbital
methods applied to small model molecules.  As an example, in
Figure 1 the vertical bars indicate the strongest electronic
excitations in ethylbenzene (as a model for polystyrene) calculated with a CNDO-S program which includes the configuration
interaction in the final state (4).  There is reasonably good
agreement with measured spectral features.  The small peak at 4.75

eV is the well known symmetry forbidden band in benzene. It is strongest in poly(2-vinylpyridine) where the nitrogen quite effectively breaks the ring symmetry (4).

The excitations of carbon 1s core electrons begin at about 285 eV. Electric dipole excitation of these atomic s states takes place to the empty p orbitals which make up the antibonding molecular orbitals. Since the 1s states are sharp and well separated from the valence orbitals, the measured spectrum maps the distribution of empty molecular orbitals (3, 4). Results for polystyrene and Lexan are given in Figure 2. As indicated by the vertical bars, which show the calculated position of the empty orbitals of ethyl-benzene (4) (computed with CNDO-S), the excitation spectrum is dominated by transitions to anti-bonding π ring orbitals which are similar in polystyrene and Lexan due to the local nature of the electronic structure. In poly(2-vinylpyridine) the initial strong peak at 285 eV is split into a doublet due to different carbon 1s electron binding energies caused by the inequivalent charge states of various carbon atoms around the pendant ring (4). Thus, electron energy loss spectroscopy can provide chemical information about atomic charge states similar to ESCA or x-ray photoelectron spectroscopy. The shift of the first peak in Lexan to higher energy may be such a chemical effect. The differences in position and intensity of the higher peaks are due to subtle differences in the antibonding orbitals of the two materials.

The σ bonds in the backbone of vinyl polymers should not be describable in terms of local states of small model molecules because of overlap of carbon atomic orbitals only 1.5Å apart. This concept can be tested in polyethylene where the least bound C-C bond band widths have been calculated to be about 3 eV (8). The energy loss function, $Im(-1/\varepsilon)$, for polyethylene is given in Figure 3 where $\varepsilon_1$ and $\varepsilon_2$, the real and imaginary parts of the dielectric function computed via a Kramers-Kronig analysis, are also given (1). Figure 4 shows the computed optical reflectivity and optical absorption coefficient which follow from the dielectric function. In the region of the fundamental absorption edge there appears to be reasonable agreement between optical and energy loss results (1) although it would be desirable to extend the comparison to higher energies as was successfully done for polystyrene, poly(2-vinylpyridine) and polymethylmethacrylate (2, 4).

Results of photoemission studies of polyethylene have shown definite evidence for wide energy bands among deep valence orbitals (9), but the nature of the fundamental absorption edge has not been resolved. Band structure calculations predict direct interband excitations to occur above 12.6 eV (8) whereas the absorption threshold is at 7.2 eV and a strong peak in $\varepsilon_2$ occurs at 9.0 eV. The momentum dependence of the absorption threshold indicates that the threshold is of excitonic origin, i.e. the excitation is localized by the strong electron-hole or configuration inter-

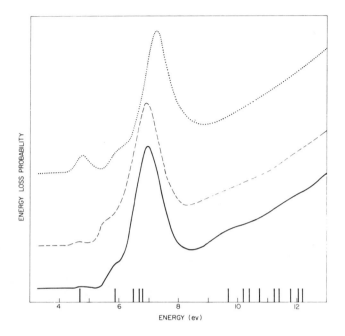

*Figure 1.   Valence electronic excitations: polystyrene (———); lexan (bisphenol-A-
(polycarbonate)) (– – –); poly(2-vinylpyridine) (· · ·). Vertical bars indicate calculated excitations of ethylbenzene.*

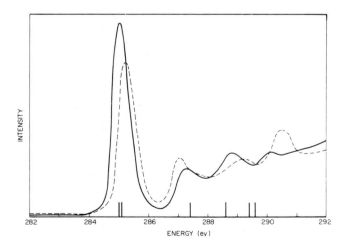

*Figure 2.   Carbon 1s core electronic excitations: polystyrene (———). Vertical bars
indicate positions of antibonding molecular orbitals of ethylbenzene with first orbital
aligned with strong peak in polystyrene Lexan (– – –).*

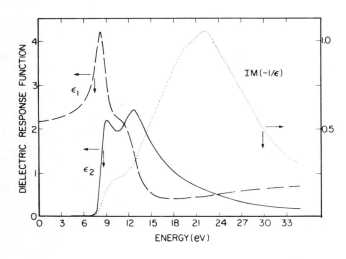

Journal of Chemical Physics

*Figure 3. Polyethylene: measured energy loss function, Im(−1/ε) (· · ·). Results of a Kramers–Kronig analysis: ε₁ (− − −); ε₂ (———) (1).*

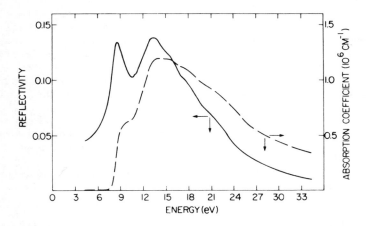

Journal of Chemical Physics

*Figure 4. Polyethylene: the normal incidence reflectivity (———) and optical absorption coefficient (− − −) computed from the dielectric function of Figure 3 (1)*

action energy, thus occurring below the original bandgap ($\underline{1}$).
This conclusion was reached by measuring the momentum dependence
of the energy loss spectra shown in Figure 3. The threshold did
not change position with momentum contrary to the large changes
expected for the direct non-vertical interband transitions required by the energy band models ($\underline{1}$). Thus, although the ground
electronic states are delocalized into relatively wide energy
bands the excited state spectrum may be localized by the electron-
hole interaction.

In fully conjugated polymers such as polyacetylene the total
bandwidth of the least bound $\pi$ orbitals is quite large, about 12
eV, and the excitation spectrum should resemble that of common
inorganic semiconductors such as Si or Ge which have similar wide
energy bands. Energy loss spectra as a function of momentum are
shown in Figure 5. Due to bond length alternation polyacetylene
is not a simple metal but rather an insulator with a fundamental
absorption edge at 1.4 eV, as measured by optical experiments ($\underline{11}$,
$\underline{12}$). This same threshold is seen in Figure 5 although our spectra
also show some residual absorption due to impurities below 0.8 eV
($\underline{6}$). As momentum is increased the onset of absorption at 1.4 eV
does not change, thus suggesting, as in the case of polyethylene,
that the threshold itself is indirect, due either to disorder or
excitonic effects ($\underline{6}$).

The existence of delocalized electronic states and wide
energy bands in polyacetylene is confirmed by the rapid dispersion to higher energy with increasing momentum of the strong peak
at 4.1 eV which is due to electronic excitations. As indicated
by Kramers-Kronig analyses of optical ($\underline{12}$) and energy loss data
($\underline{6}$) this energy loss peak is associated with a zero crossing of
$\varepsilon_1$. When the dielectric function goes through zero the response
of the solid to an infinitesimal charge fluctuation is finite and
collective oscillations of the electron gas occur which are
called plasmons. In simple metals where the valence electronic
states are completely delocalized, plasmons are the dominant
excitation. And, due to the extent of the energy bands the
plasmons exhibit positive dispersion ($\underline{6}$). In fact, the magnitude
of the plasmon dispersion in $(CH)_x$ is similar to that measured in
graphite in which the electronic states are well known to be
delocalized. Thus, above the fundamental absorption edge, the
excitation spectrum indicates the significance of wide energy
bands similar to those in inorganic solids.

In polymers with localized electronic states such as polystyrene, which also contains strong energy loss peaks due to $\pi$
electronic excitations as seen in Figure 1, positive dispersion
is not observed ($\underline{13}$). These results are shown in Figure 6. As
momentum is increased, the strong peak at 7 eV due to $\pi \rightarrow \pi*$
excitations of the benzene rings in polystyrene shows very little
momentum dependence. Since the bandwidth of valence and conduction bands due to overlap of $\pi$ orbitals on adjacent phenyl rings
is < 0.1 eV little momentum dependence to the non-vertical

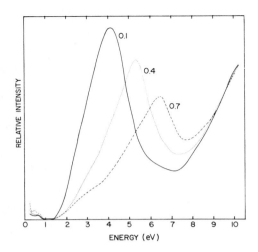

*Figure 5. Momentum dependence of the energy loss probability of polyacetylene, $(CH)_x$ ((————) q = 0.1 Å$^{-1}$; (· · ·) q = 0.4 Å$^{-1}$; (– – –) q = 0.7 Å$^{-1}$)*

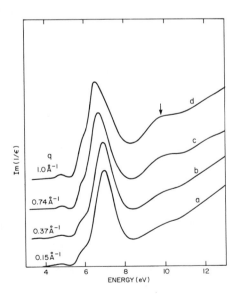

Journal of Chemical Physics

*Figure 6. Momentum dependence of the energy loss probability of polystyrene: Curve a: q = 0.15 Å$^{-1}$, very similar to Figure 1 data; Curve b: q = 0.37 Å$^{-1}$; Curve c: q = 0.74 Å$^{-1}$; Curve d: q = 1.0 Å$^{-1}$; arrow indicates optically forbidden excitation (13)*

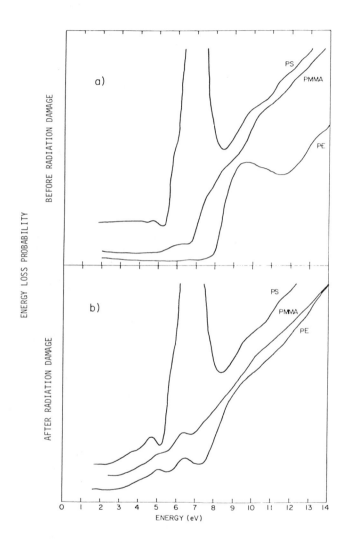

*Figure 7.    Energy loss spectra of polystyrene (PS), poly (2-vinylpyridine) (PVP), and polyethylene (PE), at* q ∼ 0 *before and after significant radiation damage: (a) dose* < 4 C/m₂²; *(b) dose 175 C/m² for PS, 88 C/m² for PE, and 51 C/m² for PMMA.*

excitations would be expected consistent with the experimental data. The very slight negative dispersion of the strong 7 eV energy loss peak was attributed to local field effects (13). The arrow indicates a feature which is much stronger at high q than at small q thus identifying it as an optically forbidden transition (13).

Hence, the momentum dependence of excitations as measured in electron energy loss spectroscopy can provide a test of the local or extended nature of electronic states in polymers. In polyace-tylene, electronic states and their excitations are clearly of an extended nature although the onset of absorption near the bandgap appears to be indirect or excitonic (6). In polyethylene while the ground states may be extended, strong excitonic effects localize excitations at the fundamental absorption edge. In pendant group polymers, both ground and excited states are local-ized (7).

All of the above spectra were taken with a total radiation dose much less than that required to produce extensive radiation damage. The effects of continued exposure to the electron beam are shown in Figure 7 (14). At the doses indicated for the bottom panel of Figure 7, the spectrum of electron-induced chromophores is similar to that produced by UV photolysis or gamma radiation (14). Specifically, electron radiation-induced peaks at 4.2 and 5.1 eV in polymethylmethacrylate as well as the 4.7 eV peak in polystyrene are also products of UV photolysis which have been assigned to unsaturated bond formation in the polymer backbone (14). In polyethylene, the electron beam cleaves C-H bonds leading to unsaturated bond formation. The strong absorption peak at 6.5 eV is clearly associated with a single isolated unsaturated bond while absorption at 5.3, 4.3, 4.0 and 3.6 eV can be associated with the formation of higher polyenyl groups.

## Literature Cited

1.  Ritsko, J.J., J. Chem. Phys., 1979, 70, 5343.

2.  Ritsko, J.J., Brillson, L.J., Bigelow, R.W. and Fabish, T.J., J. Chem. Phys., 1978, 69, 3931.

3.  Ritsko, J.J., Japan Jour. of Appl. Phys., 1978, 17-S2, 231.

4.  Ritsko, J.J. and Bigelow, R.W., J. Chem. Phys., 1978, 69, 4162.

5.  Walker, I.C., Chem. Soc. Rev., 1974, 3, 467.

6.  Ritsko, J.J., Mele, E.J., Heeger, A.J., MacDiarmid, A.G. and Ozaki, M., Phys. Rev. Lett., 1980, 44, 1351.

7.  Duke, C.B., Salaneck, W.R., Fabish, T.J., Ritsko, J.J., Thomas, H.R. and Paton, A., Phys. Rev. B, 1978, 16, 5717.

8.    McCubbin, W.L., in "Electronic Structure of Polymers and Molecular Crystals", ed. by J.M. Andre and J. Ladik (Plenum, NY 1975), p. 171.

9.    Delhalle, J., Andre, J.M., Delhalle, S., Pireaux, J.J., Caudano, R. and Verbist, J.J., J. Chem. Phys., 1974, 60, 595.

10.   Grant, P.M. and Batra, I.P., Sol. St. Comm., 1979, 29, 225.

11.   Fincher, C.R., Peebles, D.L., Heeger, A.J., Druy, M.A., Matsumura, Y., MacDiarmid, A.G., Shirakawa, H., Ikeda, S., Sol. St. Comm., 1978, 27, 489.

12.   Fincher, C.R., Ozaki, M., Tanaka, M., Peebles, D., Lauchlan, L., Heeger, A.J., MacDiarmid, A.G., Phys. Rev. B., 1979, 20, 1589.

13.   Ritsko, J.J., J. Chem. Phys., 1979, 70, 4656.

14.   Ritsko, J.J., Ann. Rep. Conf. on Elect. Ins. and Dielectric Phen., Nat. Acad. Sciences, Wash., D.C. 1978, p.109.

RECEIVED December 22, 1980.

# Exciton States and Exciton Transport in Molecular Crystals

R. SILBEY

Department of Chemistry, Massachusetts Institute of Technology, Cambridge, MA 02139

R. MUNN

Department of Chemistry, UMIST, Manchester, M60 1QD, United Kingdom

Electronic transport in molecular crystals is a complicated phenomenon for two reasons. The first is the complexity of the vibrations. In molecular crystals the phonons include molecular modes and translational and vibrational lattice modes, each with its characteristic frequency, bandwidth, and mechanism of electron-phonon coupling, again with different strengths. The second reason why transport is complicated is the absence of any clear ordering of the different parameters. The electronic bandwidths may range from being larger than most phonon frequencies and bandwidths (for charge carriers) to being smaller than either (for triplet excitons). The electron-phonon coupling energy may be large or small compared with electronic and vibrational energies, and if large may cause the electronic bandwidth to narrow rapidly with increasing temperature, so changing the parameter ordering.

Early transport theories were restricted in scope and did not reflect these complications. As well as treating only a single phonon band, for simplicity, the theories would assume a particular parameter ordering and a transport mechanism (1-11). Transport has been described by the mean-square particle displacement as a function of time, so permitting study of the clothing of the particle by phonons and the development of diffusive motion. These approaches also reveal the change from hopping to band motion as the temperature is lowered in systems with strong electron-phonon coupling.

However, the available theories have still been restricted to selected parameter orderings. In particular, it has been assumed in theories of exciton transport that the exciton bandwidth is narrower than the phonon bandwidth, and this assumption has been carried over to theories of carrier transport. In fact, carrier bandwidths may well be much larger than phonon bandwidths at low temperatures, becoming smaller than phonon bandwidths as the temperature is raised, owing to polaron band narrowing

0097-6156/81/0162-0045$05.00/0

effects. There is therefore a need for a theory which can treat both wide and narrow electronic bands in molecular crystals with either strong or weak electron-phonon coupling. We have already discussed how such a theory may be developed, (12) and here we describe the theory in detail.

## Principles

We start from the following model Hamiltonian, in which we take h=1:

$$H = \sum_n \epsilon a_n^+ a_n + \sum_{n,m} J_{nm} a_n^+ a_n$$

$$+ \sum_q \omega_q (b_q^+ b_q + 1/2) + N^{-1/2} \sum_{nq} g_q^n \omega_q (b_q^+ + b_q) a_n^+ a_n \quad (2.1)$$

Here the operators $a_n^+$ and $a_n$ create and destroy an electronic excitation (exciton or charge carrier) of energy $\omega$ at site n, while the operators $b_q$ and $b_q^+$ create and destroy a phonon of frequency $\omega_q$ and wavevector $q$. The quantity $J_{nm} = J_{mn}$ is the transfer integral between sites n and m. The last term is the electron-phonon coupling term, of magnitude determined by the dimensionales parameters $g_q^n = g_q e^{iq \cdot R_n}$, where $R_n$ is the position vector of site n. The coupling is local (diagonal) in excitation site. Such a coupling can arise from molecular distortion in the excited or ionized state, in which case $g_q$ and $\omega_q$ are expected to be almost independent of $q$, or from vibrational fluctuations in the exciton site shift or charge carrier polarization energy, in which case $g_q$ and $\omega_q$ may vary more markedly with q.

If the electron phonon coupling is weak ($|g_q| \ll 1$ for all $q$), the Hamiltonian may be used directly to study transport, as in section III. If the coupling is strong, the Hamiltonian is transformed to yield a weaker coupling as in section IV. In either case, the Hamiltonian is written

$$H = H_{ex} + H_{ph} + N^{-1/2} \sum_{k,q} V_{kq} a_k^+ a_q \quad (2.2)$$

where $H_{ex}$ and $H_{ph}$ are operators diagonal in excitation wavevector k and phonon wavevector $q$, and $V_{kq}$ is a function of phonon operators.

Transport is studied in terms of the mean-square displacement of an excitation created at the origin at time zero. At long times, the rate of change of this quantity gives the diffusion coefficient; we shall not consider the transport before it becomes diffusive. The mean-square displacement is obtained from the excitation density matrix, which is the full density matrix, of the coupled system integrated over all phonon states. Exact formal expressions for this quantity can be obtained, but more tractable expressions follow if only terms up to second order in the $V_{kq}$ are retained. After further neglect of small quantities, the diffusion coefficient can be expressed as (12).

$$D = \langle\langle v_k^2 \Gamma_{kk}^{-1} + \gamma_{kk} \rangle\rangle \qquad (2.3)$$

where the double angle brackets denote a thermal average over excitation states of energy $E_k$ and $v_k$ is the velocity $\nabla_k E_k$. The $\Gamma_{kk}$ are the rates of scattering out of state $k$, so that the first term in Eq. (2.3) has the usual form for band transport, and the $\gamma_{kk}$ have the form of hopping rates.

Both $\Gamma_{kk}$ and $\gamma_{kk}$ are obtained from quantities

$$W_{kk';qs} = \int_0^\infty d\tau \left\{ \langle V_{sk'} V_{kq}(\tau) \rangle e^{i(E_s - E_{k'})\tau} + \langle V_{sk'}(\tau) V_{kq} \rangle e^{-i(E_k - E_q)\tau} \right\} \qquad (2.4)$$

The single angle brackets denote an average over phonon states, $V_{kq}(\tau) = e^{iH_{ph}\tau} V_{kq} e^{-H_{ph}\tau}$, and $s = k' - k + q$. The scattering rates are given by

$$\Gamma_{kk} = N^{-1} \sum_q W_{qq;kk} \qquad (2.5)$$

and the hopping rates are given by

$$\gamma_{kk} = \left( \frac{1}{2} \frac{d^2}{dk^2} - \frac{d^2}{dK^2} \right) \mathrm{Re}\, N^{-1} \sum_q W_{q,q+K;k,k+K} \Big|_{K=0} \qquad (2.6)$$

(or various equivalent forms (6, 12). Derivation of the diffusion coefficient thus requires evaluation of the correlation functions $\langle V_{sk'} V_{kq}(\tau) \rangle$, of the integrals in Eq. (2.4), of the sums in Eqs. (2.5) and (2.6), and of the thermal average in Eq. (2.3).

Untransformed Coupling

General Results. If the electron-phonon coupling is sufficiently weak, the last term in Eq. (2.1) can be treated as a perturbation. Then in Eq. (2.2) the operator $V_{kq}$ is given by

$$V_{kq} = g_{k-q}\, k-q\, (b_{k-q} + b_{-k+q}^+). \qquad (3.1)$$

The required correlation functions are of the form

$$\langle V_{k+K,q+K} V_{qk}(\tau) \rangle = |g_{k-q}|^2 \omega_{k-q}^2 \times$$

$$[(n_{k-q}+1)e^{-i\omega_{k-q}\tau} + n_{k-q} e^{i\omega_{k-q}\tau}] \qquad (3.2)$$

where we have used the results $g_{-q} = g_q^*$ and $\omega_{-q} = \omega_q$. In Eq. (3.2)

$n_{k-q}$ is the thermal equilibrium number of phonons in mode k-q, given by

$$n_{k-q} = n(\omega_{n-q}) = (e^{\beta\omega k-q}-1)^{-1} \qquad (3.3)$$

where $\beta = 1/k_B T$ the Boltzman constant.

Optical Phonons. For a narrow optical phonon band we can set $g_q=g$, $\omega_q=\omega$, and $n_q=n$ for all q. The scattering rates are then

$$\Gamma_{kk}^{op} = 2\pi g^2 \omega^2 [(n+1)N_{ex}(E_k-\omega)+nN_{ex}(E_k+\omega)], \qquad (3.4)$$

where $N_{ex}(E)$ is the excitation density of states. In general the resulting diffusion coefficient is difficult to obtain except by numerical means. However for wide parabolic excitation bands such that $B>>\omega$, we can set

$$\Gamma_{kk}^{op} \approx 2\pi g^2 \omega^2 (2n+1)N_{ex}(E_k) \qquad (3.5)$$

where m* is the effective mass. Conventional procedures (13) then lead to

$$D/a^2 = [2\pi k_B T/(m*)^5]^{\frac{1}{2}}/g^2\omega^2(2n+1) \qquad (3.7)$$

where $a$ is the intermolecular distance. Since $m*\sim 1/B$, then $D\sim B^{5/2}/\omega^2$. The mobility $\mu=e\beta D$ varies as $T^{-1/2}$ at low temperatures such that $k_B T<<\omega$, changing eventually to $T^{-3/2}$ when $k_B T>>\omega$.

Acoustic Phonons. We assume that acoustic phonons can be adequately described by a Debye spectrum cut-off frequency $\omega_D$, and the electron-phonon coupling is given by the deformation potential approximation

$$g_q = (A/\omega_q)^{\frac{1}{2}}. \qquad (3.8)$$

The scattering rate is then

$$\Gamma_{kk}^{ac} = 2\pi A N^{-1} \sum_q \omega_q \{(n_q+1)\delta(E_{k-q}-E_k+\omega_q)+$$

$$n_q \delta(E_{k-q}-E_k-\omega_q)\} \qquad (3.9)$$

In some limits, results can be obtained more directly. In the conventional semiconductor limit $B>>k_B T>>\omega_D$, we have $n_q\approx k_B T/\omega_q>>1$, and $|E_{k-q}-E_k|>>\omega_q$ except for a few sets of wavevectors. These results yield

$$\Gamma^{ac}_{kk} = 4\pi A k_B T N_{ex}(E_k) \tag{3.10}$$

$$D/a^2 = [\pi/k_B T(m^*)^5]^{\frac{1}{2}}/A \tag{3.11}$$

so that $D \sim B^{5/2}/A \sim B^{3/2}$ if we assume A is proportional to B. Eq. (3.11) gives the standard $T^{-3/2}$ temperature dependence for the mobility (13).

In molecular crystals one may also require the limit $k_B T \gg \omega_D \gg B$. Then in Eq. (3.9) nonzero contributions arise only for frequencies $\omega_q \leq B$. Taking on average $|E_{k-q} - E_k| \sim \frac{1}{2}B$ yields for all k,

$$\Gamma^{ac} \sim 4\pi A k_B T N_{ph}(\frac{1}{2}B), \tag{3.12}$$

where for a Debye spectrum $N_{ph}(\omega) = 3\omega^2/\omega_D^3$. Taking similarly $\langle v_k^2 \rangle \sim (\frac{1}{2}Ba)^2$, we obtain

$$D/a^2 = \omega_D^3/12\pi A k_B T, \tag{3.13}$$

so that $D \sim \omega_D^3/A \sim \omega_D^3/B$ (with the previous assumption $A \propto B$) and the mobility varies as $T^{-2}$. The narrow excitation band in this limit greatly reduces the number of allowed one-phonon scattering processes, so that the diffusion coefficient may be large, as the ratio $\omega_D/B$ indicates.

Finally, in the limit $k_B T \gg B \gg \omega_D$, the scattering rates are given by Eq. (3.10) but the thermal averages in D have to be taken over a narrow excitation band. We take $v_k^2 \sim (\frac{1}{2}Ba)^2$ and $N_{ex}(E_k) \sim 1/B$, obtaining

$$D/a^2 = B^3/(16\pi A k_B T) \tag{3.14}$$

so that the mobility varies as $T^{-2}$. This result accords with the narrow excitation band treatments of Glarum (14) and Friedman (15).

## Transformed Coupling

Transformation. The transformation of the Hamiltonian (2.1) which yields a weak residual excitation-phonon coupling even when the $g_q$ are large has been discussed several times (4, 7, 16, 17). It produces a uniform shift in the excitation energy levels and a displacement in the equilibrium position of the phonons corresponding to the formation of a polaron. Since the transfer interactions $J_{nm}$ compete with this tendency to form a localized state, the optimum transformation should be determined variation-ally (6, 17). However, for present purposes we use the full clothing transformation which is exact for J=0 and yields the correct untransformed results for large J and weak coupling. The results are qualitatively similar to those which would be obtained with the full variational transformation, but are simplified by the absence of the temperature-dependent variational parameters.

After the transformation, the excitation part of the Hamiltonian (2.2) is

$$H_{ex} = \sum_k (\varepsilon - N^{-1} \sum_q \omega_q |g_q|^2 + \sum_n \tilde{J}_h e^{i\mathbf{k}\cdot\mathbf{R}_n}) \; a_k^+ a_k \qquad (4.1)$$

where $\mathbf{R}_n$ is a lattice vector and

$$\tilde{J}_h = J_{n+h,n} <\theta_{n+h}^+ \theta_n> \qquad (4.2)$$

$$\theta_n = \exp[N^{1/2} \sum_q (g_q^n)^* (b_q^+ - b_{-q})] \qquad (4.3)$$

$$<\theta_{n+h}^+ \theta_n> = \exp[-N^{-1} \sum_q (2n_q+1) |g_q|^2 (1-\cos q\cdot\mathbf{R}_h)] \qquad (4.4)$$

The phonon part is

$$H_{ph} = \sum_q \omega_q (b_q^+ b_q + 1/2) \qquad (4.5)$$

The residual coupling is described by the operators

$$V_{kq} = N^{-1/2} \sum_{n,h} J_h e^{i\mathbf{k}\cdot\mathbf{R}_n} e^{i(\mathbf{k}-\mathbf{q})\cdot\mathbf{R}_n} <[\theta_{n+h}^+ \theta_n - <\theta_{n+h}^+ \theta_n>]> \qquad (4.6)$$

The excitation part and the coupling are temperature dependent through the thermal averages in Eqs. (4.2) and (4.6); this partition of the Hamiltonian ensures the correct thermal equilibrium behavior.

### Diffusion Constant

When the diffusion constant is evaluated we find

$$D/a^2 = \frac{1}{\pi^{\frac{1}{2}}[I_0(y)-1]} \frac{(\tilde{B}^2 + \Gamma^2)}{(\tilde{B}^2 + 2\Gamma^2)^{\frac{1}{2}}}$$

$$+ \frac{1}{4}\pi^{\frac{1}{2}}[I_0(y)-1] \frac{\tilde{B}^2}{(2\tilde{B}^2 + \Gamma^2)^{\frac{1}{2}}} e^{-\frac{1}{4}\beta^2 \tilde{B}^2/(2\tilde{B}^2+\Gamma^2)} \qquad (4.6)$$

Although the derivation of this expression has used a number of simplifying assumptions, there are none about the relative sizes of B and $\Gamma$.

The first term in Eq. (4.6) represents the band contribution, which falls with increasing temperature owing to the increase in $I_0(y)$. The corresponding decrease in B via Eqs. (4.2) and (4.4) contributes to the fall in this contribution unless $\Gamma \gg B$. The second term in Eq. (4.6) represents the hopping contribution, which increases with increasing temperature through $I_0(y)$ and the exponential factor. However, once B becomes much less than $\Gamma$, the exponential factor becomes constant, and the decrease in $B^2$ outweighs the increase in $I_0(y)$, so that eventually the second term also decreases with increasing temperature. The occurrence of this decrease is discussed in more detail below.

In the limit $\Gamma \gg \tilde{B}$ assumed in most previous work, we find

$$D/a^2 = \frac{\Gamma}{(2\pi)^{\frac{1}{2}}[I_0(y)-1]} + \frac{\pi^{\frac{1}{2}}[I_0(y)-1]\tilde{B}^2}{4\Gamma} \qquad (4.7)$$

This is essentially of the form derived earlier (4). We note that in Eq. (4.7) $\Gamma$, or equivalently $\Delta$, cannot tend to zero (which would make D infinite), because the equation is valid only for $\Gamma \gg \tilde{B}$. On the other hand, $\tilde{B}$ can tend to zero, in which case only the band term remains. As the velocities $v_k^2$ in Eq. (2.3) tend to zero, so do the scattering rates $\Gamma_{kk}$; both factors vary as $\tilde{B}^2$, so that their ratio tends to a constant.

At sufficiently high temperatures, y becomes large, so that $I_0(y)-1$ can be replaced by its asymptotic value $e^y/(2\pi y)^{\frac{1}{2}}$, while $\Gamma \cong \Delta y^2$. The hopping term in Eq. (4.7) then dominates, and can be written as

$$D/a^2 = \left(\frac{B^2}{8\Delta g^2 2^{\frac{1}{2}}}\right) \sinh \frac{\beta\omega}{2} \exp(-2g^2 \tanh \frac{\beta\omega}{4}) \qquad (4.8)$$

In the opposite limit $B \gg \Gamma$, we obtain the new result

$$D/a^2 = \frac{\tilde{B}}{\pi^{\frac{1}{2}}[I_0(y)-1]} + \frac{(\pi/2)^{\frac{1}{2}}}{4} [I_0(y)-1] \, Be^{-\beta^2 \tilde{B}^2/8}$$

Here $\Gamma$ does not occur and so can tend to zero, being already assumed much smaller than B. If then B tends to zero, the diffusion coefficient vanishes, as expected. Vibrational dispersion (nonzero $\Delta$) is therefore not essential to obtain diffusive motion and finite transport coefficients in this limit, but to show this has required a sufficiently general theory.

## Discussion

In this paper we have shown how electronic transport in molecular crystals can be treated much more generally than hitherto. The present treatment avoids (i) the assumption that vibrational relaxation is fast compared with excitation transfer

($\Gamma \gg \tilde{B}$, as usually assumed in polaron and exciton transport theories), and (ii) the assumption that the excitation-phonon coupling is weak ($g\omega n \ll B$, as usually assumed in conventional semiconductor transport theories). In particular, such a generalization is necessary to treat charge-carrier transport in aromatic hydrocarbon crystals. In these systems, the carrier bandwidths are of the order of 1000 cm$^{-1}$ compared with phonon bandwidths of no more than a few tens of cm$^{-1}$, but the electron-phonon coupling need not be weak. This feature has also been recognized in recent specialized theories of carrier transport in anthracene and naphthalene, (21) and different regimes of exciton-phonon coupling have been discussed.

However, most of the detailed results in this paper have been derived under simplifying assumptions to permit easier algebraic and numerical analysis. Only local linear excitation-phonon coupling has been treated, and anisotropy has been ignored. The fully clothed transformation given by Eqs. 4.1 - 4.3 has been used rather than the proper variational transformation (6, 17) which requires separate numerical evaluation (although this could readily be incorporated once evaluated). The excitation and phonon densities of states in section IV have been taken as Gaussian, and the parameters have been chosen to satisfy $\beta\Delta \ll 1$ and $B/\omega < 1$.

With these restrictions on the validity of the final results in section IV, it is not appropriate to attempt a quantitative fit to the behavior of any specific substance, especially for charge-carriers. In any case, information on the parameter values is rather sparse. Adequate phonon dispersion curves and densities of states are fairly readily available or obtainable by calculations, but there are rather few exciton and charge-carrier band structures available, and these are mostly several years old and hence do not incorporate recent advances in quantum chemistry. The different types of electron-phonon coupling have been identified, but there are few attempts to deduce accurate coupling parameters. Exceptions are the use of semi-empirical calculations to obtain couplings to molecular vibrations (23) and the analysis of spectra to obtain exciton-phonon couplings (24, 25). However, even though a quantitative fit to experiment is inappropriate, we can indicate how the qualitative features of our results may relate to experimental data. We consider only charge-carrier mobilities, which have been measured as a function of temperature for enough substances to begin to reveal some kind of pattern.

Schein (26) has emphasized the tendency for mobilities in molecular crystals to lie in the range 0.1-10 cm$^2$ V$^{-1}$ s$^{-1}$ and to vary with temperature at $T^{-n}$ with typically $0 < N < 2$. Most of the crystals studied are aromatics and heteroaromatics for which relatively wide carrier bands are expected. Calculations show that although generally lower mobilities are obtained as the electron-phonon coupling strength g increases, this trend is least marked for the widest bands with $B/\omega=1$.  Mobilities in the required range are obtained at $1/\beta\omega=2$.  (room temperature if $\omega=100$ $cm^{-1}$) for $g^2=0.1-1$.  Also, the diffusion coefficient for $B/\omega=1$ depends only weakly on temperature for much of the range investigated, except for weak and very strong coupling.  Since the mobility contains the extra factor $T^{-1}$, these dependences imply a variation $T^{-n}$ with n close to 1, noticeably different from the standard narrow-band result $T^{-2}$ but in qualitative agreement with the observed trend.

It has been remarked (27) that the smallest mobilities tend to show an activated behavior, intermediate ones the weak temperature dependence just discussed, and larger ones a more rapid decrease like $T^{-n}$ with n=2.5-3.  As noted in section IV, activated behavior requires very strong electron-phonon coupling, but it is not obvious why this might be present in the very low mobility crystals studied, electrons in orthorhombic sulphur (28) and β-nitrogen (29) and holes in γ-oxygen (29).  At the opposite extreme, the hole and electron mobilities in the ab plane of durene vary as $T^{-2.5}$, attaining values as high as 50 $cm^2$ $V^{-1}$ $s^{-1}$ below 150K and falling to 5 and 8 $cm^2 V^{-1} s^{-1}$, respectively, at room temperature (30).  Durene has the same crystal structure as anthracene, but the peripheral methyl groups might be expected to reduce the carrier bandwidths in the ab plane.  However, in the present treatment a smaller bandwidth does not necessarily imply a smaller mobility if the electron-phonon coupling is weaker too.  Weaker coupling also implies a wider region in which band motion dominates with its rather strong temperature dependence.  Thus the behavior of durene could be explained in terms of particularly weak coupling, but any microscopic reason for such weak coupling remains to be found.

The present model is therefore able to interpret the observed rough correlation between the magnitude of carrier mobilities and their temperature dependence consistently as a dependence of both quantities on the electron-phonon coupling strength.  This interpretation appears physically reasonable, but the omission of nonlocal coupling from the model should be borne in mind.  Such coupling is central to theories (21, 31) of the almost temperature-independent electron mobilities in the c' direction of anthracene and naphthalene and their deuterated forms and of the rapid increase in the mobility for naphthalene and deuteronaphthalene at low temperatures (32, 33).  The present methods can be extended to treat nonlocal coupling and strong anisotropy, which appears to play a major role in anthracene and naphthalene (21).

The problem with the theory of electronic transport in molecular crystals has been to deduce the transport, given a model Hamiltonian containing what one considers to be the essential physical interactions.  Since several interactions may be comparable in size, simple perturbative methods fail.  The method (12) adopted here yields a rather direct solution to the problem.

Given the clothing transformation, one has then to evaluate the correlation functions, which can be regarded as the central factors in the transport theory.

## Abstract

A theory of exciton-phonon coupling is presented and the consequences of this coupling for spectral line shapes and exciton transport are discussed. The theory is valid for arbitrary phonon and exciton bandwidths and for arbitrary exciton phonon coupling strengths. The dependence of the diffusion constant on temperature and the other parameters is analyzed.

## Literature Cited

1.   Haken, H. and Strobl, G. in "The Triplet State," edited by A.B. Zahlan (Cambridge University Press, 1968); Z. Physik, 1973, 262, 185.

2.   Haken, H. and Reineker, P., Z. Physik, 1972, 249, 253; Reineker, P. and Haken, H., Z. Physik, 1972, 250, 300; Reineker, P. and Kuhne, R., Z. Physik, 1975, B22, 193, 201; Phys. Letters, 1978, 69A, 133; Kuhne, R. and Reineker, P., Phys. Stat. Solid., 1978, b89, 131.

3.   Kenkre, V.M. and Knox, R.S., Phys. Rev., 1974, B9, 5279; Phys. Rev Letters, 1974, 33, 803; J. Luminescence, 1976, 12/13, 187; Kenkre, V.M., Phys. Rev., 1975, B11, 1741; 1975, B12, 2150.

4.   Grover, M. and Silbey, R., J. Chem. Phys., 1971, 54, 4843.

5.   Abram, I. and Silbey, R., J. Chem. Phys., 1975, 63, 2317.

6.   Yarkony, D. and Silbey, R., J. Chem. Phys., 1977, 67, 5818.

7.   Blumen, A. and Silbey, R., J. Chem. Phys., 1978, 69, 3589.

8.   Munn, R.W., J. Chem. Phys., 1973, 58, 3230

9.   Munn, R.W., Chem. Phys., 1974, 6, 469.

10.  Munn, R.W. and Silbey, R., J. Chem. Phys., 1978, 68, 2439.

11.  Aslangul, C. and Kottis, Ph., Phys. Rev., 1974, B10, 4364, 1976, B13, 5544, 1978, B18, 4462.

12.  Munn, R.W. and Silbey, R., Mol. Cryst. Liq. Cryst., in press.

13.  Ziman, J.M., "Electrons and Phonons (Oxford University Press, 1960).

14. Glarum, S.H., J. Phys. Chem. Solids, 1963, 24, 1577. Our A corresponds to Clarum's $\gamma\hbar\omega_0$, apart from numerical factors.

15. Friedman, L., Phys. Rev., 1965, 140, A 1649. Our A corresponds to Friedman's $a^2(j')^2/Mv_s^2$.

16. Grover, M.K, and Silbey, R., J. Chem. Phys., 1970, 53, 2099.

17. Yarkony, D. and Silbey, R., J. Chem. Phys., 1976, 65, 1042.

18. Gosar, P., Phys. Rev., 1971, B3, 1991.

19. Holstein, T., Ann. Phys., 1959, 8, 343.

20. Munn, R.W. and Siebrand, W., J. Chem. Phys., 1970, 52, 47.

21. Sumi, H., Solid State Comm., 1978, 28, 309; J. Chem. Phys., 1979, 70, 3775.

22. Clarke, M.D., Craig, D.P. and Dissado, L.A., Mol. Cryst. Liq. Cryst., 1978, 44, 309.

23. Duke, C.B., Lipari, N.O. and Pietronero, L., Chem. Phys. Letters., 1975, 30, 415; Lipari, N.O., Duke, C.B. and Pietronero, L., J. Chem. Phys., 1976, 65, 1165.

24. Vilfan, I., Phys. Stat. Solidi., 1976, b78, k131.

25. Port, H., Rund, D., Small, G.J. and Yakhot, V., Chem. Phys., 1979, 39, 175.

26. Schein, L.B., Phys. Rev., 1977, B15, 1024.

27. Roberts, G.G., Apsley, N. and Munn, R.W., Physics Reports, to be published.

28. Gibbons, D.J. and Spear, W.E., J. Phys. Chem. Solids, 1966, 27, 1917; Ghosh, P.K. and Spear, W.E., J. Phys. C: Solid State Phys., 1968, 1, 1347.

29. Loveland, R.J., LeComber, P.G. and Spear, W.E., Phys. Rev., 1972, B6, 3121.

30. Burshtein, Z. and Williams, D.F., Phys. Rev., 1977, B15, 5769.

31. Vilfan, I., Lecture Notes in Physics, 1977, 65, 629.

32. Schein, L.B., Chem. Phys. Letters, 1977, 48, 571; Schein,
    L.B., Duke, C.B. and McGhie, A.R., Phys. Rev. Letters, 1978,
    40, 197; Schein, L.B. and McGhie, A.R., Phys. Rev., 1979,
    B19, 0000.

33. Karl, N., Sci. Papers Inst. Org. Phys. Chem. Tech. Univ.
    Wroclaw, 1978, 16, 43.

RECEIVED December 22, 1980.

# Percolation of Molecular Excitons

RAOUL KOPELMAN

Department of Chemistry, University of Michigan, Ann Arbor, MI 48109

## Band Theory, Localization and Percolation

Basic aspects of solid state theory face their critical tests in disordered organic solids. In the February 1980 issue of Physics Today the title of the article by Duke and Schein is: "Organic Solids: is energy band theory enough?". They show the difficulties facing not only the traditional electron-band theory but also the traditional electron hopping theory when compared with experiments. Energy (exciton, excitation) transport in organic solids may be equivalent to electron transport both in its importance and in its theoretical significance. Again the traditional band and hopping models are suspect. Two important concepts in modern theories of semiconductors (including amorphous ones) are Anderson localization (1) and percolation (2, 3) (both approaches seem to have originated at Bell Labs at about the same time, i.e. 1957). These two approaches have often been represented as irreconcilable, even though there is no basic contradiction between the two.

Anderson <u>delocalization</u> refers to the formation, at zero temperature, of a band (electron or exciton), resulting in metallic conduction. The hallmark of metallic (band) conduction is a <u>negative</u> temperature coefficient. Our experimental results show that this is not the case here (see below).

A <u>thermally assisted hopping</u> process is based on localized states. The localization could be due to Anderson localization or self-trapping (e.g. small polarons) or some other factor. This process gives a typical "non-metallic" conduction (or migration), exhibiting a positive temperature coefficient. The details of such a hopping process may be described by a percolation model. The latter is based on the clusters within which there is efficient hopping for a given time-scale. The point on the order parameter axis at which the average cluster size rises sharply is the critical <u>percolation point</u> and may define a <u>percolation transition</u> (2).

We give below a brief overview of percolation concepts (II)

and of the formalism of excitation percolation (III), which leads
to scaling and critical behavior. Experimental results are given
(IV) and shown to fully support the theoretical predictions
(III). A physical picture emerges (V), involving partial delocalization ("islands of coherence") over microdomains, while
efficient migration takes place over larger domains (which dominate the percolation transition behavior). An epilogue (VI) and
"Experimental" section (VII) conclude this paper.

## What is Exciton Percolation?

An example of macroscopic percolation is the flow of matter
through the random but fixed channels in a porous glass filter.
An example of microscopic percolation is the migration of a gas
through a solid (4). The nominal distinction between diffusion
and percolation is that the latter implies migration through a
microscopically heterogeneous medium. We are interested here in
the migration of excitons, e.g. localized Frenkel excitons
(tightly bound electron-hole pairs).
Percolation has also been viewed as an agglomeration of
small clusters into large aggregates ("infinite clusters") (2).
Interestingly enough this idea has been applied to spin clusters,
i.e. magnetic phase transitions, long before it has been applied
to polymers (2). Gelation and vulcanization have been recently
treated with a site-bond percolation approach (5, 6), and a
related model has been applied to super-cooled water (7). These
are "static" percolation problems that are related to static
critical phenomena (8) rather than to dynamic (transport) critical phenomena. The latter deal with electrical conduction and
with energy transport - the topic of interest here.
The migration of Frenkel excitons is dynamic in nature but
can still be related to the presence of large clusters or aggregates of the appropriate molecules (often called guest, donor or
carrier molecules). The larger the average cluster, the higher
the exciton migration efficiency. Simultaneously it has to be
realized that a given time scale defines an average migration
range which conceptually is closely related to an average cluster
size. The resulting migration clusters ("dynamic clusters") can
be treated mathematically like the conventional ("static") clusters, which do not depend on time. The exciton migration probes
are "internal" to the dynamic cluster and thus there is no
requirement that the excitons reach the surface of the crystal or
crystallite. This is somewhat analogous to the AC conductivity of
composite materials (e.g. silver loaded teflon). The emphasis is
thus on a 3-dimensional cluster (a "percolation volume" rather
than a percolation path).

## Excitation Percolation

Here we limit ourselves to excitation percolation in very

simple disordered molecular materials. These bear some resemblance to natural photosynthetic antenna systems and may have some bearing on future photovoltaic materials (9).

A typical "energy semiconductor" contains three kinds of molecules or molecular units: Energy donors, energy traps and "inert" host. The host's electronic energy levels are too high to be excited. The energy levels of the donor are significantly lower and thus are excitable. The energy traps have even lower energy levels, thus enabling efficient quenching of the donor excitations. The systems of interest exhibit multiple donor-donor energy transfer eventually resulting in the trapping of the excitation. The energy trapping may initiate some "photochemistry" (actually, "excitochemistry"), as in photosynthesis, or, on the contrary, may inhibit photochemistry by offering a more effective route for radiative or radiationless decay of the excitation.

One of the most interesting aspects of energy transport is the excitation percolation transition (2, 9) and its similarity (10) to magnetic phase transitions and other critical phenomena (2, 8). In its simplest form the problem is one of connectivity. In a binary system, made only of hosts and donors, the question is: can the excitation travel from one side of the material to the other? The implicit assumption is that there are excitation-transfer-bonds only between two donors that are "close enough", where "close enough" has a practical aspect (e.g. defined by the excitation transfer probability or time). Obviously, if there is a succession of excitation-bonds from one edge of the material to the other, one has "percolation", i.e. a connected chain of donors forming an excitation conduit. We note that the excitation-bonds seldom correspond to real chemical bonds; rather more often they correspond to van-der-Walls type bonds and most often they correspond to a dipole-dipole or equivalent quantum-mechanical interaction.

For a system where the donors are distributed at random it can be shown mathematically that there exists a critical concentration $C_c$ (also called "percolation concentration") below which the percolation (edge-to-edge connectivity) has a probability of zero and above which the percolation probability ($P$) rises sharply with donor concentration ($C$). A mathematical relation (2), for a substitutionally disordered binary lattice, is:

$$P_\infty \propto |C-C_c|^\beta \qquad\qquad C > C_c \qquad\qquad (1)$$

where the exponent $\beta$ is about 0.13 for a 2-dimensional topology and 0.35 for a 3-dimensional one (2, 10). We note that $0^{0.13} = 0$ (for $C=C_c$), $0.001^{0.13} = 0.41$, $0.01^{0.13} = 0.55$ and $0.1^{0.13} = 0.74$.

For ternary systems (host, donor, trap), which are usually of interest, the question is no longer whether the excitation will make it from one edge of the material to the other, but whether the excitation will make it to a near-by (or not so near-by) trap.

A relevant concept is that of a donor cluster, which is defined as a set of donors connected by bonds. The <u>average cluster size</u> ($I_{AV}$) will be infinite above the critical concentration but finite below it. This average cluster size is mathematically related to a critical exponent $\gamma$ (which is about 2.2 - 2.4 in 2-dimensions (<u>10</u>)):

$$I_{AV} \propto |c-c_c|^{-\gamma} \qquad\qquad c < c_c \qquad\qquad\qquad (2)$$

We note that $0.1^{-2.2} = 1.6 \times 10^2$, $0.01^{-2.2} = 2.5 \times 10^4$, $0.001^{-2.2} = 4.0 \times 10^6$ and $0^{-2.2} = \infty$ (for $c=c_c$). The general expression for the migration efficiency (trapping probability P) of an excitation is quite complex (<u>9</u>). However, there are some simple concentration limits, (<u>10</u>, <u>11</u>)

$$P = P_\infty \qquad\qquad\qquad c >> c_c \qquad\qquad\qquad (3)$$

and

$$P \propto I_{AV} , \qquad\qquad\qquad c << c_c \qquad\qquad\qquad (4)$$

together with the special relationship (<u>7</u>):

$$P = S^{1/\delta} \qquad\qquad\qquad c=c_c , \; S<<1 \qquad\qquad (5)$$

where $\delta = \gamma/\beta + 1$. Combining, respectively, eqs. 1 and 3 and eqs. 2 and 4, one gets, respectively:

$$P \propto |c-c_c|^\beta \qquad\qquad\qquad c >> c_c \qquad\qquad\qquad (6)$$

and

$$P \propto |c-c_c|^{-\gamma} \qquad\qquad\qquad c << c_c \qquad\qquad\qquad (7)$$

The last three equations account for the critical behavior (<u>2</u>, <u>8</u>, <u>10</u>) of the energy transport in these systems.

## Scaling Experiments

Extended experimental investigations were performed on the naphthalene triplet exciton, the first system to exhibit a critical transition (<u>9</u>). The energy transport was measured for sets of concentration ranges (C) of $C_{10}H_8/C_{10}D_8$, where each set is distinguished by a given supertrap (acceptor), supertrap concentration S (relative to $C_{10}H_8$) and temperature (Fig. 1). The six curves of Fig. 1 are replotted against the reduced concentration $C/C_c$ (Fig. 2), resulting in a striking universal behavior. Replotting again these data in conventional logarithmic fashion (Fig. 3) yields the critical exponents both above and below the critical concentration. This results in the experimental expon-

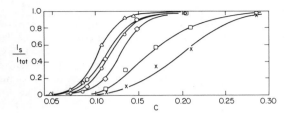

Figure 1.   Donor concentration dependence (16) of the energy transport measure $I_s I_{tot} = I_s/(I_s + I_d)$, where $I_s$ is the acceptor ("supertrap") phosphorescence (O—O) and $I_d$ is that of $C_{10}H_8$, for Series A (x-trap, $S \approx 10^{-4}$; ($\Diamond$) 1.7 K, ($\bigcirc$) 4.2 K); Series B (BMN, $S = 10^{-3}$; ($\bigcirc$) 1.7 K, ($\triangle$) 4.2 K), and Series C (BMN, $S = 10^{-4}$; ($\times$) 1.7 K, ($\square$) 4.2 K). The lines are visual guides. BMN is betamethyl-naphthalene.

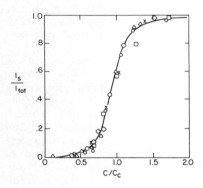

Figure 2.   Universal energy transport curve. The data points and designations are the same as in Figure 1. For each family of data points $C_c$ was derived from Figure 1 via Equation 7, using $P = I_s/I_{tot}$.

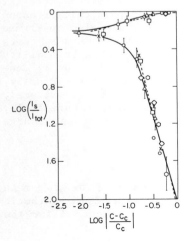

Figure 3.   "Scaled" energy transport curve. The data points and designations are the same as in Figures 1 and 2. Error bars were added to a few points to indicate experimental uncertainties. The dashed lines are least-squares fits to the experimental data, giving $\gamma = 2.1 \pm 0.2$ and $\beta = 0.13 \pm 0.05$.

ents:  $\beta$ = 0.13 ± 0.05 and $\gamma$ = 2.1 ± 0.2, in excellent agreement
with the mathematical exponents (0.13 and 2.2) for 2-dimensional
lattices (10).    An  apparently  unexpected  result  is  the  wide
scaling region, i.e. the validity of the critical exponent approach over a wide range of concentrations, even "far away" from
the critical point ($C_c$).  However, in contrast to other critical
phenomena, the percolation-transition (2) exhibits scaling relationships over a wide range of the order parameter (concentration) with non-classical exponents (2), as has been argued theoretically  (12)  and  demonstrated  by  computer  simulations  (13).
This means that eqs.  1 and 2 should be valid over a large $C$-
region, and likewise eqs. 6 and 7.   The latter is in excellent
agreement with experiment (Fig. 3).

## Physical Interpretation

Migration  clusters  may  be  defined  by  intermediate-range
interactions (11).    These give rise to intermediate-range bonds
and clusters ("secondary clusters").   Excitation delocalization
will usually happen within "primary clusters", defined by near
neighbor interactions.
The  exciton  delocalization  over  primary  clusters  has  been
demonstrated  by  investigations  of  cluster  spectra  (11).    Thus
within  the  intermediate-range  ("secondary")  clusters  there  are
excitation jumps ("hops") among the delocalized primary cluster
states.   Among the intermediate-range clusters occasional "leaks"
may  occur  via  longer-range  interactions.    The  latter  define
"tertiary"  clusters.   We point out that the scaling results are
consistent with a leaky cluster model (11) where the excitation is
not absolutely confined within a secondary cluster, but is confined within a "tertiary cluster" (a set of weakly communicating
secondary clusters).   It has actually been shown (11, 14) that (1)
raising the temperature, (2) lengthening the migration time, or
(3)  increasing the supertrap concentration, shifts the critical
point to a lower $C_{10}H_8$ concentration.   This is physically reasonable for a hopping model.   Points (1) and (3) are also confirmed
by the data in Fig. 1.   We also note that the variation of $C_c$ with
S and T (Fig. 1) is inconsistent with an Anderson–Mott mobility
edge (11) interpretation of our experimental results but is consistent with a hopping (dynamic) percolation mechanism.   Finally,
we note that the lifetime of the triplet excitation is extremely
long (3 sec for naphthalene (11)).    Thus we can, in principle,
claim an analogy between Frenkel exciton transport and charge
transport.

## Epilogue

We believe that excitation percolation will be a common
occurrence in many complex heterogeneous systems, both synthetic
and natural.   Excitation percolation has been proposed for the

primary photosynthetic step, consisting of excitation migration
on a quasilattice of chlorophyll-a (9, 15). Whether the percolation model will also be useful for the description of charge
transport in such systems remains an open question.

## Experimental

Crystal samples used for the triplet experiments were prepared from potassium fused, zone refined $C_{10}D_8$ (Merck, Sharp &
Dohme, 98% isotopic purity) and a "$C_{10}H_8$/BMN Standard." The
"standard" was made by mixing 5.0 mg zone refined BMN with 5.00 g
potassium-fused, zone-refined $C_{10}H_8$. This method allowed us to
hold the relative supertrap concentration S constant at say $10^{-3}$.
The relative supertrap concentration in this series of samples
did not vary by more than 30%. Single crystals were cut along the
ab (cleavage) plane, mounted in a sample holder "cage" and
immersed in liquid helium. Triplet decay measurements were made
by shuttering a 1600 W Hanovia xenon lamp (filtered by a
$NiSO_4/COSO_4$ solution filter) with a Uniblitz programmable shut-
ter. Spectrally resolved $C_{10}H_8$ 0-0 phosphorescence (at 21208 cm$^{-1}$) was measured with an ITT F-4013 phototube mounted in a Products
for Research housing (cooled to -25$^{\circ}$C). Data were collected with
a PAR model 1120 discriminator and model 1110 photon counter using
a DEC SCl-11 microcomputer. Specially written software permitted
collection of time resolved data in a multiscaling mode for
averaging a number of decay curves. The time resolution was
limited by the closing speed of our shutter system to about 4 ms.

## Acknowledgements

Work supported by NSF Grant DMR800679.

## Abstract

Exciton migration experiments on isotopic naphthalene crystals reveal a universal behavior with reduced donor concentration, irrespective of temperature, acceptor species or acceptor
concentration, giving a very wide scaling region with critical
exponents $\gamma = 2.1 \pm 0.2$ and $\beta = 0.13 \pm 0.05$. This is consistent
with a 2-dim. dynamic exciton percolation model, but not with an
Anderson-Mott mobility edge, down to 1.7K. The physical model
assumes a hierarchy of clusters: (1) Primary, short-range donor
clusters ("static-clusters") within which the excitation is delocalized. Among the primary clusters the excitation hops rather
quickly, provided that such primary clusters form a (2) Secondary, intermediate-range cluster ("dynamic cluster"), within which
the excitation migrates fast compared to the time-scale of the
experiment. Occasional leaks may occur among the secondary
clusters, resulting in (3) Tertiary, long-range clusters. The
hopping-percolation process is thermally assisted. Analogies

between long lived Frenkel exciton transport and charge transport are discussed.

## Literature Cited

1.  Anderson, P.W., _Rev. Mod. Phys._, 1978, 50, 191.

2.  Stauffer, D., _Phys. Rep._, 1979, 54, 1.

3.  Pollak, M. and Knotek, M.L., _J. Non-Cryst. Solids_, 1979, 32, 141.

4.  Brandt, W.W., _J. Chem. Phys._, 1975, 63, 5162.

5.  Hoshen, J., Klymko, P. and Kopelman, R., _J. Stat. Phys._, 1979, 21, 583.

6.  Coniglio, A., Stanley, H.E. and Klein, W., _Phys. Rev. Lett._, 1979, 42, 518.

7.  Stanley, H.E., _J. Phys. A._, 1979, 12, L329.

8.  Stanley, H.E., "Introduction to Phase Transitions and Critical Phenomena" (2nd. ed.), Oxford Univ. Press, London 1980.

9.  Kopelman, R., Topics in Applied Physics 15: "Radiationless Processes in Molecules and Condensed Phases", Ed. F.K. Fong, J. Springer, Berlin (1976), p. 297.

10. Hoshen, J., Kopelman, R. and Monberg, E.M., _J. Stat. Phys._, 1978, 19, 219.

11. Kopelman, R. and Francis, A.H., "Laser Spectroscopy of Ions and Molecules in Solids", Eds. W.M. Yen and P.M. Selzer, J. Springer, Berlin (in press).

12. Stauffer, D. and Coniglio, A., preprint (1980).

13. Kopelman, R., Monberg, E.M., Newhouse, J.S. and Ochs, F.W., _J. Luminescence_, 1979, 18/19, 41.

14. Ahlgren, D.C., Monberg, E.M. and Kopelman, R., _Chem. Phys. Lett._, 1979, 64, 122.

15. Swenberg, C.E., Geacintov, N.E. and Breton, J., _Photochem. Photobiol._, 1979, 28, 999.

16. Ahlgren, D.C., Ph.D. Thesis, The University of Michigan (1979).

RECEIVED April 15, 1981.

# Localization of Electronic States in Polymers and Molecular Solids

C. B. DUKE

Xerox Webster Research Center, Xerox Square-114, Rochester, NY 14644

The nature of the electronic states associated with injected charges and optical excitations in polymeric and molecular solids has been a subject of renewed interest in recent years both because of the increasing use of such materials in electronic devices and because a question has arisen concerning the appropriateness of traditional one-electron energy band theory for the description of these states (1, 2, 3). In particular, it has been suggested that under many circumstances charges injected into polymeric and molecular materials are more analogous to localized molecular ions in solution than to the extended "Bloch" states which are thought to be associated with electronic motion in metals and common covalent semiconductors (1, 2). Consequently, a suspicion has been growing that semiconducting organic solids should not be regarded as "organic semiconductors" in the traditional sense, in spite of an extensive early literature in which energy-band models were applied to interpret measurements of their optical and transport properties (4).

The purpose of this paper is the presentation of a brief overview of recent literature in which new models of electronic states in polymers and molecular solids have been proposed (1, 2, 3, 5-16). Since localized (e.g., molecular-ion) states seem prevalent in these materials, I indicate in Sec. II the physical phenomena which lead to localization. Sec. III is devoted to the description of a model which permits the quantitative analysis of the localized-extended character of electronic states and to the indication of the results of spectroscopic determinations of the parameters in this model for various classes of polymeric and molecular materials. I conclude with the mention in Sec. IV of an important practical application of these concepts and models: The contact charge exchange properties of insulating polymers (7, 17, 18, 19).

0097-6156/81/0162-0065$05.00/0

## Relaxation and Fluctuations

When a charge (or optically-excited exciton) is injected into a molecular solid, it induces changes in the electronic charge density and atomic positions both on the molecular site which it occupies (intramolecular relaxation) and on neighboring molecular sites (intermolecular relaxation). This phenomenon is called "relaxation" and leads to a lowering of the energy of the composite system of added charge plus molecular solid by an amount called the "relaxation energy," $E_r$. The relaxation energy is formally defined for an injected charge (e.g., molecular ion) as the difference between the ground state energy of the ion and the Hartree-Fock molecular orbital eigenvalue (in the canonical basis) which corresponds to the free-ion state. Both the intramolecular and intermolecular contributions to the relaxation energy can be large for molecular solids: I.e., $E_r$(intra) $\cong$ $E_r$(inter) $\cong$ 1-2eV ([1], [2], [5]). Therefore the energies of molecular anions (cations) in condensed molecular media are about 2-4eV lower (higher) than the corresponding free-molecule orbitals. Detailed models of the various contributions to the relaxation energy are given elsewhere ([2], [5], [8]).

An important feature of the intermolecular contributions to the relaxation energy is their dependence on the local atomic structure in the vicinity of the injected ion or exciton. For example, a surface ion exhibits only about 60% of the bulk relaxation energy because of the absence of polarizable neighboring species outside the condensed phase ([1], [6]). This variation in the value of the molecular-ion site energy at the surface relative to the bulk led to the prediction ([1]) and subsequent observation ([20]) of localized surface states for molecular films (specifically, anthracene). More generally, local variations in the composition and structure of polymers and molecular glasses cause spatial fluctuations in the site energies of molecular ions and excitons in these materials. For intrinsic molecular-ion states, the magnitude of these fluctuations may be inferred via analyses of the widths of valence-electron photoemission lines ([2], [6], [8]) or of contact charge exchange spectra ([2], [6], [18], [19]). Typically one finds $E_r \cong$ 0.5-1.0eV. The importance of these fluctuations is their creation of localized molecular-ion states in polymers and glasses ([1], [2], [3]). In particular, intrinsic localized molecular-ion states are thought to be involved in the contact charge exchange behavior of pendant group polymers ([7], [17], [18], [19]) and extrinsic localized molecular-ion states dominate the transport properties of molecularly doped polymers ([21], [22], [23]).

Another significant source of variations in the local site energies of molecular ions and excitons in condensed media is the modulation of these energies by the thermal vibrations either of the medium (e.g., acoustical phonons and librons) or of the molecular ion (exciton) itself (intramolecular vibrations). A model Hamiltonian which incorporates electronic interactions with

the various branches of the vibrational spectrum was proposed by
Rice et al., (10) and subsequently was extended and applied to
provide an interpretation of UV absorption and photoemission
spectra (8, 12, 24).  Three fundamentally different phenomena
occur:  the modulation of local molecular-ion energies by vibrations characteristic of the medium in which ion is embedded (8,
9), their modulation by modes internal to the ion itself (9, 10,
24), and the modulation of the site-to-site electron transfer
integrals by vibrations characteristic of the medium (3, 9, 10).
Each of these phenomena affects photoemission, UV absorption, and
transport in its own unique fashion so that a complete description
of any of these measurements must encompass all three.  A comparison of the nature and consequences of the three effects has been
given by Duke (3, 9).

## Localization

The local or extended nature of molecular-ion (or exciton)
states in molecular solids is determined by a competition between
fluctuations in the local site energies of these states (which
tend to localize them) and the hopping integrals for inter-site
excitation transfer (which tend to delocalize them).  In order to
define this fluctuation-induced localization concept more precisely, consider the model defined by the one-electron
Hamiltonian

$$H = \sum_{n,\alpha} \varepsilon_n(\alpha) a^+_{n,\alpha} a_{n,\alpha} + \sum_{\substack{n \neq n' \\ \alpha, \alpha}} V^{\alpha\alpha'}_{nn'} a^+_{n'\alpha'} a_{n\alpha}, \qquad (1)$$

in which n is a site index;  $\alpha$ designates the molecular orbital;
$V_{nn'}$ designate the intermolecular hopping integrals;  and $a_{n\alpha}$ is
the annihilation operator for an electron occupying the orbital $\alpha$
at the site n.
Both the molecular ion energies (i.e., site energies, $\varepsilon_n$)
and the hopping integrals, $V_{nn'}$, form distributions because of the
local variations in composition and structure.  While a proper
analysis of the resulting model is complicated (25, 26), the
qualitative features of interest to us can be defined in terms of
a mean site energy $\bar{\varepsilon}$, the rms deviation from this mean

$$\Delta \equiv \sqrt{<(\varepsilon_n - \bar{\varepsilon})^2>_{AV}} \qquad (2)$$

and a mean hopping integral, $\bar{V}$.  Variations in the site energies
from the mean (described by $\Delta$ ) are referred to as "diagonal
disorder" whereas analogous variations of the hopping integrals
are called "off-diagonal disorder."  Similarily, if these variations are caused by local time-independent fluctuations in composition or structure one speaks of "static disorder" while if

they are generated by thermal vibrations one employs the term
"dynamic disorder." Whereas static diagonal and off-diagonal
disorder can occur in polymers, only dynamic disorder should be
present in molecular crystals. A model properly embodying dynamic disorder is more complicated than Eq. (1) (3, 8, 9). The
important point for our present purpose, however, is the recognition that all sources of disorder must be considered in order to
determine the localized or extended character of the electronic
states associated with charges (or excitons) injected into polymers and molecular materials, and that these sources of disorder
are described by choosing suitable distributions of the $\varepsilon_n$ and
$V_{nn}$, parameters in Eq. (1) or an appropriate generalization
thereof.

In terms of the simplification of Eq. (1) defined by considering only nearest-neighbor hopping integrals (i.e., neglecting long-range hops), it can be shown (25, 26) that injected
charges are localized, i.e., they form molecular cations or
anions within the solid, if

$$\Delta > cz\bar{V} \tag{3}$$

in which z is the coordination number of the (presumably identical) molecular sites and c is a dimensionless number of the order
of unity which depends both on the connectivity (i.e., dimension-
ality) of the molecular system and on the extent of off-diagonal
disorder. Typically $c \cong 2.5$ for (isotropic) three dimensional
systems, 1.5 for two dimensional systems, and zero for one-
dimensional systems. Inequality (3) is believed to be satisfied
in most polymers: In this case charges injected into the polymer
form local, molecular anions and cations, respectively, rather
than extended mobile states like those characteristic of crystalline covalent semiconductors. Consequently, under these circumstances the electrons and holes are more accurately visualized
as ions in solution which move, however, by carrier hopping rather
than ionic diffusion. On the other hand, static diagonal disorder
should be zero in bulk organic crystals, so that inequality (3) is
not necessarily satisfied and extended band-like states are not
ruled out. In any case, if $\Delta$ and $\bar{V}$ are known, an assessment of the
localization of the injected charges can be made.

Different classes of polymers and molecular solids exhibit
systematically distinct ranges of values of $\Delta$ and $\bar{V}$. For typical
pendant-group and molecularly doped polymers, $0.1eV \leq \Delta \leq 1eV$,
$\bar{V} \leq 0.1eV$ for motion along the polymeric backbone of pendant group
polymers, and otherwise $\bar{V} \ll 0.1eV$ (1, 8). Therefore inequality
(3) is clearly satisfied for these materials, so injected charges
form intrinsic (pendant-group polymers) or extrinsic (molecular-
ly-doped polymers) local molecular-ion states (1, 2, 3, 6, 7, 8).
A similar situation seems to prevail for molecular glasses,
although the parameter values are not yet firmly established in
this case. For bulk molecular crystals in the absence of defects

the static disorder contribution to $\Delta$ vanishes so one expects $\Delta$ = c'$\kappa$T in which c' is a constant.  This observation led to the prediction (1) and subsequent observation (3, 27, 28) of a transition with decreasing temperature from hopping motion between localized states to band motion in extended states in naphthalene and deuterated naphthalene.  Finally, for segregated-stack linear chain charge transfer salts (e.g., TTF-TCNQ) and non-saturated-backbone polymers (e.g., poly(sulfur nitride) and polyacetylene) $\Delta \leq \bar{V}$ so that transport via delocalized states is possible provided the electronic motion is not strictly one dimensional.  In some cases, however, the quasi-one-dimensional nature of the materials strongly suggests the occurrence of Fermi-surface instabilities (10-15, 29).  For example, in polyacetylene Peierl's instabilities, which manifest themselves as bond-alternating backbone structures in undoped materials, seem to predominate (15, 16).  It further has been suggested (13, 14, 30) that these instabilities generate rather different phenomena upon doping than those which occur in "traditional" one-electron semiconductors.

We conclude, therefore, that although the optical and transport properties of organic polymers and molecular solids may appear analogous to those of typical covalent semiconductors, the materials should not be thought of as "organic semiconductors" in the traditional one-electron sense.  Most commonly (i.e., when $\Delta \gg \bar{V}$) they are more accurately regarded as Fermi glasses characterized by localized intrinsic and/or extrinsic molecular ion states with energies near the Fermi energy.  Sometimes (e.g., in the case of polyacetylene) they are best considered to be organic metals which have become semiconductors because of Fermi-surface instabilities.  Occasionally, as in the case of ultrapure naphthalene and deuteronaphthalene at T $\leq$ 100K, organic crystals do seem to exhibit traditional one-electron energy-band transport.  This situation is, however, the exception rather than the rule.

## Contact Charge Exchange

An important practical problem which requires for its solution the concepts developed in the preceeding section is that of contact charge exchange between insulators.  Long viewed in terms of the traditional semiconductor model (31, 32), charge injection and motion in high polymers have successfully eluded quantitative interpretation until the recognition in recent years (5, 6, 7, 18, 19) that the polymeric materials involved are best described as Fermi glasses.  Specifically, the localization of injected charges predicted by the Fermi-glass model is required both to understand the good insulating properties of polymeric materials (32) and to interpret the non-equilibrium features of contact charge exchange (7, 18, 19, 33, 34).

The basic construct of the Fermi-glass model is a spatially averaged density of donor and acceptor states which is a function

of energy.  This state density either may be calculated from the
microscopic models noted above ($\underline{2}$, $\underline{6}$, $\underline{7}$, $\underline{8}$, $\underline{9}$) or may be determined phenomenologically via photoemission ($\underline{8}$) or metal-polymer
contact charge exchange ($\underline{6}$, $\underline{8}$, $\underline{18}$, $\underline{19}$) measurements.  Once the
densities of state are known for two insulators, their steady-
state contact charge exchange can be predicted ($\underline{7}$).  A phenomenological version of this Fermi-glass model has been shown to
predict successfully the contact charge exchange of various copolymers of polystyrene and poly(methyl methacrylate), and of
carbon-black pigmented polymers ($\underline{7}$).  The model also has been used
to interpret quantitatively photoemission and UV absorption data
for polystyrene and poly(2-vinyl pyridine) ($\underline{2}$, $\underline{8}$), and to
calculate hopping activation energies in poly(2-vinyl pyridine)
($\underline{35}$).  Consequently, it constitutes the first successful attempt
to provide a simple, unified theory of optical absorption, charge
injection, and charge transport in high polymers.

## Abstract

     The extended or localized character of charges injected into
solids is determined by a competition between fluctuations in the
energies of these charges from one site to another and their
probability of extending over more than one site.  We illustrate
for trinitrofluorenone and triphenylamine the fact that injected
charges can exhibit widely different transport behaviors in condensed molecular systems characterized by similar geometrical and
electronic structures.  A model of charge localization in condensed molecular solids is developed.  This model embodies distributions of site energies and of electronic overlap integrals.
A procedure for estimating these distributions using spectroscopic data is indicated and applied to a variety of polymers and
molecular solids.  The model predicts the systematic occurrence
of localized states in certain types of materials (e.g., aromatic
pendant-group polymers like polystyrene) but not in others (e.g.,
crystalline, non-saturated-backbone polymers like polyacetylene).  Fermi-surface instabilities occur, however, in linear
polymer systems which exhibit extended states.  A survey is given
of the varied behaviors expected for localized and extended-state
polymeric materials.  As an example of their utility, the results
of the analysis are applied to an important practical problem:
the description of contact charge exchange between polymeric
insulators.

## Literature Cited

1.   Duke, C.B., Surface Sci., 1978, 70, 674.

2.   Duke, C.B., Mol. Cryst. Liq. Cryst., 1979, 50, 63.

3.   Duke, C.B. and Schein, L.B., Physics Today, 1980, 33(2), 42.

4.  Meier, H., "Organic Semiconductors", (Verlag Chemie, Weinheim, 1974).

5.  Duke, C.B. and Fabish, T.J., Phys. Rev. Lett., 1976, 37, 1075.

6.  Duke, C.B., Fabish, T.J. and Paton, A., Chem. Phys. Lett., 1977, 49, 133.

7.  Duke, C.B. and Fabish, T.J., J. Appl. Phys., 1978, 49, 315.

8.  Duke, C.B., Salaneck, W.R., Fabish, T.J., Ritsko, J.J., Thomas, H.R. and Paton, A., Phys. Rev. B., 1978, 18, 5717.

9.  Duke, C.B., in "Tunneling in Biological Systems", B. Chance et al., eds. (Academic Press, New York, 1979), pp. 31-66.

10. Rice, M.J., Duke, C.B. and Lipari, N.O., Solid State Commun., 1975, 17, 1089.

11. Madhukar, A., Chem. Phys. Lett., 1974, 27, 606.

12. Rice, M.J., Solid State Commun., 1977, 21, 757; 1978, 25, 1083.

13. Rice, M.J., Phys. Lett., 1979, 71A, 152.

14. Su, W.P., Schrieffer, J.R. and Heeger, A.J., Phys. Rev. Lett., 1979, 42, 1678.

15. Kertesz, M., Koller, J. and Azman, A., J. Chem. Phys., 1977, 67, 1180; 1978, 68, 2779.

16. Duke, C.B., Paton, A., Salaneck, W.R., Thomas, H.R., Plummer, E.W., Heeger, A.J. and MacDiarmid, A.G., Chem. Phys. Lett., 1978, 59, 146.

17. Fabish, T.J., Saltsburg, H.M. and Hair, M.L., J. Appl. Phys., 1976, 47, 930; 1976, 47, 940.

18. Fabish, T.J. and Duke, C.B., J. Appl. Phys., 1977, 48, 4256; in "Polymer Surfaces", D.T. Clark and W.J. Feast, eds. (Wiley-Interscience Chichester, 1978), pp. 109-119.

19. Duke, C.B., J. Vac. Sci. Technol., 1978, 15, 157.

20. Salaneck, W.R., Phys. Rev. Lett., 1978, 40, 60.

21. Hoegl, H., J. Phys. Chem., 1965, 69, 755.

22. Gill, W.D., _J. Appl. Phys._, 1972, 43, 5033; in "Photo-
    conductivity and Related Phenomena", J. Mort and D. Pai,
    eds. (Elsevier, Amsterdam, 1976), pp. 303-334.

23. Pfister, G., _Phys. Rev. B._, 1977, 16, 3676.

24. Duke, C.B., in "Synthesis and Properties of Low-Dimensional
    Materials", J.E. Miller and A.J. Epstein, eds. (New York
    Academy of Sciences, New York, 1978), pp. 166-178.

25. Anderson, P.W., _Rev. Mod. Phys._, 1978, 50, 191.

26. Weaire, D. and Srivastave, V., in "Amorphous and Liquid
    Semiconductors", W.E. Spear, Ed. (G.G. Stevenson, Ltd.,
    Dundee, 1977), pp. 286-290.

27. Schein, L.B., Duke, C.B. and McGhie, A.R., _Phys. Rev. Lett._,
    1978, 40, 197.

28. Schein, L.B. and McGhie, A.R., _Phys. Rev. B._, 1979, 20, 1631.

29. Chaikin, P.M., in "Synthesis and Properties of Low Dimen-
    sional Materials", J.S. Miller and A.J. Epstein, eds. (New
    York Academy of Sciences, New York, 1978), pp. 128-144.

30. Epstein, A.J. and Miller, J.S., _Solid State Commun._, 1978,
    27, 325.

31. Harper, W.R., "Contact and Frictional Electrification",
    (Clarendon Press, Oxford, 1967), pp. 185-222; 245-268.

32. Wintle, H.F., _Trans. IEEE._, 1977, EI-12, 97.

33. Ruckdeschel, F.R. and Hunter, L.P., _J. Appl. Phys._, 1975,
    46, 4416.

34. Lowell, J., _J. Phys. D: Appl. Phys._, 1976, 9, 1571; 1979, 12,
    1541.

35. Duke, C.B. and Meyer, R.J., _Bull Am. Phys. Soc._, 1979, 24,
    309; _Phys. Rev._, (submitted).

RECEIVED December 22, 1980.

# Electronic Structure of Some Simple Polymers and of Highly Conducting and Biopolymers

J. LADIK, S. SUHAI, and M. SEEL

University Erlangen-Nürnberg, D-8520 Erlangen, West Germany, and Department of Applied Mathematics, University of Waterloo, Waterloo, Ontario, Canada, N2L 3G1

## Method

The _ab initio_ SCF LCAO crystal orbital (CO) method (which applies a non-local exchange and keeps all the occurring three- and four center integrals if the number of neighbours to be taken into account has been chosen) was developed about twelve years ago (1). In this theory one has to solve the generalized eigenvalue equation

$$\underline{\underline{F}}(\vec{k})\underline{d}_i(\vec{k}) = \varepsilon_i(\vec{k})\underline{\underline{S}}(\vec{k})\underline{d}_i(\vec{k}) \tag{1}$$

where

$$\underline{\underline{F}}(k) = \sum_{\vec{q}} e^{ik\vec{R}_{\vec{q}}} \underline{\underline{F}}(\vec{q}) \quad , \quad \underline{\underline{S}}(\vec{k}) = \sum_{\vec{q}} e^{ik\vec{R}_{\vec{q}}} \underline{\underline{S}}(\vec{q}) \tag{2}$$

$$(\underline{\underline{S}}(\vec{q}))_{r,s} = < \chi_r^{\vec{0}} | \chi_s^{\vec{q}} > \tag{3a}$$

$$(\underline{\underline{F}}(q))_{r,s} = < \chi_r^{\vec{0}} | \hat{F} | \chi_s^{\vec{q}} > = < \chi_r^{\vec{0}} | \hat{H}^N | \chi_s^{\vec{q}} > + \sum_{u,v} \sum_{\vec{q}_1} \sum_{\vec{q}_2}$$

$$P(\vec{q}_1 - \vec{q}_2)_{u,v} \times (< \chi_r^{\vec{0}}(1)\chi_u^{\vec{q}_1}(2) | \frac{1}{r_{12}} | \chi_s^{\vec{q}}(1)\chi_v^{\vec{q}_2}(2) > - \tfrac{1}{2} < \chi_r^{\vec{0}}(1)$$

$$\chi_u^{\vec{q}_1}(2) | \frac{1}{r_{12}} | \chi_v^{\vec{q}_2}(1) \chi_s^q(2) >) \tag{3b}$$

Here $\chi_s^{\vec{q}}$ is the s-th AO in the unit cell characterized by the position vector $\vec{R}_q$ and the generalized charge-bond order matrix element is given by the expression (1):

$$P(\vec{q}_1 - \vec{q}_2)_{u,v} = \frac{1}{\omega} \int_\omega \sum_{i=1}^{n^*} d_{i,u}^*(\vec{k})d_{i,v}(\vec{k}) \exp ik(\vec{R}_{q_1} - \vec{R}_{q_2}) \, d\vec{k}, \tag{4}$$

where $\omega$ stands for the volume of the first Brillouin zone in reciprocal space, $\vec{k}$ is the crystal momentum and n* denotes the number of filled bands.

It was easy to show that we can formulate the method also in the case of a combined symmetry operation (for instance helix operation = translation + rotation) instead of simple translation (2). In this case $\vec{k}$ is defined on the combined symmetry operation and from going from one cell to the next one, one has (1) to put the nuclei in the positions required by the symmetry operation and (2) one has to rotate accordingly also the basis set.

Finally it should be mentioned that efficient techniques have been developed to handle long range electrostatic effects in polymers using a modified multipole expansion (3).

## Polyacetylenes and Polydiacetylenes

The described method has been applied to cis-transoid

$( =CH \diagdown \diagup CH= )$ and trans-cisoid
$\phantom{( =CH}CH=CH$

$( -CH \diagdown\diagdown \diagup\diagup CH- )$ polyacetylene and to polydiacetylene (ideal
$\phantom{( -CH}CH-CH$
acetylene structure). For the polyacetylenes the geometry established by Itoh et al (4) and for polydiacetylene the geometry published by Chance et al (5) has been used. For the calculations an STO-3G basis set (6) has been applied (using an STO-4G basis set no significant change has been found in the band structures). For the polyacetylene sixth neighbour's interactions and for the polydiacetylene fifth neighbours interactions have been taken into account. In Table 1 we show the conduction and valence bands of these systems.

As we can see from the Table all the three chains (and this is the case also for further two other polyacetylene and three polydiacetylene chains (7) which also have been calculated) have broad valence and conduction bands with widths between 4.4 and 6.5 eV-s. Comparing the band structure of the two polyacetylene chains we can find that the position of the bands and their widths is not very strongly influenced by the different geometries. This is again the case if we compare the here not described band structures of the further polyacetylene and polydiacetylene chains. On the other hand the position of the valence and conduction bands and the widths of the valence bands of the polydiacetylene chains is more different from those of the poly-acetylene chains. To conclude we can say that due to the broad valence and conduction bands of these systems (which mean rather large hole and electron mobilities,respectively) one can expect that if doped with electron acceptors or donors these systems will become good conductors, which is, as it is experimentally estab-

Table I

The Conduction and Valence Band of cis-transoid and trans-cisoid Polyacetylenes and of Polydiacetylene (Ideal Acetylene Structure) in eV's

| | Valence Band | | | Conduction Band | | |
|---|---|---|---|---|---|---|
| | $\varepsilon_{min}$ | $\varepsilon_{max}$ | $\delta\varepsilon^{\dagger}$ | $\varepsilon_{min}$ | $\varepsilon_{max}$ | $\delta\varepsilon$ |
| Polyacetylene | | | | | | |
| c-t | -9.75 | -4.52 | 5.23 | 3.30 | 9.77 | 6.47 |
| t-c | -9.64 | -4.56 | 5.08 | 3.65 | 9.71 | 6.06 |
| Polydiacetylene | -8.08 | -3.65 | 4.43 | 1.78 | 8.04 | 6.26 |

† Bandwidth

lished, the case. In this respect one should mention that a $(CH_2)_x$ or a $(SiH)_x$ chain has also broad valence and conduction bands (8). Since these chains do not contain $\pi$ electrons, one would need more powerful acceptors (like alkali metal ions) to take out electrons from the valence band. In the authors' opinion this could be done and in this way these chains could become also good conductors (since acceptors never take out a full electronic charge per bond, but only a small fraction of it, the chances that in this way the $\sigma$ bonds of the chain will be broken is very small).

Turning now to larger systems the ab inito SCF LCAO CO method has been applied also to neutral TCNQ and TTF stacks (using the geometry which they have in the mixed crystal (9)) and to the stacked DNA bases applying the same geometry as in a single chain of the in vivo stable B form of DNA (10). The STO-3G basis has been applied also for these calculations and since the stacking distance in all these systems is over 3Å (3.47Å for TTF, 3.17Å for TCNQ and 3.36Å for the homopolynucleotides) only second neighbour's interactions have been taken into account. In Table II we present the results for the valence and conduction bands of these systems. In the case of the homopolynucleotides we give also (in parenthesis) valence and conduction bands corrected for long range correlation using the electronic polaron model (11).

In the case of the TTF and TCNQ stacks, it is interesting to note that the valence band of TTF (from which in the mixed crystal the charge transfer occurs) and the conduction band of TCNQ (to which the charge is transferred) are comparatively broad ($\sim$0.3 eV and $\sim$ 1.2 eV, respectively), while the conduction band of TTF and the valence band of TCNQ are very narrow. The position of the

Table II

The Valence and Conduction Bands of TCNQ and TTF Stacks and of the
Four Homopolynucleotides (in eV-s)

|  | Valence Band | | | Conduction Band | | |
|---|---|---|---|---|---|---|
|  | $\varepsilon_{min}$ | $\varepsilon_{max}$ | $\delta\varepsilon$ | $\varepsilon_{min}$ | $\varepsilon_{max}$ | $\delta\varepsilon$ |
| TTF | −3.80 | −3.50 | 0.30 | 8.51 | 8.59 | 0.08 |
| TCNQ | −7.25 | −7.16 | 0.09 | −0.49 | 0.69 | 1.18 |
| Poly- | −9.67 | −9.11 | 0.56 | 1.54 | 2.78 | 1.24 |
| cytosine | (−9.19) | (−8.67) | (0.52) | (1.12) | (2.27) | (1.15) |
| Poly- | −9.58 | −9.15 | 0.43 | 3.00 | 3.93 | 0.93 |
| thymine | (−9.22) | (−8.81) | (0.41) | (2.67) | (3.56) | (0.89) |
| Poly- | −9.18 | −8.77 | 0.41 | 2.82 | 3.78 | 0.96 |
| adenine | (−8.82) | (−8.41) | (0.41) | (2.48) | (3.41) | (0.93) |
| Poly- | −8.03 | −7.40 | 0.63 | 3.68 | 4.73 | 1.05 |
| guanine | (−7.65) | (−7.06) | (0.59) | (3.35) | (4.36) | (1.01) |

bands (high-lying valence band of TTF and low-lying conduction
band of TCNQ) favors also strongly the charge transfer from TTF to
TCNQ.

Comparing the widths of the valence bands and of the conduction bands of the four homopolynucleotides one finds that they
have about the same widths as the valence band of TTF and the
conduction band of TCNQ. This means that if one would dope them
with good electron acceptors or donors, one would expect a rather
large conductivity also in these systems. One should point out
that though real DNA is aperiodic, the homopolynucleotides (which
contain also the sugar phosphate part neglected in these calculations, but which most probably would not effect too much the
valence and conduction bands, because both are $\pi$ bands) have been
synthetized in the laboratory. According to our knowledge no
transport property measurements of doped homopolynucleotides have
been performed. The numbers in parenthesis indicate that long
range correlation decreases the bandwidths and gaps only moder-
ately (in contrary to simple metals and ionic crystals (11)).

The Aperiodicity Problem; the $(SN)_x$ − $(SNH)_x$ System. We have
reported previously (12) an ab inito SCF LCAO CO band structure
calculation on the $(SN)_x$ chain using the experimental geometry
(13) and a doublefbasis set (14). Though this calculation treated
$(SN)_x$ only as a one-dimensional system, rather good agreement
with experiment has been achieved for the effective mass and
density of states at the Fermi level ($m^*(E_F) = 1.71m_e$, exp: $2.0m_e$;
$\rho(E_F) = 0.17$ (eVspin mol)$^{-1}$, exp:0.18) and with the amount of
charge transferred from S to N(0.4e, exp: 0.3-0.4e).

In the last years 4-8 mol percent hydrogen impurities have been found in $(SN)_x$ at IBM, San Jose (15). One of the most probable site of H bonding is the N atom in the $(SN)_x$ units. To investigate the effect of randomly distributed H atoms on the band structure of $(SN)_x$ we have performed a single site one-band coherent potential approximation (CPA) calculation for the $(SN)_x$ - $(SN)_x$ mixed system assuming different concentrations of hydro-
$\overset{|}{H}$
gen impurities (16). For this purpose we have executed besides the $(SN)_x$ chain also a band structure calculation for an $(SN)_x$
$\overset{|}{H}$
chain. Since the density of states curves of the two systems are very different, no constant self energy ($\Sigma$) could be applied, but one had to use a k and energy dependent $\Sigma(k,E)$. An appropriate formalism has been worked out (for the details see (16)) and the coupled Dyson and CPA equations were solved in an iterative way until self consistency has been reached.

According to the results obtained already at 3 percent H spikes and dips occur in the density of states curve of the mixed system. At larger concentrations of impurities due to clustering effect of the impurity, a part of the gaps disappear (16). Finally, it should be mentioned that with the increase of the impurity concentration, as the calculations show, the density of states at the Fermi level increases and therefore we would expect an increase of the transition temperature between the supercon- ducting and normal states. (This experiment, as far as we know, has not been performed yet). We can conclude from this CPA calculation that aperiodicity (disorder) has a rather serious effect on the band structure of polymers.

This finding was supported by an investigation of the mixed glycine-alanine system (the corresponding homopolypeptides have very similar band structures) taking 1,000-10,000 units of the mixed chain of given sequences. In the course of this study (17), the Huckel matrices of the long chains (the diagonal elements of these matrices were fixed on the basis of the positions of the SCF LCAO bands of the homogeneous chains and the off-diagonal $\beta_{ij}$ parameters have been determined with the help of the band widths) were brought to an upper triangular form with the aid of sub- sequent Gaussian transformations. Using Dean's negative eigen- value theorem (18) from the diagonal elements of the transformed matrix one can obtain directly the density of states of the mixed chains.

The results which were obtained show that depending on the sequence and concentration of the impurity units (alanine units in a glycine chain) the band structure of polyglycine is very strongly effected. If we performed self consistent calculations between the different units (cluster calculations to take into account interunit self consistency are in progress) probably this effect would be smaller, but most probably still rather import- ant. Such calculations are in progress.

## Concluding Remarks

Besides the mentioned aperiodicity problem the treatment of correlation in the ground state of a polymer presents the most formidable problem. If one has a polymer with completely filled valence and conduction bands, one can Fourier transform the delocalized Bloch orbitals into localized Wannier functions and use these (instead of the MO-s of the polymer units) for a quantum chemical treatment of the short range correlation in a subunit taking only excitations in the subunit or between the reference unit and a few neighbouring units. With the aid of the Wannier functions then one can perform a Moeller-Plesset perturbation theory (PT), or for instance, a coupled electron pair approxima- tion (CEPA) (18), or a coupled cluster expansion (19) calcula- tion. The long range correlation then can be approximated with the help of the already mentioned electronic polaron model (11).

The situation is much more difficult in the case of partially filled valence band(s). Since in this case we cannot easily obtain localized orbitals (only if a commensurable part, e.g. 1/2, 1/4, etc. part of the band is filled which is usually not the case), one usually falls back to homogeneous or inhomogeneous gas methods which are rather questionable in the case of organic polymers with comparatively small densities and large density gradients (if there are hetero atoms).

A promising possibility for the treatment of correlation in metallic polymers is to try to subdivide the partially filled band (and a few neighbouring filled and unfilled bands) into regions each of which is characterized by a single representative level (this discretization of a continuum is familiar in the theory of scattering processes, but has not been applied for the correla- tion problem). This subdivision can be achieved by studying the density of states curve of the band and investigating the change with k of the spatial distribution of the partial charge due to the band under concentration. (Using these two criteria a division of the valence and conduction bands of polycytosine has been already performed (21)). After achieving a more or less consistent division of the band, one can attribute to each region a weight factor integrating the density of states curve over the given region (and taking the square root of this quantity). One can then perform excitations between the discrete levels repre- senting these regions and multiplying each matrix element with the corresponding weight factors. These quantities can be sub- stituted finally into the expressions of any size consistent method for the treatment of correlation (Moeller-Plesset P.T., CEPA, CPMET, etc.). The investigation of the method (the depen- dence of the results on the subdivision of the band) and its application to metallic polymers is in progress.

## Acknowledgements

The authors would like to express their gratitude to Professors T.C. Collins, J. Cizek, F. Martino and G. Del Re for the many interesting stimulating discussions. One of us (J.L.) should like to acknowledge very much the warm hospitality of Professor J. Cizek and of the Department of Applied Mathematics, University of Waterloo, extended to him during his visit in Waterloo.

## Abstract

For the calculation of the Hartree-Fock band structures of polymers a method has been developed including non-local exchange and full-self consistency. It is applicable also in the case of a combined symmetry (e.g., helix).

The method has been applied to polyene, polyacetylene and polydiacetylene chains, to formamide chains (both hydrogen-bonded and stacked). Applications have been done also to TCNQ and TTF stacks, to $(SN)_x$ and to periodic DNA models (the four homopolynucleotides), to the sugar phosphate chain of DNA and to different periodic protein models (homopolypeptides). All these systems have relatively broad valence and conduction bands (bandwidths around or larger than 0.5eV) according to our results.

In the final part of the lecture methods to treat the electronic correlation and the aperiodicity of polymers will be outlined. For the latter case CPA results for a $(SN)_x$ $-(SN)_x$ mixed system will be discussed and extensions to aperiodic biopolymers (which are in progress) will be indicated.

## Literature Cited

1. Del Re, G., Ladik, J. and Biczo, G., Phys. Rev., 1967, 155, 997; Andre, J.-M., Gouverneur, L. and Leroy, G., Int. J. Quant. Chem., 1967, 1, 427 and 451.

2. Blumen, A., Merkel, C., Phys. Stat Sol., 1977, B83, 425; Ladik, J., "Excited States in Quantum Chemistry," eds. C.A. Nicolaides and D.R. Beck, D. Reidel Publ. Co., Dordrecht-Boston-London, 1979, p. 495.

3. Piela, L., Delhalle, J., Int. J. Quant. Chem., 1978, 13, 605; Andre, J.M., Fripiat, J.G., Demanet, C., Bredas, J.L. and Delhalle, J., Int. J. Quantum Chem., S12, (accepted); Suhai, S. and Seel, M., to be published.

4. Itoh, T., Shirayawa, H. and Ikeda, S., J. Polymer Sci., Polymer Chem. Ed., 1974, 12, 11.

5. Chance, R.R., Baughman, R.H., Muller, H. and Eckhardt, C.I., J. Chem. Phys., 1977, 67, 3616.

PHOTON, ELECTRON, AND ION PROBES

6. Hehre, W., Steward, R.F. and Pople, J.A., J. Chem. Phys., 1969, 51, 2657.

7. Suhai, S., to be published.

8. Suhai, S., to be published.

9. Kistenmacher, T.J., Phillips, T.E. and Cowan, D.O., Acta Cryst., 1974, 33, 763.

10. Arnott, S., Dover, S.D. and Wonacott, A.J., Acta Cryst., 1969, B25, 2192.

11. Kunz, A.B., Phys. Rev., 1972, B6, 606; Collins, T.C., "Electronic Structure of Polymers and Molecular Crystals," J.-M. Andre and J. Ladik eds., Plenum Press, New York-London, 1975, p. 405.

12. Suhai, S. and Ladik, J., Solid State Comm., 1977, 22, 227.

13. Mikulski, C.M., Russo, P.J., Saran, M.S., Madoiarmid, A.G., Garito, A.F., Heeger, A.J., J. Am. Chem. Soc., 1975, 97, 6358.

14. Roos, B., Siegbahn, T., Theor. Chim. Acta, 1970, 17, 1209.

15. Smith, R.D., Wyatt, J.R., Weber, D., DeCorps, J.J., and Saalfed, F.E., Inorg. Chem., 1978, 17, 1639.

16. Seel, M., Collins, T.C., Martino, F., Rai, D.K. and Ladik, J., Phys. Rev., 1978, B18, 6460.

17. Seel, M., Chem. Phys., 1979, 43, 103.

18. Dean, P., Rev. Mat. Phys., 1972, 44, 127.

19. Meyer, W., J. Chem. Phys., 1972, 58, 1017; Kutzelnigg, W., J. Chem. Phys., 1975, 62, 1225.

20. Cizek, J., J. Chem. Phys., 1969, 36, 4256; Paldus, J. and Cizek, J., Adv. Quant. Chem., 1975, 9, 105.

21. Ladik, J., "Quantum Theory of Polymers," J.-M. Andre, J. Delhalle, J. Ladik and G. Leroy editors, Springer-Verlag, Berlin-Heidelberg-New York (in press).

RECEIVED December 22, 1980.

# Spectroscopic Studies of Polydiacetylenes

D. BLOOR

Department of Physics, Queen Mary College, Mile End Road,
London E1 4NS, United Kingdom

Conjugated polymers have been studied for many years principally as potential semiconductors. In 1970 the field was seen to be expanding, but still in its early stages and lacking a unifying theory (1). The absence of such a theory was not due to lack of effort by theorists, who had developed several alternative approaches to explain the occurrence of an energy-gap in conjugated polymers (2-9). The main problem was the amorphous structure of the conjugated polymers then available. Thus, it was not possible to distinguish intrinsic from extrinsic properties, for example the paramagnetism observed in all conjugated polymers had been attributed to an intrinsic spin-alternation (10), bond-alternation defects (11) and to a number of other causes (12).

Interest in conjugated polymers waned somewhat in the early part of the last decade but has re-emerged following the discovery of metallic conductivity and superconductivity in $(SN)_x$ (13) and metallic conductivity in doped $(CH)_x$ (14). The latter has been shown not to be unique (15, 16) indicating the generality of high conductivity in doped conjugated polymers and the possibility of the discovery of technologically applicable materials. However, though $(CH)_x$ is partially crystalline the morphology is still complex and the observed properties have been interpreted in terms of either the fundamental properties of the $(CH)_x$ polymer (14, 17), or the disordered and inhomogeneous nature of the samples (18). Thus, despite the emergence of sophisticated theories (17, 19, 20) the basic problem of interpretation imposed by sample morphology remains.

One approach to the determination of intrinsic properties, which has been utilized since the earliest interest in conjugated polymers, is to study the properties of related oligomers, as in the preceding paper (21). It is, however, also possible to study model macromolecules, the polydiacetylene. The existence of solid state polymerization in diacetylene monomers has a long history (22, 23, 24, 25), but it was not thoroughly studied until about ten years ago (26). It was then established that the polymerization could proceed to completion by a homogeneous solid state reaction to give a polymer single crystal as the product (27, 28). The reaction is shown schematically in Figure 1, typical end-groups (R, R'), which lead to the formation of ordered polymer chains, are listed in Table I. As indicated in the table numerous diacetylene compounds have now been synthesized, polymerized and their physical properties studied; this work has been extensively reviewed (29, 30, 31, 32). The crystals obtained display the intrinsic properties of the polymer since defect levels are sufficiently low to be a minor perturbation.

Table I

Substituent groups for the polydiacetylenes which are referred to in the text by the listed abbreviations

| R | R' | Abbreviation |
|---|---|---|
| $-CH_2OSO_2C_6H_4CH_3$ | Symmetric | TS |
| $-CH_2OCONHC_6H_5$ | " | PU |
| $-(CH_2)_3OCONHC_6H_5$ | " | PUDD |
| $-(CH_2)_4OCONHC_6H_5$ | " | TCDU |
| $-(CH_2)_4OCONHC_2H_5$ | " | ETCD |
| $-(CH_2)_4OCONHCH(CH_3)_2$ | " | IPUDO |
| $-(CH_2)_nOCONHCH_2COOZ$ | " | NZCMU |
| $-CH_2NC_{12}H_8$ | " | DCH |
| $-(CH_2)_2OH$ | " | OD |
| $-(CH_2)_3OH$ | " | DD |
| $-COO(CH_2)_8CH_3$ | " | CAP |
| $-CH_2OH$ | $-CH_3$ | 1OH |
| $-(CH_2)_9CH_3$ | $-(CH_2)_8COOH$ | TCDA |
| $-(CH_2)_9CH_3$ | $-(CH_2)_8COO^-.Li^+$ | Li-TCDA |
| $-C_6H_4OCO(CH_2)_3OCOC_6H_4-$ | $-(Cyclic)$ | BPG |

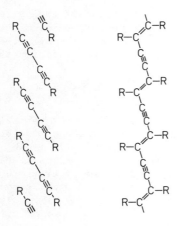

*Figure 1. Solid-state polymerization of diacetylenes shown schematically: (left) an array of monomer molecules in the crystal lattice; (right) the resulting polydiacetylene chain.*

The range of environments to which the polymer chains may be subject varies widely. In the crystalline state the structure of the lattice is strongly influenced by the bulky side-groups, which provide the arrangement of monomer molecules necessary for solid state reactivity. Thus, polymer molecules in partially or fully polymerized crystals of different monomers are subjected to different static distortions. Dramatic changes in both static and dynamic distortions can occur when the crystal lattice undergoes a phase transition. Such changes can also be produced when the polymer is prepared in a disordered form, e.g. in solution (33), as colloidal particles (34), by deformation of single crystals (35) and by extraction of the polymer from partially polymerized crystals (36). Studies of crystals can, therefore, provide information about fundamental properties and their dependence on chain structure while the less perfect samples can be used to study the modification of physical properties by disorder.

These ideas will be illustrated by discussions of the optical and ESR spectra of a range of polydiacetylenes. The individual materials mentioned have different side-groups which are listed, together with the abbreviations used subsequently to identify the materials, in Table I.

## Optical Spectroscopy

The polarized, metallic reflectivity of a fully polymerized polydiacetylene crystal shows immediately its intrinsic anisotropy. A typical reflection spectrum, for poly TS (37), is shown in Figure 2, the optical constants obtained by Kramers-Kronig analysis show that the absorption maximum occurs near 2 eV (16,000 cm$^{-1}$) (38, 39). This electronic excitation is the only strong feature of the spectrum over the range 1.65 to 10 eV (38). The absorption is highly anisotropic with an absorption tensor with its principal axes parallel to the polymer chain and perpendicular to the chain, first in the plane of the chain and secondly normal to the plane of the chain. The values of the principal absorption coefficients for poly TS at 2 eV are $9 \times 10^5$, $2 \times 10^3$ and less than 5 cm$^{-1}$. Essentially similar results have been obtained for a number of other polymers, e.g. TCDU, ETCD, DCH, 4BCMU and IPUDO (40, 41, 42, 43). The temperature dependence of the absorption energy is small, see Figure 3, as a consequence of the very small thermal expansivities of the crystals (44). The polymers with urethane sidegroups have crystallographic phase transitions involving reorientation of the sidegroups and related large shifts in absorption energy. Films of nBCMU polymer cast from solution have large continuous shifts in absorption energy (42). These indicate the onset of considerable disorder at high temperatures, as discussed later.

The absorption was initially interpreted as an interband transition since the absorption profile was of the correct form for a van Hove singularity in one dimension:

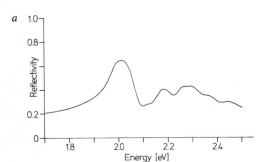

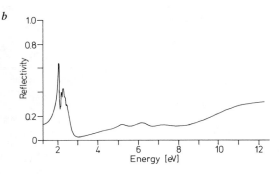

**Physics Status Solidi**

*Figure 2.   Reflection spectrum of poly TS crystal at room temperature for light polarized parallel to the polymer chains: (a) detail of the reflection in the visible and (b) spectrum over the range 1.6 to 12 eV (38).*

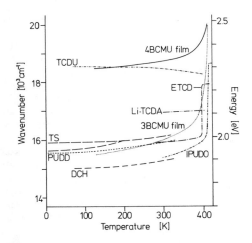

*Figure 3.   Temperature dependence of the intense optical absorption for a number of polydiacetylenes. With the exception of the nBCMU solution cast films, all samples were crystalline.*

$$\omega^2 \varepsilon_2 \propto (h\omega - E_g)^{-\frac{1}{2}} \tag{1}$$

where $E_g$ is the gap energy. However, the absence of marked photoconductivity at 2 eV (45, 46) cast doubt on this inter-pretation. The original suggestion (37) that the absorption is due to an exciton is now widely accepted. Theoretical models indicate the existence of an excitonic state which can steal intensity from the interband transition (47, 48, 49). The line shape can also be described by an asymetric Lorenztian (38)

$$I(\omega) = C \left[ \frac{\Gamma + A(\omega - \omega_o)}{(\omega - \omega_o)^2 + \Gamma^2} \right] \tag{2}$$

where C is a constant, $\Gamma$ the linewidth parameter and A the asymmetry parameter. This form was originally deduced for exci-tons with coupling to low frequency phonon modes (50). The presence of such low frequency sidebands in polydiacetylenes has been confirmed by comparison of the absorption profile with the pure electronic profile obtained from resonant Raman scattering experiments (51). The results of the Raman experiments are also characteristic of an excitonic state. Exciton surface polaritons have been observed in poly TS (52). Thus the weight of evidence is in favor of assignment of an excitonic state to the intense optical absorption.

However, as shown in Figure 3 the absorption can occur over a wide energy range. A somewhat larger range of absorption energies is found for polymer chains in partially polymerized crystals (53), see Figure 4. This, and the generally larger temperature dependence, reflects the greater variation in environment possible in partially polymerized crystals as compared with fully polymerized crystals. At low temperatures the small changes in linewidth and position can be described by coupling with a Debye phonon spectrum (54) and at higher temperatures by the response of the polymer chains to the monomer lattice expansion (51). The parameters relating absorption energy to polymer chain deforma-tion were taken from the results obtained for poly TS samples in extension (55) and under hydrostatic pressure (56). This inter-pretation is based on the occurrence of polymer chains with a large number of repeat units, of the order 100 or more, even at low levels of conversion to polymer. Evidence supporting this is provided by molecular weight measurements for radiation polymer-ized samples (33), and by the small optical linewidths observed for partially polymerized samples at low temperatures which are comparable with those of fully polymerized samples (38). It should be noted that the shifts in absorption energy shown in Figure 4 could not readily be produced by external forces.

The spectra of fully and partially polymerized polydiacetylenes can be described in terms of the response of the polymer chain to its environment. The most important contribution is taken to be the static (time averaged) deformation of the polymer backbone. The dynamic (thermal) deformations are considered to give rise to the homogeneous (temperature dependent) linewidths. Large inhomogeneous linewidths can also be observed in highly strained or disordered systems. The static deformation primarily depends on the side-groups, which determine the lattice packing of the monomer crystal and, therefore, to a large extent the polymer crystal structure.

The influence of the side-groups will be much less in solution but, until recently, there have been few studies of polydiacetylene solutions since most fully polymerized materials are insoluble. However, a number of soluble polydiacetylenes have now been identified (33, 34, 57, 58, 59, 60). In the case of nBCMU polymers hydrogen-bonding between the side-groups of a single polymer chain can produce a rigid-planar conformation in solution (57), which has an absorption energy close to that of single crystals. All other solutions have a broad absorption spectrum, as shown in Figure 5, which is independent of the side-group. The electronic origin of this absorption does not occur at the absorption maximum but, as shown by second derivative and fluorescence spectra (34), lies near 2.5 eV ($\sim$20,000 cm$^{-1}$). Colloidal polymer powders have been obtained, which retain the solution conformation; these have a narrower linewidth with the electronic absorption at 2.5 eV (20,160 cm$^{-1}$), as shown in Figure 6.

The interpretation of these results is based on the fact that the polymer chains retain an extended form in solution. This has been shown by viscosity measurements (33) and the observation of distinct spectral features for the colloidal polymer. If chain deformation primarily produced a distribution of short conjugation sequences, as has been suggested (57), this spectrum would have a featureless profile; the alternative of a single well defined fold length does not appear reasonable. The range of spectral profiles observed can be explained simply in terms of differences in linewidth. Figure 7 shows calculated profiles using the lineshape of Equation 2 and including three phonon sidebands, further details of these calculations are given elsewhere (34). Linewidth factors of 100, 300, $\sim$800 and more than 1,200 cm$^{-1}$ give profiles similar to those of poly TS as single crystals of 4K and 300K, colloidal particles and in solution respectively. In the latter two cases the linewidths must also contain an inhomogeneous broadening factor, this been simply treated as a contribution to the linewidth factor. These factors appear reasonable for the different environments involved. Thus, we conclude that even in solution the time averaged structure of the extended polymer backbone is the main factor determining the

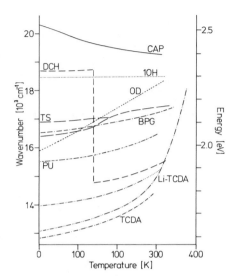

*Figure 4. Temperature dependence of the intense optical absorption for poly-diacetylene chains in their monomer matrices at typical concentrations of 1%; all samples were crystalline.*

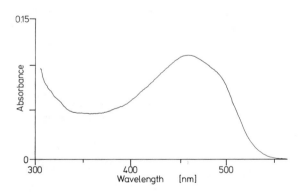

*Figure 5. Absorption spectrum of poly TS in chloroform. The rising absorption at 300 nm is due to side-group absorption, principally to the monomer, which was present at about 100 times the concentration of the polymer.*

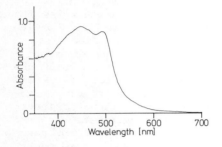

*Figure 6. Absorption spectrum of poly TS colloidal powder precipitated from chloroform solution. Similar results are obtained for acetone and nitrobenzene (34).*

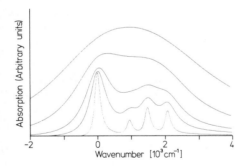

*Figure 7. Model absorption profiles calculated from Equation 2 with $A = 0.08$ and vibrational sideband frequencies of 950, 1480, 2080 $cm^{-1}$. The integrated intensities are not normalized, resulting in a vertical displacement as the linewidth parameter takes the values 100, 300, 500, 1,000, and 2,000 $cm^{-1}$, respectively.*

absorption energy, while the dynamic perturbations determine the
linewidth.

The influence of the side-groups can also be reduced in the
solid state. If the crystal structure becomes disordered there is
a general tendency for the absorption energy to increase. This
may occur in the solid either due to a phase transition, for
example in the high temperature disordered phases of Li-TCDS (60)
as indicated in Figure 4, or by extraction of the polymer from the
monomer lattice at low conversion or a combination of both, as for
nBCMU films, Figure 3. For extracted polymer unless there is
strong side-group interaction, the absorption energy moves towards the solution value (59). Typical absorption energies for
extracted polymers are 2.26 eV (18,250 $cm^{-1}$) for OD, 2.29 eV
(18,420 $cm^{-1}$) for DD and 2.31 eV (18,620 $cm^{-1}$) for TCDA.

These results show that the absorption energy of a polydiacetylene extended chains in solution is 2.5 eV and that the
lower values observed for crystals, polymer extract, films and
rigid-planar conformation in solution reflect the structure imposed on the polymer by the intra- and inter-molecular interactions of the side-groups. While reasonable agreement has been
obtained in some cases, using experimentally determined parameters, the details of the relationship between absorption energy
and structure have yet to be worked out.

Progress in this direction is hindered by the limited number
of polymer structures available. In addition, the largest shifts
in excitation energy occur in situations where X-ray structural
information is not readily obtained, i.e. in disordered polymers
and partially polymerized crystals (see Figures 3 and 4). Even
when structural information is available it may be inadequate,
e.g. the splitting observed for TS (Figures 3 and 4) occurs for
distortions which are not detected by structural analysis. In the
case of TCDU the X-ray structure shows a polybutatriene chain,
$\leftarrow RC = C = C = CR \rightarrow_x$, which has prompted the assignment of a high
absorption energy to this structure (40, 41, 42, 43). However,
the structure shows a number of other unusual bond lengths (61).
This suggests a forced computer fit of a slightly disordered
structure, an effect which has been observed for other polydiacetylenes (62). Thus, in view of the essentially continuous spread
of experimental absorption energies over the range 1.6 to 2.5 eV,
this assignment is open to question. Further doubt is caused by
the fact that the polybutatriene chain should have a somewhat
lower excitation energy than the yne-ene structure of Figure 1
(51). In addition recent calculations suggest that the polybutatriene structure does not have a stable energy minimum, in the
absence of side-group interactions (63). Raman studies of colloidal polymer particles show that the disordered chains have
bond lengths close to an ideal yne-ene structure (34). This
suggests that a more detailed analysis of the excitation energies
of deformed yne-ene chains is capable of explaining the experimental data.

Electron Spin Resonance Spectroscopy

The existence in conjugated polymers of paramagnetic centers, with g values close to the free electron value and having a narrow spectral linewidth, has been known for many years ($\underline{64}$, $\underline{65}$). It was, therefore, natural to investigate the ESR spectrum of polydiacetylenes. The earliest studies revealed a similar but very weak ESR signal ($\underline{37}$, $\underline{66}$). More detailed studies revealed that the signal strength was correlated with sample history and was very weak in the best single-crystal samples ($\underline{67}$, $\underline{68}$, $\underline{69}$). It was, therefore, concluded that the paramagnetic centers in polydiacetylenes were associated with defects and, unlike the optical absorption discussed above, were not an intrinsic property of the ideal polymer chain. However, the spin concentration for disordered samples was only an order of magnitude larger than that in good single crystals ($\underline{69}$) showing that only a small fraction of all the defects were paramagnetic.

Subsequent studies were primarily concerned with the identification of reactive species present during polymerization ($\underline{68}$, $\underline{70-74}$). The emergence of (CH)$_x$, the rediscovery of its ESR spectrum ($\underline{75}$) and the interpretation using the soliton model ($\underline{17}$) caused renewed interest in the paramagnetic centers in fully polymerized polydiacetylenes. Initial studies were concerned with the possibility of doping polydiacetylene samples. In general this was unsuccessful because either the dopant could not diffuse into the single crystal samples or it reacted with the side-groups. However, extracted poly 10H samples were found to absorb large quantities of iodine with a consequent large increase in the g $\sim$2 ESR signal ($\underline{76}$, $\underline{77}$). The broad signal in doped samples can be attributed to radicals formed at static defects, but on illumination with visible light a narrow spectral component, similar to the stable found in (CH)$_x$, appeared in both doped and undoped samples.

The structure of 10H monomer is shown in Figure 8 ($\underline{78}$). Adjacent rows of monomer are linked by hydrogen bonds and though the packing is close to the optimum for solid state reactivity failure of the weaker Van der Waals bonds leads to break up of the crystal before polymerization is complete. Small quantities of fibrous polymer with a greenish luster can be obtained by dissolving the monomer lattice at low levels of conversion. These samples give a well defined powder X-ray diffraction pattern indicating a high degree of crystallinity. Electron micrographs show fine fibers, up to 25 nm in diameter, extended in the polymer chain direction with fiber bundles with lengths greater than 20$\mu$m ($\underline{79}$). The g $\sim$2 ESR signal of such samples has a linewidth of about 10 gauss ($10^{-3}$T) and is of low intensity. This broad component increases in intensity on iodine doping. When an undoped sample is illuminated an additional narrow component, with intensity related to the irradiation intensity, is observed. Typical spectra taken at different temperatures are shown in Figure 9 to 11.

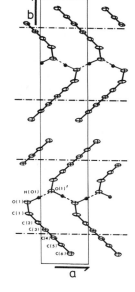

*Figure 8. Crystal structure of 10H monomer projected on to the ab-plane. Hydrogen bonds between H(01) and O(1)′ are shown dashed. Polymer chains grow with their axes along the chain lines.*

The narrow spectral component was found to be metastable down to 4.2K. The increase of signal under illumination and the subsequent decay was reproducible for repeated illumination. The rate of growth and decay was temperature dependent, as shown in Figure 12. These curves cannot be described by a single exponential factor and, although the data is insufficient for an accurate Arhenius plot, the activation energy appears to be low, $\stackrel{\sim}{<}0.1$ eV. With a filtered light source the maximum photo-sensitivity at 4.2K was found to occur at about 700 nm ( $\sim$1.75 eV, $\sim$ 14,000 $cm^{-1}$ ) but to extend to beyond 800 nm at room temperature. The intensity of the photo-induced signal was found to depend on sample volume showing that it is a bulk effect, and not a surface effect. Transmission spectra of polymer films, Figure 13, have an absorption onset near 800 nm with an absorption coefficient low enough for volume irradiation of a bulk sample down to about 650 nm.

The ESR spectra were fitted by two Lorenztian components with linewidths at room temperature of 9.4 and 2.4 gauss. The narrow component is displaced about 2 gauss to higher field relative to the broad component. The fitting revealed weak hyperfine structure associated with the broad spectral component. This, together with the increased signal strength observed on doping, suggest that a localized defect is responsible for this spectral component. The broad component is also more strongly coupled to the lattice, typical saturation results are shown in Figure 14. At 4.2K the differences in spin-lattice relaxation times become more pronounced, the narrow component being spin polarized at all except the lowest microwave powers ( $\sim$ 2 x $10^{-8}$ watt).

Three possible explanations for these observations are (a) a photo-chemical process, (b) a mobile defect, similar to the $(CH)_x$ 'soliton' and (c) photo-excited charge-transfer. The first of these can be eliminated since photochemistry even in lOH monomer requires u-v irradiation and the radicals produced in irradiated monomer (80) and related matrix isolated species (81) have spectra with strong hyperfine structure.

The spectra reported here and those found for trans-$(CH)_x$ have comparable linewidths, see Table 2, and lineshapes. It may, therefore, be appropriate to extend the model developed for $(CH)_x$ to polydiacetylenes. The model of a mobile bond-alternation defect has been used in general terms to explain the earlier observations on polydiacetylenes (66, 67, 68, 69). A detailed theory was not formulated until methods evolved in the analysis of solitons were applied to bond-alternation defects in $CH_x$ (85, 17). In this model the bond-alternation is not abrupt, as originally proposed by Pople and Walmsley (11), but extends over several polymer repeat units as shown in Figure 15. The creation energy is small, 0.4 eV, and the energy for motion along the polymer chain very small, $\sim$ 2 meV, so that a stable concentration of mobile centers with a delocalized spin is predicted, as required to explain the experimental data.

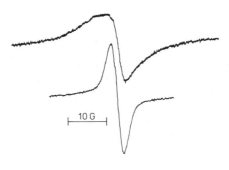

*Figure 9.    The g ~ 2 ESR signals ob-*
*served for extracted 10H polymer at*
*300 K: (*upper *spectrum) after dark stor-*
*age with a gain of 10 × 10³; (*lower *spec-*
*trum) during illumination with a gain of*
*4 × 10³. The microwave power was the*
*same for both spectra.*

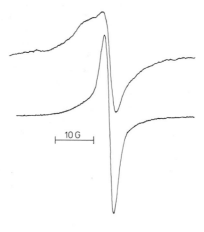

*Figure 10.    The same spectra as in Fig-*
*ure 9 but with a sample temperature of*
*105 K: (*upper *spectrum) after overnight*
*dark storage at < 100 K, gain 5 × 10³,*
*power attenuation 20 dB; (*lower *spec-*
*trum) after illumination, gain 2.5 × 10³,*
*power attenuation 35 dB (0 dB ≡ 20*
*mW).*

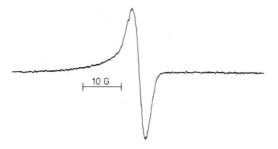

*Figure 11.    The g ~ 2 ESR signal observed for extracted 10H polymer at 4.2 K*
*after illumination (gain 2.5 × 10³, power attenuation 60 dB (0 dB = 20 mW))*

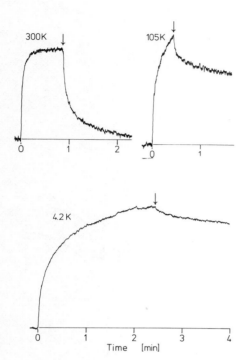

*Figure 12.   Growth and decay of the photoinduced ESR signal in 10H polymer extracts at the different temperatures indicated. Illumination commences at time zero and terminates at the time indicated by the arrow. The spectral range used was from about 570 to 700 nm.*

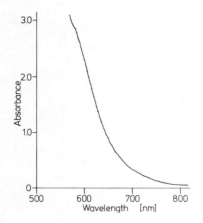

*Figure 13.   Transmission spectrum at room temperature of a poly 10H extract about 0.5-mm thick*

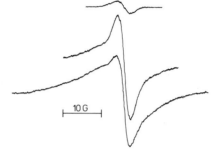

*Figure 14. Signal saturation at 105 K after sample illumination. Spectra were recorded at constant gain with power attenuations of 50 dB (upper spectrum); 20 dB (middle spectrum); and 10 db (lower spectrum) (0 dB ≡ 20 mW).*

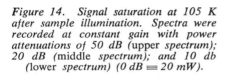

*Figure 15. Extended (soliton) bond-alternation defects in polyacetylene (upper) and polydiacetylene (lower) chains. The regions in which the bond lengths change gradually, and over which the unpaired electron is delocalized, are shown dashed.*

Table II

ESR linewidths for g ∿ 2 species in poly 1OH extracts and polyacetylene at different temperatures, all values in gauss.

| Material | Reference | Linewidth 300K | 105K | 4.2K |
|----------|-----------|--------|------|------|
| poly 1OH | This work | $9.4^a$ $2.4^b$ | $3.\bar{2}^b$ | $3.\bar{7}^b$ |
| $(CH)_x$ | 82 | $8.0^c$ $1.4^d$ | – – | – – |
| $(CH)_x$ | 75 | $1.43^d$ | – | $4.67^d$ |
| $(CH)_x$ | 84 | $4.1^d$ | – | $6.7^d$ |
| $(CD)$ | 84 | $1.9^d$ | – | $3.1^d$ |

a  Broad component
b  Light induced component
c  cis-form
d  trans-form

The application of this model to polydiacetylenes (86) leads to different results since the chain structure must be different on either side of the defect, as shown in Figure 15. Bond-alternation defect must, therefore, be created in pairs at either end of a polybutatriene sequence. The creation of the new chain structure will require considerable energy and will be hindered by bulky, interacting side-groups. Thus, they are liable to be few in number, and, in view of the higher energy intrinsic and the relative instability of the polybutatriene structure, will be metastable. The occurrence of such defects will be favored in partially crystalline materials with small side-groups such as poly 1OH. Population of such species via the polymer absorption tail would be possible if the electron distribution in the excited states was distorted towards the polybutatriene form in the excited state, as suggested by model calculations (87). Thus, in principle this model can explain the experimental data for poly 1OH.

The third possibility has been suggested previously to explain a very narrow, 1.4 gauss linewidth, component observed in amorphous poly TS extracts (69). Charge transfer excitation is possible between conjugated polymer chains of different lengths and different ionization potentials (88). For long chains these differences could be small. Again disordered samples are necessary to allow for chain contact, which is also favored for small side-groups. However, the brief discussions of this model available in the literature (89) do not enable it to be tested adequately against the experimental data for poly 1OH.

## Conclusions

The intrinsic optical absorption of polydiacetylenes can occur over a wide energy range depending on the polymer chain environment. "Free" chains in solution adopt a conformation with an yne-ene bonding and an absorption energy of 2.5 eV. Lower values occur in situations where side-group interactions play an important role. Disorder in the polymer appears mainly to affect the absorption linewidth. A similarly weak impact on polymer properties is seen in the small increases in paramagnetism of disordered samples relative to that of single crystals. These small effects reflect the rigidity of the polyacetylene chains, which retain in chain extended form even when disordered.

However, disorder appears to be crucial in the occurrence of photoexcited paramagnetic centers since irradiation of single crystal samples, e.g. poly TS, has no effect. The energy levels in the low energy absorption tail of the disordered samples play an important part. It is not clear if these derive from excitonic or interband transitions but it should be noted that in principle any distortion of the polymer chain can lead to defect states in the optical gap (90) and that the weak absorption tail in crystals is a major factor determining photoconduction (91). Thus, it appears that a basis exists for the correlation of structure and the intrinsic absorption spectra but that further efforts are required to obtain a better understanding of defect states and their spectroscopic properties.

## Acknowledgements

I would like to thank my past and present colleagues in the Polymer Group at Queen Mary College who have assisted in the experiments described here, in particular Dr. D. N. Batchelder, Dr. G. C. Stevens, Dr. F. H. Preston, Dr. R. J. Kennedy, Dr. B.

Tieke, C. L. Hubble and D. J. Ando.  My thanks are also due to Dr. H. Sixl, Dr. A. Karpfen, Prof. H. Bassler, Prof.  G. Wegner, Dr. R. R. Chance, Dr. A. R. Bishop and Dr. R. J.  Young for the communication of results prior to publication.  This work was supported by the Science Research Council and the ESR measurements were made possible by the hospitality of Prof. H. C. Wolf of the University of Stuttgart.

## Abstract

The complex morphology of most conjugated polymers prevents direct measurement, and hinders understanding, of their intrinsic properties.  Such problems do not occur for the polydiacetylenes, which can be prepared as macroscopic single crystals.  Since polydiacetylenes can also be obtained in less perfect forms the effects of disorder can be studied.  Spectroscopic techniques have been widely used to study the intrinsic properties of polydiacetylenes and their modification by disorder.  The results obtained by optical and electron spin resonance spectroscopy are discussed.

## Literature Cited

1.    Rembaum, A. J. Polymer Sci., Part C. No. 29, 1980, p.  157.

2.    Longuet-Higgins, H.C. and Salem, L., Proc. Roy. Soc., 1959, A251, 172.

3.    Pople, J.A. and Walmsley, S.H., Trans. Faraday Soc., 1962, 58, 441.

4.    Fukutome, H., Prog. Th. Phys., 1948, 40, p. 998 and 1227.

5.    Misurkin, I.A. and Ovchinnikov, A.A., Theor. Chim. Acta., 1969, 13, 115.

6.    Cazes, D., Salem. L. and Tric, C., J. Polymer Sci, Part c. No. 29, 1970, p. 109.

7.    Partridge, R.H., Int. J. Quant. Chem., 1972, 6, 167.

8.    Pugh, D., Mol. Phys., 1973, 26, 1297.

9.    Ovchinnikov, A.A., Ukrainskii, I.I. and Kventsel, G.V., Sov. Phys. Uspekhi, 1973, 15, 575.

10.   Berlin, A.A., Vinogradov, G.A. and Ovchinnikov, A.A., Int. J. Quant. Chem., 1972, 6, 263.

11.   Pople, J.A. and Walmsley, S.H., Mol. Phys., 1962, 5, 15.

12.  Nechstein, M., J. Polym. Sci., Part C. No. 4, 1965, p. 1367.

13.  Labes, M.M., Love, P. and Nichols, F., Chem. Rev., 1979, 79,
     1.

14.  MacDiarmid, A.G. and Heeger, A.J., "Molecular Metals," Ed.
     W.E. Hatfield, Plenum, New York, 1979, p. 161.

15.  Ivory, D.M., Miller, G.G., Sowa, J.M., Shacklette, L.W.,
     Chance, R.R. and Baughman, R.H., J. Chem. Phys., 1979, 71,
     1506.

16.  Kanazawa, K.K., Diaz, A.F., Geiss, R.H., Gill, W.D., Kwak,
     J.F.,Logan, J.A., Rabolt, J.F. and Street, G.B., JCS Chem.
     Comm., 1979, 854.

17.  Su, W.P., Schrieffer, J.R. and Heeger, A.J., Phys. Rev.
     Lett., 1979, 42, 1698.

18.  Tomkiewicz, Y., Schultz, T.D., Broom, H.B., Clarke, T.C.
     and Street, G.B., Phys. Rev. Lett., 1979, 43, 1532.

19.  Berlin, A.A., Vinogradov, G.A. and Ovchinnikov, A.A., J.
     Macromol. Sci. Chem., 1977, A11, 1701.

20.  Andre, J.M. and Ladik, J., "Recent Advances in the Quantum
     Theory of Polymers," Ed. J.M. Andre, J-L. Bredas, J.
     Delhalle, J. Ladik, G. Leroy and Moser, C., Springer Verlag,
     Berlin, 1980 and other contributions to this publication.

21.  See for example Karplus, M. in this volume.

22.  Strauss, F., Kollek, L. and Heyn, W., Ber. Deutch. Chem.
     Gas, 1930, 63, 1868.

23.  Bohlmann, F., Chem. Ber., 1951, 84, 785.

24.  Black, H.K. and Weedon, B.C., J. Chem. Soc., 1953, 1785.

25.  Hirshfeld, F.L. and Schmidt, G.H.J., J. Polym. Sci., 1964
     A2, 2181.

26.  Wegner, G., Z. Naturforsch, 1969, 24b, 824.

27.  Hadicke, E., Mez, E.C., Krauch, C.H., Wegner, G. and Kaiser,
     J., Agnew Chem. Int. Ed., 1971, 10, 266.

28.  Kaiser, J., Wegner, G. and Fischer, E.W., Israel J. Chem.,
     1972, 10, 157.

29. Wegner, G., "Molecular Metals," Ed. W.E. Hatfield, Plenum Press, New York, 1979, p. 209.

30. Baughman, R.H. and Chance, R.R., Ann. N.Y. Acad. Sci., 1978, 313, 705.

31. Bloor, D., "Recent Advances in the Quantum Theory of Polymers," Ed. J.M. Andre, J-L. Ladik, J-L. Bredas, J. Delhalle, J. Ladik, G. Leroy and C. Moser, Springer-Verlag, Berlin, 1980, p. 14.

32. Enkelmann, V., loc. cit. p. 1.

33. Patel, G.N. and Walsh, E.K., J. Polym. Sci. Polym. Lett., 1979, 17, 203.

34. Bloor, D., Batchelder, D.N., Ando, D.J., Reed, R.T. and Young, R.J., to be published.

35. Young, R.J., Reed, R.T., Batchelder, D.N. and Bloor, D., J. Polym. Sci., in press.

36. Bloor, D., Koski, L., Stevens, G.C., Preston, F.H. and Ando, D.J., J. Mater. Sci., 1975, 10, 1678 and 1689.

37. Bloor, D., Ando, D.J., Preston, F.H. and Stevens, G.C., Chem. Phys. Lett., 1974, 24, 407.

38. Bloor, D. and Preston, F.H., Phys. Stat. Sol., 1976, (a)37, 427 and 607.

39. Reimer, B., Bassler, H., Hesse, J. and Weiser, G., Phys. Stat. Sol., 1976, (b)73, 709.

40. Muller, H., Eckhardt, C.J., Chance, R.R. and Baughman, R.H., Chem. Phys. Lett., 1977, 50, 22.

41. Hood, R.J., Muller, H., Eckhardt, C.J., Chance, R.R. and Yee, K.C., Chem. Phys. Lett., 1978, 54, 295.

42. Chance, R.R., Patel, C.N. and Witt, J.D., J. Chem. Phys., 1979, 71, 206.

43. Enkelmann, V. and Lando, J.B. Acta. Cryst., 1978, B34, 2352.

44. Batchelder, D.N., J. Polym. Sci. Polym. Phys., 1976, 14, 1235.

45. Reimer, B. and Bassler, H., Phys. Stat Sol., 1975, (a)32, 435.

46. Chance, R.R., Baughman, R.H., J. Chem. Phys., 1976, 64, 3889.

47. Cade, N.A., Chem. Phys. Lett., 1978, 53, 45.

48. Philpott, M.R., Chem Phys. Lett., 1977, 50, 18.

49. Yarkony, D.R., Chem. Phys., 1978, 33, 171.

50. Toyazawa, Y., Prog. Theor. Phys., 1958, 20, 53.

51. Bloor, D. and Hubble, C.L., to be published.

52. Philpott, M.R. and Swalen, J.D., J. Chem. Phys., 1978, 69, 2912.

53. Bloor, D. and Hubble, C.L., Chem. Phys. Lett., 1978, 56, 89.

54. Burke, F.P. and Small, G.J., Chem. Phys., 1974, 5, 198.

55. Batchelder, D.N. and Bloor, D., J. Phys. C. Sol. State Phys., 1978, 11, L629.

56. Cottle, A.C., Lewis, W.F. and Batchelder, D.N., J. Phys. C. Sol. State Phys., 1978, 11, 605.

57. Patel, G.N., Chance, R.R. and Witt, J.D., J. Polym. Sci. Polym. Lett., 1978, 16, 607 and J. Chem. Phys., 1979, 70, 4387.

58. Patel, G.N., J. Polym. Sci. Polym. Phys., 1979, 17, 1591.

59. Chance, R.R., Macromol., 1980, 13, 396.

60. Tieke, B., Thesis, University of Freiburg (1978).

61. Enkelmann, V. and Lando, J.B., Acta Cryst., 1978, B34, 2352.

62. Fisher, D.A., Batchelder, D.N. and Hursthouse, M.B., private communication.

63. Karpfen, A., private communication.

64. Berlin, A.A., J. Polym. Sci., 1961, 55, 621.

65. Pohl, H.A. and Chartoff, R.P., J. Polym. Sci., A3, 1964, 2, 2787.

66. Baughman, R.H., Exarhos, G.J. and Risen, W.M. Jr., J. Polym. Sci. Polym. Phys., 1974, 12, 2189.

67. Stevens, G.C. and Bloor, D., J. Polym. Sci. Polym. Phys., 1975, 13, 2411.

68. Eichele, H., Schwoerer, M., Huber, R. and Bloor, D., Chem. Phys. Lett., 1976, 42, 342.

69. Stevens, G.C. and Bloor, D., Phys. Stat. Sol., 1978, (a)45, 483; 1978, (a)46, 141 and 619.

70. Bubeck, C., Sixl, H. and Wolf, H.C., Chem. Phys., 1978, 32, 231.

71. Huber, R., Schwoerer, M., Bubeck, C. and Sizl, H., Chem. Phys. Lett., 1978, 53, 35.

72. Hori, Y. and Kispert, L.D., J. Chem. Phys., 1978, 69, 3826 and Am. Chem. Soc., 1979, 101, 3173.

73. Bubeck, C., Sixl, H., Bloor, D. and Wegner, G., Chem. Phys. Lett., 1979, 63, 574.

74. Bubeck, C., Sixl, H. and Neumann, W., Chem. Phys., in press.

75. Goldberg, I.B., Crowe, H.R., Newman, P.R., Heeger, A.J. and MacDiarmid, A.G., J. Chem. Phys., 1979, 70, 1132.

76. Bloor, D., Hubble, C.L. and Ando, D.J., "Molecular Metals," Ed. W.E. Hatfield, Plenum Press, New York, 1979, p. 243.

77. Hubble, C.L. and Bloor, D., to be published.

78. Fisher, D.A., Batchelder, D.N. and Hursthouse, M.B., Acta. Cryst., 1978, B34, 2365.

79. Young, R.J., private communication.

80. Bloor, D., unpublished data.

81. Sullivan, P.J. and Koski, W.S., J. Am. Chem. Soc., 1963, 85, 384.

82. Bernier, P., Rolland, M., Galtier, M., Montaner, A., Regis, M., Candille, M., Benoit, C., Aldissi, M., Linaya, C., Schule, F., Sledz, J., Fabre, J.M. and Giral, L., J. de Phys. Lett., 1979, 40, L297.

83. Snow, A., Brant, P., Weber, D. and Yang, N.L., J. Polym. Sci. Polym. Lett., 1979, 17, 263.

84. Schwoerer, M., Lauterbach, U., Muller, W. and Wegner, G., Chem. Phys. Lett., 1980, 69, 359.

85. Rice, M.J., Phys. Lett., 1979, 71A, 152.

86. Bishop, A.R., Solid State Comm., 1980, 33, 955.

87. Boudreaux, D.S., Chem. Phys. Lett., 1976, 38, 341.

88. Salem, L., "The Molecular Orbital Theory of Conjugated Systems," W.A. Benjamin Inc., New York, 1966, p. 154.

89. Penkovskii, V.V. and Kruglyak, Yu.A., J. Struct. Chem., 1969, 10, 378.

90. Kventsel, G.F. and Kruglyak, Yu.A., Theor. Chim. Acta., 1968, 12, 1.

91. Siddiqui, A.S. and Wilson, E.G., J. Phys. Sol. St. Phys., 1979, 12, 4237 and Siddiqui, A.S., J. Phys. Sol. St. Phys.

RECEIVED February 17, 1981.

# Bond Length Alternation and Forbidden Energy Gap in Conjugated Periodic Polymers

M. KERTESZ

Central Research Institute for Chemistry, Hungarian Academy of Sciences, 1525 Budapest, Pf. 17, Hungary

J. KOLLER and A. AZMAN

B. Kidric Chemical Research Institute, 61001 Ljubljana, P.p. 380, Yugoslavia

The Nuclear Distortion Problem

In recent years study of quasi one-dimensional (1D) systems has become an increasingly important field of solid state physics and chemistry (1). One basic idea here is the Peierls' theorem (2) stating that equidistant nuclear configurations of strictly 1D chains with partly filled energy bands (metallic spectrum) are unstable towards nuclear distortions leading to forbidden energy gaps in the one-particle spectrum. There is another instability of a quite different kind (3) present in the Hartree-Fock-type wavefunction of metallic (and sometimes also other) systems: introduction of certain extra variational freedom (usually formally increasing the lattice period) may lead to occurrence of symmetry breaking (SB) solutions such as charge-density-waves (CDW) and spin-density-waves (SDW). Performing Hartree-Fock (HF) type electronic band structure calculations on a number of 1D polymers (4, 5, 6) it has been observed that occurrence of CDW solutions points towards energetical advantageousness of nuclear distortions. Quite recently Calais, Pulay (7) and others have drawn attention to the study of this coupling mechanism. We examine the relation of these broken symmetry HF wavefunctions and the behavior of the potential energy curve of the nuclei around the equidistant configuration.

The degree of bond length alternation is closely related to the magnitude of the forbidden energy gap, $E_g$, which in turn affects electrical, optical and other physical properties of the system. Within the one-electron picture (Huckel-model) one can show, that $E_g = 2 \mid \beta_{short} - \beta_{long} \mid$ where $\beta_{short}$ and $\beta_{long}$ are the resonance integrals corresponding to the short and long bonds in an alternating polyene, respectively. The origin of the $E_g$ in pure polyacetylene, $(CH)_x$, has been discussed repeatedly (8) and we present some theoretical evidence at Hartree-Fock level, that nonzero experimental $E_g$ is compatible with the alternating model. A related problem is found at the polydiacetylenes (PDA's) where the two differently bond-alternated structures have different absorption.

0097-6156/81/0162-0105$05.00/0

## Forces Acting On The Nuclei

The results of a study of the Huckel Hamiltonian enabled the following classification of the shape of the total energy per unit cell, $E_t$, as function of the asymmetric distortion, $\Delta$, of the chain: (i) $E_t \sim x \, |\Delta| \, \ln \, |\Delta| + y\Delta^2$ (Peierls case); (ii) $E_t \sim -x'\Delta^2 + y'\Delta^2$ (heteroatomic chain), (iii) $E_t \sim \pm \, x''\Delta + y''\Delta^2$ (bond-alternating case). The corresponding Huckel Hamiltonians may be illustrated as: (i) $\underline{\beta_\alpha \beta_\alpha \beta}$; (ii) $\underline{\beta_\alpha \beta_{\alpha'} \beta}$; and (iii) $\underline{\beta_\alpha \beta'_{\alpha'} \beta}$. Case (i) is labile, case (iii) predicts necessarily bond length alternation, while in case (ii) the stability of the equidistant configuration ($\Delta = 0$) depends on the sign of $y' - x'$.

Without discussing more sophisticated Hamiltonians, explicitly including Coulomb interaction let us discuss the effect of charge density waves (CDW) heuristically. Suppose that there is a CDW-type electronic distribution advantegous for some or other reason in the chain and its effect is felt by the nuclei through Coulomb-forces. In order to take into account this effect within the Huckel model we have to change the $\alpha$'s and/or $\beta$'s. Traditional CDW corresponds to alternating + and − charges on neighboring atoms and consequently neighboring $\alpha$ values change also in an alternating pattern. The alternation of the resonance integrals, $\beta$, correspond to an off-diagonal CDW where charges alternate in the bond region (off-diagonal CDW or bond-order alternation wave, BOAW). As a consequence within the Huckel model, in case of BOAW, bond length distortion is predicted while in case of diagonal CDW this depends on the relation of x' and y'.

For a qualitative discussion the Hellmann-Feynman theorem may be used. Then, the forces acting on the nuclei are determined solely by the classical electrostatic forces due to the charge distribution of the electrons and nuclei. In order to see if the x component (along the chain) of the force acting on nucleus i, $F_{i,x}$ is zero or not, it is sufficient to check whether the charge distribution is invariant under the $x \to -x$ transformation or not. Since in equidistant linear arrangement the field of the other nuclei is invariant, it is necessary to consider the electrons only. From this consideration solely follows that nonzero Hellmann-Feynman forces occur in the BOAW case but not in the diagonal CDW case. This conclusion is also valid for zigzag chains if the nuclei are in planes of mirror symmetry orthogonal to the x-axis, like in equidistant polyacetylene, $(CH)_x$ and polymethyneimine, $(HCN)_x$. From the technical point of view it is important to note, that symmetry violating incorrect integral approximations used in the literature ([9]) lead in several cases incorrectly to BOAW and, consequently, to wrong potential energy curve around $\Delta = 0$.

## Actual Calculations  $(H)_x$, $(C)_x$, $(CH)_x$ and $(HCN)_x$ :

The main results of the calculations at <u>ab initio</u> Hartree-Fock level may be summarized, that the equidistant hydrogen, carbon and $(CH)_x$ chains have, besides the metallic solution another one with lower total energy possessing a CDW. The latter is of the BOAW type in all three cases, and consequently the corresponding total energy curve starts linearly as a function of $\Delta$ as in case (iii). Therefore the most stable nuclear configuration corresponds to an alternating structure in all three cases. Polymethneimine is different:  here the density matrix is of diagonal CDW-type, and this corresponds to case (ii). Actually the alternating geometry is slightly more favorable (<u>10</u>), but this conclusion may depend e.g. on the basis set used.

## The Energy Gap In Trans $(CH)_x$ :

From the above results it follows that the most stable geometry is an alternating one at HF level with nonzero $E_g$. Its actual value, however, is too large in comparison with the optical absorption maximum observed below 2 eV (<u>9</u>). This problem of the theoretical overestimation of the energy gap is a known drawback of the <u>ab initio</u> Hartree-Fock method (<u>11</u>) and serves as one of the main motivations to go beyond the HF description. Unfortunately, as we have recently observed (<u>12</u>), the simplest such description (the spin-unrestricted HF method) does not lower the gaps at all for the polymers in question in contrast to the suggestion of Misurkin and Ovchinnikov (<u>13</u>). The situation is complicated by the difficulty in separating the excitations creating free electrons and holes from the bound electron-hole quasi-particles (excitons). Only the former correspond to the $E_g$, while the creation of the energy of the latter, $E_{ex}$, lies within the gap. $E_g - E_{ex}$ may be even a large fraction of $E_g$, and the experiments which may distinguish between $E_{ex}$ and $E_g$ (photoconduction, photovoltaic effect) are difficult to carry out to yield unambiguous result. One of us has recently carried out an exciton calculation using a Pariser-Parr-Pople (PPP) model (<u>14</u>) working out exciton formalism in the intermediate binding (localized on, but also delocalized over several (CH) units). According to this as well as other exciton calculations (<u>15</u>), the 2 eV peak in $(CH)_x$ may very well be due to excitons. In case of the polydiacetylenes the difference of the photoconduction onset and the absorption threshold is $\sim 0.5$ eV, therefore in that case the exciton picture seems well established (<u>16</u>).

## The Blue Shift In Polydiacetylenes:

The structure of some polydiacetylene (PDA) single crystal polymers which closely resemble $(CH)_x$ are known due to existence of large single crystal samples. The two possible structures of the main chain are the acetylenic-type, $(= RC - C \equiv C - RC=)_n$, AT, and the butatriene-type $-(RC = C = C = RC)_n$, BT structures. There is a significant blue shift in the 2 eV range optical absorption spectrum of these materials of the order of 0.3 eV going from AT to BT structure. We believe, that this phenomenon is a consequence of the interaction of the electrons with the static lattice of the nuclei in the main chain. Its understanding is desirable as the experimental facts are much better known for DPA than for $(CH)_x$, although the alternation problem here is principally different from that of $(CH)_x$ since both structures have already a non-zero gap, and probably side chains are fixing the AT or BT type structure.

We have carried out crystal orbital calculations using the Huckel-model (as Wilson (17) did) as well as the more refined self-consistent PPP model. Next table summarizes a few results (in eV). As is apparent, the Huckel values are small and a big red shift is obtained. The PPP model with uniform $\beta$'s (but repulsion integrals taken according to alternating geometry) predicts blue shift and $E_g$ in rough agreement with experiment. True PPP model with alternating $\beta$'s gives very large gaps and a small blue shift.

Table I

| Methods | Acetylene | Butatriene | Shift |
|---|---|---|---|
| Experiment | 2.0 | 2.3 | +0.3 |
| Huckel [a] | 1.487 | 0.578 [b] | -0.909 |
| PPP (all $\beta = -2.39$eV) | 1.843 | 2.129 | +0.286 |
| PPP [a] | 5.855 | 5.905 | +0.050 |

a) Using values depending on bond length ala L. Salem (2);
b) Wilson value (17).

These computational findings may be rationalized by taking into account the exchange term in PPP. Within the first neighbor approximation $E_g = |2\beta_s^* - \beta_d^* - \beta_t|$ (17), where $\beta_i^* = \beta_i - 1/2 P_i \gamma_i$. The i index refers to the single (s), double (d) and triple (t) bond in the AT structure, and similarly for BT. The $-1/2 P_i \gamma_i$ term comes from exchange. In the Huckel model $\gamma_i = 0$. When we take all $\beta_i = -2.39$ eV only exchange determines the gap. The alternating effect of the $\beta$ resonance term enforces the alter-

nation of the exchange terms in the SCF procedure and this results
in the very large gaps of the third row in the table.

This result raises a question on the nature of the excited
state of an alternating conjugated polymer (18). It can be shown,
that if the excited state contains a collective shift of the
charge-density-wave (all $\pi$ electrons) over a finite region of N $\simeq$
10-20 $\pi$-electrons, then (a) this excitation is strongly coupled
to photons (large transition moment, $\mu \sim CN$ exp (-dN), derived in
analogy to Mulliken's charge-transfer theory (19)), (b) exhibits
the right shift for PDA's as given in the third row of Table 1,
even if corrected for the energy of the shift of the CDW over N
atoms, and (c) explains qualitatively some of the optical properties of PDA's (the nonlinear band tailing due to laser light (20),
the absence of a van Hove singularity in the spectrum (21), the
large third order susceptibilities (22), peculiarities of the
Raman spectra (16, 23)). These collective states involving a
shifted charge-density-wave plus electron hole are strongly coupled to photons, but the pure shifted CDW is not; its transition
moment does not contain a multiplicative factor of N (18). It is
worth mentioning that the importance of multiexcitations in the
spectra of finite polymers has been realized earlier (23).

This work was partly supported by the RC of Slovenia, by the
Hungarian Academy of Sciences, by the Division of Sponsored
Research as well as by the National Science Foundation (grant CHE
7906129). We are indebted to Dr. T.J. Fabish for the opportunity
to present our results at the 1980 ACS Houston Meeting.

## Abstract

The structural and electronic properties of semiconducting
conjugated polymers as polyacetylene (equivalent terminology:
polyene, $/CH/_x$), polydiacetylene /PDA/, and poly /p-phenylene/
are closely interrelated. Peierls theorem on nonexistence of
one-dimensional metals predicts e.g. the structure of trans-$/CH/_x$
to contain alternatingly short double- and long single-bonds,
with a nonzero forbidden energy gap, Eg. The equidistant model
with Eg=0 (metal) is, accordingly, energetically unstable. We
summarize herein the theoretical work done on some conjugated
polymers using combined solid state physical and quantum chemical
methods, as ab initio and semi empirical Hartree-Fock crystal
orbital calculations. We offer a classification of situations
which may occur for equidistant chains based on Hellmann-Feynman
forces acting on the nuclei: i/ Peierls case, ii/ diagonal charge
density wave (CDW) with zero forces, and iii/ non-diagonal CDW
with necessarily implied bond-distortion. According to the results of actual calculations $/CH/_x$ belongs to case ii/. The
corresponding energy gaps are discussed with reference to electronic correlation effects including the possibility of exciton-
formation. We attempt to elucidate the blue-shift of PDA's, which
is observed going from acetylene to butatriene structure. In-

formation concerning dopant- /CH/$_x$ interaction is obtained using protonation potential maps.

Literature Cited

1. See e.g. in "Quasi One-Dimensional Conductors", Proceedings Dubrovnik 1978, Lecture Notes Phys. 96, Ed. S. Barisic et al. Springer Verlag, 1979.

2. Peierls, R., "Quantum Theory of Solids", Oxford, 1955; Longuet-Higgins, H.C. and Salem, L., Proc. Roy. Soc., 1959, A251, 772.

3. Paldus, J. and Cizek, J., J. Chem. Phys., 1970, 52, 2919.

4. Kertesz, M., Koller, J. and Azman, A., Theoret. Chim. Acta (Berlin), 1976, 41, 89; Kertesz, M., Koller, J. and Azman, A., J. Chem. Phys., 1977, 67, 1180.

5. Kertesz, M., Koller, J. and Azman, A., J. Chem. Phys., 1978, 68, 2779.

6. Karpfen, A. and Petkov, J., Solid State Commun., 1979, 29, 329.

7. Calais, J.-L., Int. J. Quantum Chem., 1977, 12, 411 and Pulay, P. private communication.

8. For a discussion of this problem see 4 as well as Kertesz, M., Koller, J. and Azman, A., "Recent Advances in the Quantum Theory of Polymers," Lecture Notes in Physics Vol. 113, Ed. J.M. Andre, J.L. Bredas, J. Delhalle, J. Ladik, G. Leroy and C. Moser, Springer V., 1980, p. 56, as well as the Proceedings of the 3rd Int. Conf. Quant. Chem., Kyoto, 1979 (Int. J. Quantum Chem., in press).

9. For review see e.g. Fabish, T.J., "Electronic Structure of Polymers," Critical Reviews in Solid State Sciences, CRC Press, 1979, 8(4), 383, and references therein.

10. Karpfen, A., Chem. Phys. Lett., 1979, 64, 299.

11. See e.g. Pantelides, S.T., Mickish, D.J. and Kunz, A.B., Phys. Rev., 1974, B10, 2602.

12. Kertesz, M., Koller, J. and Azman, A., Phys. Rev., 1979, B19, 2034.

13. Misurkin, A. and Ovchinnikov, A.A., Uspekhi Khimii., 1977, 46, 1835.

14. Kertesz, M., Chem. Phys., 1979, 44, 349.

15. Duke, C.B., Paton, A., Salaneck, W.R., Thomas, H.R., Plummer, E.W., Heeger, A.J. and MacDiarmid, A.G., Chem. Phys. Lett., 1978, 59, 146.

16. For review of data and original references see e.g. Bloor, D., in "Quantum Theory of Polymers," CECAM Workshop, Namur Belgium, 7-11 Feb., 1979, Ed. J.M. Andre, J.L. Bredas, J. Delhalle, J. Ladik, G. Leroy and C. Moser, Lecture Notes in Physics 113, Springer Verlag, 1980, p. 14; see also D. Bloor, this volume.

17. Wilson, G., J. Phys. C; Solid State Phys., 1975, 8, 727.

18. Kertesz, M. (to be published).

19. Mulliken, R.S. and Person, W.B., "Molecular Complexes," Wiley, N.Y., 1969.

20. Lequime, M. and Hermann, J.P., Chem. Phys., 1977, 26, 431.

21. Shirakawa, H., Iko, T. and Ikeda, S., Makromol. Chemie., 1978, 179, 1565.

22. Sauteret, C., Hermann, J.P., Frey, R., Pradere, F., Ducuing, J., Banghman, R.H. and Chance, R.R., Phys. Rev. Lett., 1976, 36, 956.

23. See e.g. Schulthen, K., Ohmine, I. and Karplus, M., J. Chem. Phys., 1976, 64, 4422.

RECEIVED February 12, 1981.

# Models of Radical Cation States in Molecules, Molecular Solids, and Polymers

C. B. DUKE

Xerox Webster Research Center, Xerox Square-114, Rochester, NY 14644

In this paper I describe the construction of a model to interpret photoemission and optical-UV absorption from molecules in both gaseous and condensed phases. In the case of isolated molecules in the gas phase, molecular-orbital theory suffices to describe such spectra, although it must be adapted to include a consideration of the intramolecular relaxation induced by the ion (photoemission) or molecular exciton (UV absorption) generated via the absorption of an incident photon (1, 2). In the case of condensed systems, like molecular glasses, molecular crystals and polymers, additional phenomena must be incorporated into a model which affords a quantitative description of photoemission spectroscopy (PES) and UV absorption (UVA). Foremost among these is the intermolecular relaxation induced by ion (PES) and exciton (UVA) states (3, 4, 5). This relaxation causes a shift to lower binding energies of the order of 1eV for ion states and 0.1eV for exciton states (6). In addition, however, this shift varies from one site to another in the bulk of molecular glasses and polymers as well as at the surfaces of all condensed molecular and polymeric materials (6, 7). These variations generate surface states for molecular crystals (6, 8), localized molecular ion states in polymers and molecular glasses (6, 7, 9), and inhomogeneous broadening of PES and UVA spectra (7, 8, 10).

I proceed by first describing in Sec. II the construction of a molecular-orbital model suitable for interpreting molecular PES and UVA spectra. Then, this model is extended to encompass intramolecular (Sec. III) and intermolecular (Sec. IV) relaxation. The paper concludes with an indication of the applications of the model to interpret PES and UVA from polyacetylene (11) and from two pendant-group polymers, polystyrene and poly(2-vinyl pyridine) (9, 10).

0097-6156/81/0162-0113$05.00/0
© 1981 American Chemical Society

## CNDO/S3 Model

The molecular orbital model which has been developed to describe PES and UVA from aromatic hydrocarbons and heterocycles is a spectroscopically parameterized CNDO model, called the CNDO/S3 model, constructed to describe these spectra for benzene (12), p-xylene (12), pyrolle (13), furan (14) and p-difluorobenzene (15). The CNDO equations for the one electron orbitals are specified by Eqs. (1b)-(4b) in Lipari and Duke (12). The parameters utilized in these equations to define the CNDO/S3 model are given in Table I.

Table 1

Parameters used to define the CNDO/S3 model. The interatomic coulomb integrals are specified in terms of their intra-atomic values $(\gamma_A, \gamma_B)$ and the distance between the atomic centers $(R_{AB})$ via $\gamma_{AB} = 14.397 \; [28.794(\gamma_A + \gamma_B)^{-1} + R_{AB}]^{-1}$.

| Atom | $I_s$ (eV) | $I_p$ (eV) | $\beta_s$ (eV) | $\beta_p$ (eV) | $\gamma$ (eV) | $\zeta$ (A$^{-1}$) |
|------|------|------|------|------|------|------|
| H | 13.60 | ----- | 10 | -- | 12.85 | 2.33 |
| C(sp$^2$) | 21.34 | 11.54 | 20 | 17 | 10.63 | 3.78 |
| C(sp$^3$) | 21.34 | 11.54 | 20 | 17 | 10.63 | 3.07 |
| N | 27.51 | 14.34 | 25 | 20 | 12.37 | 3.03 |
| O | 35.50 | 17.91 | 31 | 26 | 13.10 | 4.32 |
| F | 43.70 | 20.89 | 39 | 33 | 15.18 | 4.25 |

The CNDO/S3 model has been applied to evaluate the electronic structure of a large number of aromatic hydrocarbons including m-xylene (12), p-xylene (12), 1,4-dis(trifluoromethyl) benzene (12), 1,2 di(p-tolyl)ethane (16), [2.2]-paracyclophane (16), pseudopara-dicyano[2.2]paracyclophane (16), pseudo-para-dichloro[2.2]paracyclophane (16), 1,1,2,2,9,9,10,10 octafluoro[2.2]paracyclophane (16), naphthalene (12), anthracene (12), napthacene (12), pentacene (12), ethylene (11, 15), tetrafluoroetylene (15), the fluorobenzenes $C_6F_{6-n}F_n, 1 < n \; 6 \leq (15)$, azulene (17), stilbene (17), diphenylbutadiene (17), diphenylhexatriene (17), diphenyloctatetraene (17), octatetraene (11), and polyacetylene (11). It also has been utilized to describe UPS and/or UV

absorption    spectra    of    7,7,8,8,-tetracyano-p-quinodimethane
("TCNQ") (18), pyrrole (13, 19), indole (13), isoindole (13),
9,10-dihydroacridine (13), 5H-dibenz[b,f]azepine (19), 10,11-di-
hydro-5H-dibenz[b,f]azepine (13, 19), free-base porphin (20),
chlorin (20), phlorin (20), phthalocyanine (19), triphenylamine
(21), N,N,-N',N'-tetramethyl-p-phenylenediamine (13), N,N'-di-
(m-tolyl)N,N,'-diphenyl-p-phenylenediamine    (13),    N,N,N',N'-
tetramethyl-4,4'-diaminobiphenyl (13), N,N'-di-(m-tolyl)-N,N'-
diphenyl-4,4' diaminobiphenyl (13), N,N'-di-(m-tolyl)-4,4' diaminobiphenyl (13), pyridine (10), 2-methyl pyridine (10), 2-
ethyl pyridine (10), 2-methyl pyridine (10), 2-ethyl pyridine
(10), and carbazole (22). Finally, although less attention has
been given the oxygen heterocycles, fairly extensive CNDO/S3
calculations have been performed for furan, benzofuran and di-
benzofuran (14, 19). We conclude, therefore, that the CNDO/S3
model has proved useful in evaluating the electronic structure of
a considerable number of aromatic hydrocarbons and heterocycles
via the interpretation of PES and UVA data obtained for such
molecules.

Two further extensions of the CNDO/S3 model also have proven
valuable. First, it has been utilized to evaluate electron-
molecular-vibration interactions in benzene (23, 24), TCNQ (25,
26), and TTF (26). A review of these calculations and their
applications to interpret the transport properties of molecular
crystals has been given elsewhere (27). In addition, the CNDO/S3
model also has been extended to encompass chalcogenide molecules,
most notably $S_8$ (28, 29), $Se_8$ (29), $S_4N_4$ (30), $As_4S_4$ (31), $As_4S_6$
(3), and $As_4Se_4$ (31).

## Relaxation In Isolated Molecules

When the absorption of a photon causes an electron to change
its quantum state (e.g., by leaving the solid in PES), the quantum
states of the other electrons in the molecule also are modified in
response to the changes in charge density associated with the
photon-stimulated transition. The total energy of the molecule
(or molecular cation in the case of PES) is lowered by this
process, which is called relaxation. Specifically, in PES the
difference in energy between the ground-state energy of the
molecular cation caused by the photoionization event and the
Hartree-Fock molecular orbital eigenvalue (in the canonical
basis) is defined to be the "relaxation energy" (1, 2, 4).

The contributions to the relaxation energy of an isolated
molecule, designated as intramolecular in character, $E_r$ (intra),
arise from two sources (4, 6). Electronic relaxations generate
energies $E_{re}$ (intra) $\simeq$ 1-2eV whereas atomic relaxations yield $E_{ra}$
(intra) $\simeq$ 0.1-0.2eV. The electronic contributions are built into
the semiempirical CNDO/S3 model by construction. Specifically,
the procedure of determining the model parameters by fitting PES
data for examplary molecules (benzene, furan, pyrrole) provides a

direct correspondence between the CNDO/S3 orbital eigenvalues of
the neutral molecules and the eigenstates of the corresponding
radical cations.

The atomic relaxations which occur upon photoionization have
been described using a composite model consisting of the CNDO/S3
description of the orbital eigenvalues and eigenfunctions together with suitable valence-force-field models of the molecular
normal modes of vibration ($\underline{27}$). Although the magnitude of the
atomic contributions to the relaxation energy are modest for
valence-electron photoionization ($E_{ra}$ (intra) $\cong 0.25eV$), atomic
relaxations in large rigid organic molecules make important modifications of the photoionization lineshapes which can be used as
"signatures" of the nature of the radical cation state. The
effect has been used, for example, to clarify the ordering of
photoinduced radical cation states in benzene, deuterobenzene,
and the fluorobenzenes ($\underline{15}$, $\underline{24}$). Electron-vibration interactions
deduced from examinations of PES data are crucial for the descrip-
tion of transport in molecular solids ($\underline{27}$) and of electron-
transfer reactions between such molecules in biological and electrochemical systems, like photosynthetic membranes ($\underline{32}$).

Another manifestation of photoinduced atomic relaxation in
molecules is the photogenerated change in geometry of composite
molecules comprised of fragments linked by single bonds. Moreover such changes are sensitive to the molecule's environment and
hence can differ between the gaseous and various condensed
states. The CNDO/S3 model has been applied to determine the
difference between radical cation geometries of aromatic amines
in the gas phase and in condensed thin films. A simple illustra-
tion of this phenomenon is afforted by the PES of triphenylamine
($\underline{21}$).

## Relaxation In Condensed Media

The photoinduced creation of an ion or exciton in condensed
molecular media causes greater relaxation than in isolated molecules because the induced charge redistribution generates electronic and atomic relaxation in the other constitutents of the
medium as well as in the excited molecule itself. The associated
intermolecular contributions to the relaxation energy are comparable in magnitude to the intramolecular ones from the molecule
itself, i.e., $E_{re}$ (inter)$\cong 1-2eV$, $E_{ra}$ (inter)$\cong 0.1eV$ ($\underline{4}$, $\underline{6}$, $\underline{9}$, $\underline{10}$).
A wide variety of models have been proposed to evaluate these
intermolecular contributions to the relaxation energy, including
ab initio models for hydrated electrons ($\underline{33}$) and hydrated ions
($\underline{34}$), microscopic models based on dipolar lattices ($\underline{35}$), and
dielectric models ranging from the classic Born model of solva-
tion to its more modern extensions ($\underline{9}$, $\underline{10}$, $\underline{36}$). The model which
we utilize to describe PES and UVA spectra is the dielectric model
developed by Duke $\underline{et}$ $\underline{al}$., ($\underline{9}$, $\underline{10}$). Specifically, this model
predicts successfully both the relaxation induced shifts and

additional widths relative to the gas phase of condensed-state
PES ionizations ($\underline{6}$, $\underline{7}$, $\underline{10}$).  Specific examples include $S_8$ ($\underline{6}$),
TCNQ ($\underline{6}$), anthracene ($\underline{8}$), ethyl benzene ($\underline{9}$, $\underline{10}$) and 2-ethyl
pyridine ($\underline{10}$).

## Applications To Polymers

The details of the applications of these models to describe
the electronic properties of polymers depend on the polymer being
considered.  In the case of polymers which consist of aromatic
pendant groups suspended from an aliphatic backbone, photoinduced
molecular cation states tend to be localized on the individual
aromatic moieties rather than extended over the entire macromole-
cule ($\underline{6}$, $\underline{7}$, $\underline{10}$).  PES and UVA from polystyrene and poly(2-vinyl
pyridine) afford excellent examples of the spectroscopy of such
localized molecular ion states ($\underline{9}$, $\underline{10}$).  Moreover, based on the
concepts developed from such spectroscopy, a quantitative phenomenological model of the steady state contact charge exchange
properties of these and related polymers has been developed ($\underline{37}$).
In the case of non-saturated-backbond polymers like poly(sulfur
nitride), polyacetylene and the polydiacetylenes, however, the
electronic excitations may be extended over a larger portion of
the macromolecular backbone, the detailed extent depending upon
the chain conformation and polymer morphology.  Trans-polyacetyl-
ene constitutes an example of such a situation, because its PES
and UVA spectra admit a simple interpretation in terms of the same
CNDO/S3 model which describes these spectra for large, even
polyenes ($\underline{11}$).  From the success of the composite CNDO/S3-
dielectric relaxation model in describing PES and UVA from such
diverse linear polymers, we conclude that this composite model,
described in detail by Duke et al. ($\underline{10}$), should be widely useful
in the quantitative interpretation of these spectra from other
polymers as well.

## Literature Cited

1.   Wittel, K. and McGlynn, S.P., Chem. Revs., 1977, 77, 745.

2.   Salaneck, W.R., preceeding paper.

3.   Duke, C.B., Salaneck, W.R., Paton, A., Liang, K.S., Lipari,
     N.O. and Zallen, R., in "Structure and Excitations of Amor-
     phous Solids", G. Lucovsky and F. Galeener, eds. (American
     Institute of Physics, New York, 1976), pp. 23-30.

4.   Duke, C.B. and Fabish, T.J., Phys. Rev. Lett., 1976, 37,
     1075.

5.   Grobman, W.D. and Koch, E.E., in "Photoemission in Solids",
     L. Ley and M. Cardona, eds. (Springer-Verlag, Berlin, 1979),
     Vol. 2, pp. 261-298.

6.  Duke, C.B., Surface Sci., 1978, 70, 671.

7.  Duke, C.B., Fabish, T.J. and Paton, A., Chem. Phys. Lett., 1977, 49, 133.

8.  Salaneck, W.R., Phys. Rev. Lett., 1978, 40, 60.

9.  Duke, C.B., Mol. Cryst. Liq. Cryst., 1979, 50, 63.

10. Duke, C.B., Salaneck, W.R., Fabish, T.J., Ritsko, J.J., Thomas, H.R. and Paton, A., Phys. Rev. B, 1978, 18, 5717.

11. Duke, C.B., Paton, A., Salaneck, W.R., Thomas, H.R., Plummer, E.W., Heeger, A.J. and MacDiarmid, A.G., Chem. Phys. Lett., 1978, 59, 146.

12. Lipari, N.O. and Duke, C.B., J. Chem. Phys., 1975, 63, 1748; 1975, 1768.

13. Duke, C.B. and Yip, K.L., Bull. Am. Phys. Soc., 1977, 22, 399.

14. Yip, K.L. and Duke, C.B., Bull. Am. Phys. Soc., 1977, 22, 399.

15. Duke, C.B., Yip, K.L., Ceasar, G.P., Potts, A.W. and Streets, D.G., J. Chem. Phys., 1977, 66, 256.

16. Duke, C.B., Lipari, N.O., Salaneck, W.R. and Schein, L.B., J. Chem. Phys., 1975, 63, 1758.

17. Yip, K.L., Lipari, N.O., Duke, C.B., Hudson, B.S. and Diamond, J., J. Chem. Phys., 1976, 64, 4020.

18. Lipari, N.O., Nielsen, P., Ritsko, J.J., Epstein, A.J. and Sandman, D., Phys. Rev. B., 1976, 14, 2229.

19. Duke, C.B., Int. J. Quant. Chem., 1979, 13, 267.

20. Yip, K.L., Duke, C.B., Salaneck, W.R., Plummer, E.W. and Loubriel, G., Chem. Phys. Lett., 1977, 49, 530.

21. Duke, C.B., Lin, J.W-p, Paton, A., Salaneck, W.R. and Yip, K.L., Chem. Phys. Lett., 1979, 61, 402.

22. Duke, C.B., J. Vac. Sci. Technol., 1978, 15, 157.

23. Duke, C.B., Lipari, N.O. and Pietronero, L., Chem. Phys. Lett., 1975, 30, 415; J. Chem. Phys., 1976, 65, 1165.

24. Duke, C.B. and Lipari, N.O., Chem. Phys., 1975, 36, 51.

25. Lipari, N.O., Duke, C.B., Bozio, R., Girlando, A., Pecile, C. and Padva, A., Chem. Phys. Lett., 1976, 44, 236.

26. Lipari, N.O., Rice, M.J., Duke, C.B., Bozio, R., Girlando, A. and Pecile, C., Int. J. Quant. Chem., 1977, 11, 583.

27. Duke, C.B., in "Synthesis and Properties of Low-Dimensional Materials", J.E. Miller and A.J. Epstein, eds. (New York Academy of Sciences, New York, 1978), pp. 166-178.

28. Salaneck, W.R., Lipari, N.O., Paton, A., Zallen, R. and Liang, K.S., Phys. Rev. B., 1975, 12, 1493.

29. Salaneck, W.R., Duke, C.B., Paton, A., Griffiths, C. and Keezer, R.C., Phys. Rev. B., 1977, 15, 1100.

30. Salaneck, W.R., Lin, J.W-p., Paton, A., Duke, C.B. and Ceasar, G.P., Phys. Rev. B., 1976, 13, 4517.

31. Salaneck, W.R., Liang, K.S., Paton, A. and Lipari, N.O., Phys. Rev. B., 1975, 12, 725.

32. Duke, C.B., in "Tunneling in Biological Systems", B. Chance et al., eds. (Academic, New York, 1979), pp. 31-65.

33. Newton, M.D., J. Phys. Chem., 1975, 79, 2795.

34. Clementi, E., "Lecture Notes in Chemistry", (Springer Verlag, Berlin, 1976), Vol. 2.

35. Lyons, L.E. and Mackie, J.C., Proc. Chem. Soc., 1962, 1962, 71.

36. Dogonadze, R.R. and Kornyshev, R., J. Chem. Soc. Farad. Trans. II, 1974, 70, 1121.

37. Duke, C.B. and Fabish, T.J., J. Appl. Phys., 1978, 49, 315.

RECEIVED December 22, 1980.

# Intermolecular Relaxation Effects in the Ultraviolet Photoelectron Spectroscopy of Molecular Solids

## Molecular-Ion States in Aromatic Pendant Group Polymers

W. R. SALANECK

Xerox Webster Research Center, Webster, NY 14580

This contribution deals with the use of ultraviolet photo-electron spectroscopy (UPS) for the study of the surface and bulk electronic structure of organic molecular and polymeric solids. In so far as is necessary, some features of the UPS of isolated model monomer molecules in the gas phase are described in order to provide a basis for an understanding of certain phenomena that occur in the corresponding condensed molecular and polymeric solids. Some features of photoelectron spectroscopy in general are outlined with an emphasis on the phenomenological interpretation of spectra for the several case studies to be reviewed. The complimentary nature of X-ray photoelectron spectroscopy (XPS or sometimes ESCA) and UPS is pointed out. The discussions presented are focused upon the experimental aspects of the UPS of insulating organic molecular and polymeric solids, but specific hardware considerations are not included. A variety of references, some of a review nature, are included, but the content is not intended to be historically complete. Examples for examination are drawn primarily from the author's own experience.

## Historical Background

The techniques of UPS (1, 2) and XPS, also known as ESCA (3, 4), are well documented in the literature (5). These two branches of photoelectron spectroscopy have been divided historically into two categories based upon technical grounds: UPS utilizing far uv photons with energies typically less than about 50 eV; with XPS utilizing the soft X-ray photon sources $MgK_{\alpha}$ (1254 eV) and $AlK_{\alpha}$ (1487 eV). Recent use of synchrotron radiation sources offering a wide range of photon energy make this distinction less appropriate (5, 6, 7). From another point of view, however, UPS deals with valence electronic structure of matter by virtue of the photon energies involved, whereas XPS can be used to study both valence and certain core-level electronic states.

0097-6156/81/0162-0121$07.25/0

ESCA or XPS of organic polymer materials has been pioneered by Clark and coworkers (8), who have worked mainly with core-level spectra. Valence levels of organic polymers have also been studied by XPS by, among others, Clark and coworkers (8) as well as Andre, Pireaux and collegues (9). The UPS of organic molecules in gas phase is very widely practiced (1, 2), but far less has been done on organic molecular solids (10, 11). Some inorganic molecular and polymeric solids have been studied by photoelectron spectroscopy (10), but are not referred to here. Very few studies of organic polymeric solids have been reported, primarily because of the incipient technical difficulties involved in obtaining undistorted data. The earliest work that involves UPS spectra of a real polymer is that of Fujichira and Inokuchi, who used phonon energies up to 11.27 eV to study the density-of-states of poly-ethylene (12). The polymer studies discussed here, however, represent the most complete combined UPS and optical absorption studies of any polymer systems to date.

## Photoelectron Spectra of Molecular Solids: Basic Issues

Isolated Molecules. Some familiarity with a few central issues in photoelectron spectroscopy (PES) is a prerequisit to the discussion of localized molecular-ion states in aromatic pendant group (APG) polymers that follows. One must be concerned with what is actually measured in PES, in relation to what one would really like to know about. The basic problem is that information is usually required on the chemical and electronic structure of neutral molecules in the ground state, whereas PES measures the (sometimes multi-) excited states of the correspond-ing molecular cations. Keeping in mind the proper concept enables at least a better phenomenological interpretation of spectra. Quantitative analysis then relies either on comparison of a given spectrum with corresponding spectra of a series of logically chosen model molecules, or comparison with appropriate model calculations. The photoemission process for molecular photo-electron spectroscopy is outlined below (13).

For an isolated molecule the basic excitation by absorption of a photon of energy $\hbar\omega$ is

$$E_T^i(N) \; + \; \hbar\omega \; = \; E_T^f(N,k), \tag{1}$$

where $E_T^i(N)$ is the total initial ground state energy of the neutral molecule containing N electrons, and $E_T^f(N,k)$ is the total energy of the N-electron system in the k final state, including the photoelectron. When $\hbar\omega$ is sufficiently large, the photo-electron is only weakly coupled to the ion containing N-1 elec-trons in the final state of the excitation process. Then the final state wave function is separable, and the energy can be partitioned into

$$E_T^f(N) = E_T^f(N-1,k) + E_K(k), \tag{2}$$

where $E_T^f(N-1,k)$ is the total energy of the N-1 electron molecular ion in the k final state, and $E_K(k)$ is the kinetic energy of the electron emitted from state k. A photoelectron spectrum consists of a plot of the number of electrons detected per unit time at an energy $E_K$, versus $E_K$, that is, $n(E_K)$. Combining Eq's (1) and (2) yields the standard operation equation in photoelectron spectroscopy,

$$E_B(k) = \hbar\omega - E_K(k), \tag{3}$$

where

$$E_B(k) = E_T^f(N-1,k) - E_T^i(N). \tag{4}$$

The $E_B(k)$ are the so-called binding energies of the electrons emitted from the state k and are determined from the peaks in $n(E_K)$ through Eq. (3). Note that k labels the possible <u>final</u> electronic configurations of the molecular ion, and specifically not initial states. This means that the peaks in $n(E_K)$ correspond to states of the molecular ion and must be related to the states of the neutral molecule through some intermediary step(s).

Phenomenological interpretation of PES spectra most commonly relies upon one-electron models of the neutral molecule. One draws a one-to-one correspondence between the major peaks in $n(E_K)$ and the one-electron molecular orbitals (MO's) (<u>19</u>). If one sets the numerical values of the energies of the peaks in $n(E_K)$ equal to the energy eigenvalues of the MO's of the neutral ground state molecule, one has the commonly employed Koopman's theorem (<u>1-6</u>). This simple equality is obviously quantitatively, and often qualitatively (<u>15</u>), insufficient. For interpretational simplicity the most popular practice is to discuss the excited states of the molecular ion in terminology that employs the one-electron basis functions (MO's) of the neutral molecule in the ground state. It is then necessary to put in "by hand" (or in some more sophisticated theoretical fashion <u>15</u>) the electronic reorganization and electron correlation effects that lead to the so-called relaxation phenomena that account for the differences between the MO's of the neutral molecule and the molecular ion. In a phenomenological view, the eigenstates of the molecular ion are not those of the neutral molecule, and this fact must be taken into account in interpreting spectra, and in comparing spectra with numerical values of model calculations. So-called "shake-up" satellites are manifestations of the multielectron processes that do not fall within the framework of one-electron models of photoemission (<u>17</u>). However, shake-up effects are many-body effects they are sometimes rationalized within the basis of a one-electron framework. A wealth of literature exists that deals with these issues (17). For the present purposes, however, we need to keep in mind the fact that intra-molecular relaxation occurs when a hole-state is created (positive charge left behind) on a given molecule in photoemission, and to extend the concept to inter-molecular relaxation in molecular solids.

Molecular Solids. Consider a Van der Waals molecular solid as one in which the molecules not only retain their identity but are held together only by weak Van der Waals forces (10). The polymers to be considered subsequently fall into this category with respect to the polymer chains. In the condensed Van der Waals molecular solid phase, in addition to the intramolecular electronic relaxation effects that occur in response to the generation of the hole-state in PES of isolated molecules in the gas phase, as discussed above, there exist intermolecular relaxation effects (10, 11, 18). Phenomenologically, the intermolecular relaxation is due to the electronic polarization of electrons in molecules surrounding the given molecule which experiences photoemission of an electron. This polarization energy, or intermolecular relaxation energy, is given to the escaping photoelectron. The corresponding peak in the $n(E_k)$ spectrum then appears at higher kinetic energy, and thus lower apparent binding energy, when photoemission is from the molecule in the condensed molecular solid state as compared with photoemission from the same level, k, when the molecule is isolated in the gaseous state. The photoionized molecule itself experiences a change in total charge, a subsequent redistribution of the remaining net charge and an ultimate accompanying distortion in the molecular geometry (21). This relaxation then results in several special characteristics of photoelectron spectra of molecular solids.

For example, molecules in the surface molecular layer of a molecular solid experience different intermolecular relaxation than their counterparts in the bulk (18). This difference is due to the existance of a smaller number of nearest neighbor molecules for a given molecule on the surface, as compared to within the bulk solid. Also, the electronic states of a insulating organic Van der Waals molecular solid are localized on the individual molecules (21, 22). This localization in space leads to localization in energy of the electronic states of the excited neutral molecule in optical absorption, and of the molecular cation in photoelectron spectroscopy. In the question of charge (electron or hole) transport in molecular solids, sufficient localization of the molecular-ion states can preclude an energy band picture for the solid (10, 11, 23). Electronic transport is governed by so-called transfer integrals between nearest-neighbor molecules which tend to delocalize electronic states in molecular crystals (21, 23). The energy localization now provides an energy difference between neighboring one-electron orbitals. The competition between these two phenomena determines the detailed nature of the charge transport in molecular solids (21, 23). Surface transport

properties may be expected to be significantly different, however, due to the smaller intermolecular relaxation on the molecular solid surface (18).

Other consequences of intermolecular relaxation are evident
in UPS. Both the intermolecular relaxation energy shift (in gas
versus solid state spectra) and the broadening of UPS lines in
molecular solids as compared with the corresponding molecular
gases, can be explained by starting with a one-electron formalism
and including the coupling of the electronic states (MO's) of an
isolated molecule to the vibrational modes of the parent molecule
as well as to the longitudinal polarization fluctuations of the
molecular solid (dielectric medium) in which the parent molecule
is imbedded (20). When cast in a dielectric constant framework,
the intermolecular relaxation energy shifts result from weak
coupling of the localized molecular electronic states to certain
excitations of the dielectric medium, while the broadening of UPS
lines in molecular solids result from strong coupling of the
electronic states of the molecule to other excitations of the
medium (20, 21, 25, 26, 27). These issues will be elucidated
further in subsequent parts of this article.

UPS Versus XPS. There are several differences between the
UPS and XPS spectra of insulating organic molecular solids. XPS
of course enables the investigation of core-levels to within
about 1400 eV of the vacuum level (3, 4). UPS is limited to
valence region and some very low binding energy core-like levels
by virtue of the lower photon energies involved (1, 2, 5, 7). Two
other features are perhaps less obvious. First, because of the
inherent line width of the typical photon sources involved, UPS is
capable of higher resolution (1, 2). X-ray monochromatization
helps this point for XPS or ESCA (6), but not to the extent of
UPS. Several discussions can be found in the literature for
details (11). The resolution inherent in UPS is not realized in
condensed molecular solids or polymeric materials, however. Line
widths of UPS peaks in condensed molecular solids exhibit widths
(full-width-at-half-maximum) around 1 eV near room temperature
(18, 20, 26, 27), even when the particular UPS peak is very narrow
for the molecule in the gas phase. There are a variety of
contributions to the UPS line widths in condensed molecular
solids. These will be addressed in the discussions of the model
molecule work on anthracene and isopropyl benzene below.

Perhaps the most useful of the differences between XPS and
UPS valence spectra are the photoelectron cross sections that
vary with energy. We mention only two extremes, XPS at about 1200
to 1400 eV and UPS in the range of about 20 to 40 eV. When dealing
with organic insulating molecular solids, we are concerned mainly
with the $C(2p)$ and $C(2s)$ atomic orbitals in as much as they
contribute to the valence spectra of the various molecules. At
about 21 eV photon energy, the $C(2p)$ cross section is over ten
times that of the $C(2s)$ cross section, while at 1200 to 1400 eV,

the C(2s) cross section is over a factor of ten larger than the
C(2p) cross section (6). Thus at UPS energies, the C(2p)-derived
states are greatly emphasized over the C(2s)-derived states,
while in XPS valence spectra, mostly the broad C(2s)-derived
states are observed. In this sense XPS and UPS are complementary
for the study of organic molecular solids. Incidentally, the np-
to-ns cross section ratios are different for different elements,
and the above scenario does not necessarily apply for other
elements.

In the examples of our work on organic molecular and polymeric solids that follow, first some contributions to the UPS line
widths in condensed molecular solids are discussed for two prototype systems, anthracene and isopropyl benzene; then the UPS of
two aromatic pendant group polymers, polystyrene and poly(2-vinyl
pyridine), are discussed and compared with some spectra concerning the simplest linear conjugated polymer, polyacetylene.

## Some Technical Aspects of the UPS of Insulating Solids

Sample preparation is of critical importance in obtaining
distortion free UPS spectra of insulating molecular solids. This
issue has been discussed at length in the literature (30). We
usually employ vapor deposition onto cold metal substrates to
prepare samples of insulating molecular solids (18, 30). In the
case of polymers, however, solution casting in an inert atmosphere with subsequent vacuum drying has been employed successfully (20). It has been determined that thin films of thickness
in the 20 to 30 Å range, are required. This thickness is
determined by the magnitude of the escape depth of the photoelectrons, which for UPS and organic insulators is in the range of
about 5 Å (18, 31). Films of thickness greater than about 50 Å
yield debilitating charging affects that affect the acquired UPS
spectra (32). If the sample is an insulator and too thick, for
every electron removed from the surface region of the sample, a
positive charge remains behind. There is then Coulomb attraction
between the positive charge and subsequent excaping photoelectrons. This attractive interaction leads to various degrees
smearing of the UPS spectra. At a thickness of about 20 to 30 Å,
however, the sample is still thick enough that photoemission from
the metal substrate does not occur, yet thin enough that sample
charging does not occur. The lack of charging can be attributed
to any or all of several possible phenomena: photoinjection of
negative charge from the substrate, since the u.v. photons have an
attenuation length greater than the sample thickness; electron
tunneling from the substrate; or direct electric-field-dependent
transport of the positive charge from the surface, through the
thickness of the film, and to the substrate.

Some XPS valence band spectra of thick insulating samples
have been obtained using a low energy electron emission source
(flood gun) to neutralize the positive surface charge. This

practice has been shown in at least one case, however, to itself
lead to distortions of the valence spectra obtained (33).   We
never employ flood gun techniques for this reason.

Another issue of critical importance is that of surface
cleanliness.   Since the photoelectron escape depths from organic
insulating materials at typical UPS energies lie in the region of
about 5 Å, surface contamination can  distort UPS spectra.   The
spectroscopy is thus carried out in an ultra high vacuum environment, with pressures of about $10^{-9}$ Torr or less, and numerous
steps are taken to assure a clean sample surface.   Probably the
most certain practice is that of in situ vapor deposition with
subsequent examination by UPS before contamination can occur.
This technique has been used to prepare both non-single-crystal-
line and crystalline samples.   In the case of the polymers to be
discussed, the films were cast upon oxidized aluminum substrates
in a flowing $N_2$ atmosphere following by vacuum drying by immediate
insertion into the UPS spectrometer.   By trial and error, the
technique was perfected to generate uniform thickness films with
negligible surface contamination.   The details are discussed in
Ref. 20 and in the Appendix to this article.

## Examples of Contributions of Intermolecular Relaxation Effects to UPS Spectra of Molecular Solids

In this section, several studies of model molecular solids
are reviewed that shed some light on the origin of the line widths
and gas-to-solid spectral energy shifts in the UPS spectra of
molecular solids.   First, examples are given of contributions to
the solid state UPS line broadening.   A following discussion deals
with the energy shift and electronic localization issues, which
are directly related.

### Surface Effects in Intermolecular Relaxation.
Using anthra-
cene molecules, the effect of the surface on the intermolecular
relaxation energy shift in gas versus molecular solid UPS spectra
has been investigated (18).   This shift is sometimes referred to
as the polarization energy shift.   In UPS studies, it is observed
that spectra of molecules in the condensed state are shifted
approximately rigidly towards higher kinetic energies (towards
lower apparent binding energies) with respect to the corresponding gas phase spectra.   The amount of the shift is called the
intermolecular relaxation energy, or polarization energy, $E_r$ (11,
18, 19, 34).   The surface monomolecular layer will experience a
polarization different from that of succeeding layers in the bulk
film, since the number of nearest neighbor molecules is different
on the surface compared with in the bulk.

Anthracene was chosen as an ideal molecule with which to look
for a surface contribution to $E_r$ for several reasons:   (a) the
first IP (photoemission peak) corresponds to a delocalized  π-
state (35), well separated in energy from the remaining IP's; (b)

the molecule is flat and stable, enabling the preparation of thin vapor-deposited films, which under the proper conditions will have the molecules lying flat on the substrate (36); and (c) optical studies had already indicated that surface excitons in anthracene are different than those in the bulk (37).

For this study the spectrometer geometry was important. The angle between the photon source and the analyzer entrance slit was fixed at 90°. However, the sample could be rotated such that the analyzer would collect electrons photoemitted nearly parallel to the sample normal or nearly perpendicular to the sample normal (i.e., nearly parallel to the surface). This latter geometry enables the observation of only the top surface molecular mono-layer when electrons of sufficiently small electron elastic mean free path, $\lambda$, are collected (38). In order to avoid any possible complication of the spectra due to s-polarization (vector $\overline{E}$ of the photon beam perpendicular to the place of incidence) and p-polarization (vector $\overline{E}$ parallel to the plane of incidence) of the light beam, the uv lamp was mounted on a port 45° off of the axis of rotation of the sample, while maintaining the 90° angle to the analyzer. Thus, the polarization effects were minimized. In addition, the data were only accumulated over a 60° range of $\theta$ (from $\theta = 20°$ to $\theta = 80°$), further minimizing any possible polariza-tion effects.

The samples were prepared as follows: A gold foil substrate was ion-sputter cleaned, and the energy scale calibration was checked by assuring that the recorded UPS kinetic-energy differ-ence between the gold Fermi level in the HeI (21.21 eV) spectrum and that in the He II (40.81 eV) spectrum was exactly 19.60 eV. Energy resolution was about 0.1 eV as judged by the width of the Fermi edge of the gold substrate. Then 99.99 % pure zone-refined anthracene, purchased from Materials Limited, was vapor deposited at the very low rate of about 40 Å/hour onto the gold foil at -100°C. Under proper conditions, (36) planar molecules lie "flat" on the substrate. At very low temperatures, the surface mobili-ties are apparently small enough to prevent nucleation and growth of needle-like crystals as, for example, observed for other planar organic molecules deposited near room temperature (31). The final sample thicknesses were always about 20 Å, as determined by attenuation of the XPS Au($4f_{7/2}$) signal using MgK$\alpha$ (1253.7 eV) radiation and a $\lambda = 22.1$Å appropriate for organic overlayers on Au substrates and MgK$\alpha$ X-rays (39). The 20 Å sample thickness was important. It was thick enough that the Au-substrate photo-emission peaks were completely absent from the UPS data, yet thin enough that no sample charging occurred, as measured by the sharpness of the cutoff of the secondary-electron distribution.

The wide-scan UPS spectrum obtained at 21.2 eV photon energy (18) (He I radiation) is shown in Fig. 1. The data are in excellent agreement with those of Grobman and Koch (11) on anthracene crystals. In Fig. 2, narrow (4 eV width) scans through IP(1) reveal a broadening on the high-kinetic-energy (lower bind-

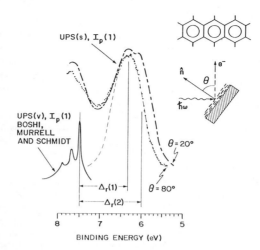

*Figure 1.   HeI UPS data for a 20-Å anthracene film on a gold substrate collected at θ = 20° (data points).   The inset defines the geometry.   The gas phase UPS data of Ref. 61 are traced below.   Note the relaxation energy shift of $E_r = 1.2$ eV between the energy scales, both of which refer to the vacuum level.*

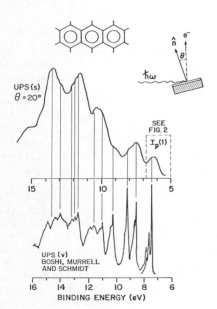

*Figure 2.   High-resolution HeI UPS angle-dependent data on $I_{po}(1)$ of anthracene.   For clarity, the θ = 20 data are approximately normalized to the peak intensity of the θ = 80° data at about 6.3-eV binding energy; IP(1) from Ref. 61 is shown for determination of the top-layer relaxation energy, $E_r(1) = 1.2 \pm 0.1$ eV and the subsequent layer value, $E_r(n > 2) = 1.5 \pm 0.1$ eV.*

ing energy) side for near normal electron exit ($\theta$ = 20$^\circ$). This
broadening is not observed in otherwise identical studies employing larger cagelike molecules where molecular dimensions are
larger than the electron elastic mean free path, $\lambda$ , and thus the
effect is not expected to occur (18). The data of Fig. 2 are
normalized to approximately equal peak intensity for display
purposes. The dashed line for the $\theta$ = 80$^\circ$ data (electron exit
nearly parallel to the surface) are expected to correspond to only
the top molecular monolayer when $\lambda \lesssim d$, the molecular thickness
(38). As an estimate of $\lambda$ for the present study, recall that
Nielsen has determined that $\lambda$ is approximately one molecular
thickness (about 3.5 Å) in tetrathiafulvalene (TTF) and tetracyanoquinodimethane (TCNQ) molecules at about 5-eV electron kinetic energy above the vacuum level (31).

Thus, using the leading (low-binding-energy) edge of the
spectrum to determine the line shape, the peak shown as a dashed
line in Fig. 2 represents IP(1) of the top molecular monolayer of
anthracene. The full width at half-maximum (FWHM) of IP(1) is 0.7
eV. This width is not due to sample charging in view of the 20-Å
mean thickness and the sharpness of the observed cutoff of the HeI
secondary-electron spectrum. Origins of this 0.7 eV contribution
to the UPS line width will be discussed below. Note that the
relaxation-energy shift, $E_r$, is 1.2 eV for IP(1), in agreement
with the results of Grobman and Koch (11) for approximately
tangential electron exit. Here $E_r$ is referred to the most intense
peak in vibrationally split IP(1) consistent with Ref. 11 rather
than the center of gravity of the vibrational envelope. This
latter case would lead to an increase in the magnitude of $E_r$ by
about 0.1 eV.

Finally, at $\theta$ = 80$^\circ$, the unscattered electrons emanate from
$\lambda(80^\circ)$ = $\lambda(\cos 80^\circ)$ = 0.17 $\lambda$ from within the solid. Since the
intermolecular spacing in anthracene is (8.7 Å per unit cell)/(3
molecules per unit cell) = 2.9 Å/molecule (40), $\lambda$ would have to
be about 26 Å in order to see a significant contribution to IP(1)
from the middle of the second anthracene layer in the $\theta$ = 80$^\circ$
data, an unreasonable number at 15 eV kinetic energy (31, 39).
Alternatively, if $\lambda$ =26 Å, then the features of the Au substrate
would show distinctively in the $\theta$ = 20$^\circ$ UPS data for 20-Å films.
Thus the XPS thickness determination alone puts an upper limit on
$\lambda$ in the UPS energy region of less than 10 Å.

The $\theta$ = 20$^\circ$ UPS data of Fig. 2 then include contributions to
IP(1) from the succeeding sublayers of the sample. Using the
dashed line as a measure of IP(1) from only the first molecular
monolayer, the $\theta$ = 20$^\circ$ data were deconvoluted with a Du Pont No.
610 curve resolver into only two peaks of FWHM = 0.7 eV, of
approximately equal intensity, and separated in energy by 0.3 eV.
This fact implies that the second and subsequent anthracene
layers have $E_r$ = 1.5 ± 0.1 eV to within the accuracy of this
experiment. A very simple analysis based upon the peak-intensity
ratio and intermolecular spacing yields a measure of the elastic

mean free path for the photoelectrons involved: $\lambda = 4$ Å at 15 eV
above the vacuum level, consistent with the 20-Å sample thickness
necessary to obscure all traces of photoemission from the Au
substrate in the UPS data, while not in the XPS data, where $\lambda = 22$
Å at about 1200 eV (36).

The surface effect discussed above is an aspect of intermolecular relaxation in molecular solids is that associated with
the depth of the molecule from the sample surface. This effect
manifests itself as an apparent line-broadening in UPS spectra,
other contributions to line-broadening also exist. These, as
well as some specific mechanisms that lead to the observed
intermolecular polarization effect will be discussed below.

Homogeneous and Inhomogeneous Contributions. Two other
contributions to UPS linewidths for molecular solids have been
articulated in a study of isopropyl benzene films at low temperatures (24). The shape and size of the isopropyl benzene molecule
prohibited the explicit observation of the surface effect discussed for anthracene. Isopropyl benzene was of interest as a
model molecule for polystyrene, however. The measurements were
carried out on condensed molecular-solid films in the temperature
range $15^{\circ}K < T < 150^{\circ}K$.

Isopropyl benzene, a schematic diagram of which is included
in Fig. 3, was utilized in this study for several reasons. The
primary motivation for the study was the acquisition of data on
the temperature dependence of UPS linewidths in molecular
solids. These data might then either confirm or disprove our
recently-proposed dielectric model of charge-induced relaxation
in these materials (20, 21). Since polystyrene (PS) and condensed
films of ethyl benzene had been studied as prototype systems in an
earlier investigation of the relaxation energy shifts in solid-
state versus gas phase photoemission spectra (20), it was appropriate to utilize an analogous model molecule in the present
examination of the fluctuation-induced widths of the solid-state
spectra. Isopropyl benzene has a lower vapor pressure than
benzene and ethyl benzene which were studied previously (20),
consequently, it lends itself to investigation over a slightly
larger temperature range. Isopropyl benzene is otherwise as
suitable a model molecule for polystyrene as ethyl benzene.

The UPS spectrum, obtained with 32 eV photons, for a 15 Å
thick vapor-deposited film of isopropyl benzene $30^{\circ}K$ is shown in
Fig. 3. Also shown in the figure is the UPS spectrum from the
clean palladium crystal used as a substrate signal, and the
isopropyl benzene UPS spectra previously obtained in the gas
phase at 21.2 eV photon energy (41).

The data on the temperature dependence of the width of the
lowest binding energy ionization (i.e., at a binding energy of
about 7.6 eV) from thicker, approximately 25 Å films of isopropyl
benzene are summarized in Fig. 4. The data have been analyzed
experimentally by assuming that the observed full-width-at-half-

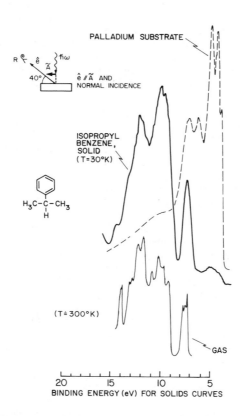

*Figure 3. UPS spectra of the clean Pd substrate of a molecular thin-film ($\sim 15$ Å) sample of isopropyl benzene in the gas phase (41).*

The photon energies are 32 eV, 32 eV, and 21.2 eV (top to bottom). The energy scale is for the solid-state spectra and should be shifted to read higher building energy by 1.5 eV for the gas-phase spectrum. The isopropyl benzene film in this case is only about 15-Å thick, and consequently the Pd d-band is not totally obliterated. The data of Figure 4 are for approximately 25-Å thick films, where the Pd signal can not be seen.

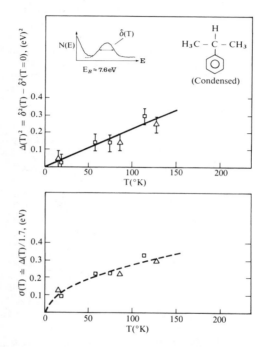

*Figure 4.    Square of the experimental temperature contribution to UPS linewidth, $\Delta(T)^2$, as a function of temperature for two totally independent runs ($\triangle$, $\square$) in the top panel.*

*The large error bars reflect the signal-to-noise ratio relative to the magnitude of the total observed linewidths. In the lower panel, a solid line is drawn for the standard deviation of a gaussian linewidth using the slope from the upper panel, where c $= 2.6$ ($\pm0.5$) $\times$ $10^{-2}T^{1/2}$ ($---$). The linewidth data points plotted in the lower panel show consistency with the $T^{1/2}$ behavior predicted by a molecular-ion model calculation (20, 21).*

maximum is given by the square root of the sum of the squares of the individual contributions to the line width, namely

$$(T) = (\sum_i \Delta_i^2)^{1/2}, \tag{5}$$

where the set of $\Delta_i$ are independent contributions to $\delta(T)$, the observed peakwidth: $\Delta_N$ is the 0.67 eV linewidth of the peak in the gas phase; $\Delta_r$ is the instrumental resolution estimated at 0.1 eV from the width of the Fermi level of the clean metal substrate; $\Delta_T = \Delta(T)$ is the T-dependent contribution that is extracted from the data; and $\Delta_o$ is the residual T-independent contribution obtained by the extrapolating to T = 0. Operationally, we plot $\delta(T)^2$ versus T, extrapolate to T = 0 and then subtract ($\Delta_r^2 + \Delta_N^2$) = 0.68 eV in order to determine $\Delta_o$. We find $\Delta_o \gtrsim 0.4$ eV, depending on the vapor deposition conditions. For example, slow, multiple exposures to total about 20 L yield $\Delta_o$ = 0.4 eV. However, a fast 20 L exposure yielded $\Delta_o \sim 0.6$ eV. This latter case probably corresponds to a greater degree of molecular disorder in the resultant sample. We have attributed the residual, T-independent factor to spatial variations in local site energies and to the corresponding inhomogeneous intermolecular relaxation (polarization) energies as described previously (18, 20, 21, 28). After determining the specific $\Delta^2$ contribution for each separate run, the data are plotted as $\Delta(T)^2 = \delta(T)^2 - \delta(T=0)^2$, in Fig. 4 for two totally independent runs. The data can be converted to the standard deviation parameter of a gaussian curve, $\sigma$, and plotted as a function of T, as also shown in Fig. 4. Then a curve that goes as $T^{1/2}$ appears to be consistent with the data as shown in Fig. 4b.

A most important aspect of these data is the temperature dependence of the width of the molecular solid photo-ionization peak. This T-dependence is of the form $\Delta(T)^2 = \Delta_o^2 + c^2 T$, which is the form which had been previously discussed, but not quantitatively observed, in connection with an X-ray photoelectron spectroscopy (XPS) study of core-level peaks of the alkali halides (25). Sample charging effects inhibited the study of the linear T-dependence of $\Delta_2(T)^2$ in the XPS case. A linear temperature dependence of $\Delta^2$ is predicted by any model in which a low-frequency vibrational mode, $\hbar\omega < kT$, is strongly coupled to the molecular ion state (21, 25, 26, 27, 42, 43, 49). The most obvious such coupling is associated the excitation of intramolecular modes upon photoionization; a phenomenon which produces structure on photoionization lines in the gas phase as well as solid state. Evaluation of the relevant coupling constants for benzene based molecular ions (43) suggests that the coupling to these modes is too weak to provide a quantitative description of the observed value of $c^2$, although we cannot definitively exclude this mechanism because we do not know the normal modes of a condensed isopropyl benzene. A more likely mechanism is the coupling of the molecular ion state to dipole active normal modes of neighboring molecules via long-range Coulomb interactions. The consequences of this interaction can be described by a generalization of the Huang-Rhys model (44, 45, 46), that was proposed to interpret the earlier XPS studies of alkali halides (25). A recent model calculation has been carried out (21) that predicts $c = 4 \times 10^{-2}$ $eV(^{o}K)^{-1/2}$ for condensed benzene, a value in good correspondence with the observed value of $c = 2.6 \pm 0.5 \times 10^{-2}$ $eV (^{o}K)^{-1/2}$ for isopropyl benzene obtained from Fig. 4. From this close correspondence we have inferred that coupling of a given isopropyl benzene molecular ion state to infra-red active vibrations of neighboring neutral isopropyl benzene molecules is the predominant cause of the observed temperature dependent linewidth.

The origin of the temperature-independent solid state broadening, $0.4$ eV $\leqslant \Delta_o < 0.6$ eV we attributed to spatial variations in the electronic contributions to the intramolecular relaxation energies in the vicinity of the surface (18, 19, 22). Similar widths (to 0.6 eV) have been observed in a variety of other contexts, including condensed thin films of $N_2$ and CO molecules (28) and the sub-monolayer adsorption of these molecules on metal surfaces (29). Interatomic Auger and electron-hole shakeup processes have been proposed, but found to be too small to account for the observed widths in these cases (28, 47, 48). On an energetic basis these latter proposed processes are expected to occur for inner valence orbitals and not for the outer-most IP studied here. Also, they contain no significant temperature dependence. By the process of elimination, spatial variations in the relaxation energies are the only extent interpretation of the temperature-independent (inhomogeneous) broadening.

The study of isopropyl benzene can be summarized as follows. The width of the UPS line corresponding to removal of the lowest binding energy $\pi$-electron is temperature dependent. This temperature dependence contributes significantly to the UPS linewidth at elevated temperatures. The fact that the width is temperature dependent indicates that the mechanism involves vibrations. Although experimentally, intramolecular and intermolecular effects could not be separated, theoretical models predict that of the effect is mostly intermolecular. The small residual linewidth observed is due to sample inhomogeneities. Presumably, an ideal single crystal thin film would exhibit the same $\Delta(T)$ but have a smaller $\Delta_o$.

Relaxation Energy Shifts and Localized Molecular-Ion States in Aromatic Pendant Group Polymers. The electronic structure of polystyrene (PS) and poly vinyl pyridine (PVP) have been studied using a variety of electronic spectroscopies and model calculations (20). Here, we review the results of the UPS and ultra violet absorption spectroscopy (UAS) portion of that study, and discuss the results in a phenomenological manner. The aim of this

study was to characterize the electronic excitations of the
polymers in terms of those of the molecular building blocks of the
polymers. To this end, the appropriate model monomer molecules
were studied in both the gas phase and the condensed molecular
solid phases. Some technical details of the photoelectron spectroscopy of the polymer films are given in the Appendix. Some of
the general practices discussed there apply to the condensed
molecular solids as well.

At this point it is worth rephrasing some of the issues of
the above discussions. The UPS spectra are a measure of the
single-particle excitation spectrum of the molecule, in so far as
removal of an electron is concerned, while UAS data are a measure
of the particle-hole excitation spectrum. In other terms, UPS
measures the molecular-ion states while UAS measures excited
states of the neutral molecule. For a molecule in isolation, in a
one-electron picture the valence electron molecular cation states
are comprised of the set of one-electron molecular orbitals
(MO's) containing one half-filled (usually non-degenerate) molecular orbital and the totality of other fully occupied orbitals,
distorted from their situation in the neutral molecule due to the
removal of an electron from the molecule in a photoelectron
transition (21). The charge density change that occurs upon the
generation of a hole-state results in a reorganization of the
total charge distribution of the molecule. The differences
between the molecular cation states measured by photoelectron
spectroscopy and the eigenstates of the neutral molecule are then
just those dictated by the redistribution of total charge (reorganization) as well as the accompanying change in electron correlation (15, 16) (neglecting relativistic effects). The energy
gained by the molecule in the excited state is termed the intramolecular relaxation energy, of which there are atomic (a) and
electronic (e) contributions (21). This energy is sometimes
called the Koopman's defect, since it is this energy which
accounts for the discrepency between Koopman's theorem and observed UPS spectra in a one-electron formalism (15, 16).

In a condensed molecular solid, however, there are also
intermolecular relaxation (polarization) effects that occur in
addition to the intramolecular effects (10, 11), as discussed
above. In fact, in any dielectric medium, the total net positive
charge density on the molecular cation induces corresponding
electronic, and ultimately atomic, distortions in the surrounding
medium.

The clearest evidence for the validity of the molecular-ion
concept is the obvious similarity between the UPS and UAS spectra
of PVP and those of its corresponding model molecule 2-ethyl
pyridine. These spectra are shown in Figs. 5 and 7, respectively.
In particular, in Fig. 5 the UPS spectra for the series of
materials pyridine(v), 2-ethyl pyridine(v), 2-ethyl pyridine(s),
and PVP are presented in order to show explicitly the relative
consequences of the 2-ethyl substituent and condensation effects

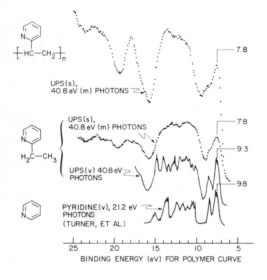

*Figure 5.    UV photoemission spectroscopy data for the series of materials pyridine (v) (2), 2-ethyl pyridine (s), and poly(2-vinyl pyridine) (s).*

*The shift of 0.5 eV between the lowest binding-energy peak for pyridine and for 2-ethyl pyridine reflects the effects of hyperconjugation induced by the ethyl group substituent. The additional shift of 1.5 eV between the vapor (v) and solid-state (s) data on 2-ethyl pyridine arises from the intermolecular relaxation effects discussed in text. The spectra are presented in the figure shifted relative to each other by an amount such that the lowest energy ionization potentials are aligned. The absolute values of these ionization potentials are indicated in the figure.*

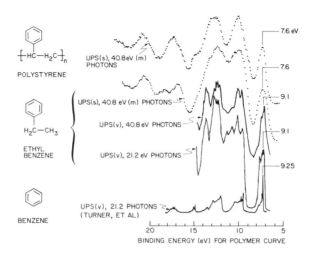

*Figure 6.  UV photoemission spectroscopy data for the series of materials ben-
zene (v) (3), 2-ethyl benzene (v), 2-ethyl benzene (s), and polystyrene (s).*

*The shift of 0.14 eV between the lowest binding-energy peak for benzene and the 2-ethyl
benzene reflects the effects of hyperconjugation induced by the ethyl group substituent.
The additional shift of 1.5 eV between the vapor (v) and solid-state (s) spectra on 2-ethyl
benzene arises from the intermolecular relaxation effects discussed in the text. The
spectra are presented in the figure shifted relative to each other by an amount such that
the lowest energy ionization potentials are aligned. The absolute values of these ionization potentials are indicated in the figure.*

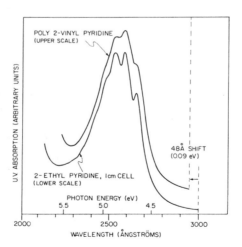

*Figure 7.  UV absorption spectra in the region of the first absorption band in pyri-
dine ($^1A_1 \rightarrow {}^1B_1, {}^1B_2$) for 2-ethyl pyridine and poly(2-vinyl pyridine).*

*The similarity of these two spectra indicates that the aromatic pyridine group determines
these lowest energy optical excitations of the polymer. The slight relaxation energy shift
is discussed in the text. The vibrational progression involving the $\eta'_{10}$ vibration of pyridine (50, 62) is clearly visible.*

into a solid on the evaluation of the spectrum of PVP from that of
its pyridine pendant-group moiety. In the above notation, v
refers to the vapor phase, while s refers to solid phase. Based
upon these spectra in conjunction with model MO calculations
(20), it is clear that the uppermost three ionization potentials
in PVP, emanating from $2b_1(\pi)$, $1a_2(\pi)$, and $11a_1(\sigma)$ states of
pyridine are preserved throughout this sequence of materials,
although the relative ordering of these states is altered by the
2-ethyl substitution (20). Similarly, the detailed resemblance
between the lowest energy ($\pi \rightarrow \pi*$) uv absorption (UAS) bands
in PVP and 2-ethyl pyridine (50) shown in Fig. 7, demonstrates
that the low-energy electronic transitions in neutral PVP are
essentially those of 2-ethyl pyridine. Consequently, we see that
the low-energy electronic excitations of PVP are essentially
those of the pyridylethylene repeat units which make up PVP.
Within the context of the uv absorption bands of substituted
polyethylenes this result is well-known in the literature on both
polystyrene (PS (51, 52), and PVP (53)). Our spectra extend these
observations to include molecular cation states (i.e., photoemission spectra) in addition to molecular exciton states. The
UPS and UAS spectra for PS analogous to those shown in Figs. 5 and
7 for PVP are given in Fig's 6 and 8 for the series benzene(v),
ethyl benzene(v), ethyl benzene(s), and PS. It is clear from
Fig's. 6 and 8 and the literature on the uv absorption properties
of substituted polyethylenes (51, 52, 53) that the low-energy
electronic excitations of these materials are properly identified
as being localized on the pendant-group moieties. We conclude,
therefore, that the molecular-ion concept is adequately based on
spectroscopic data in the case of aromatic substituted polyethylenes.

    Figures 5-8 show further that the two features of the
condensed phase spectra which differ from their gas-phase counterparts are the energies and widths of the individual ionization
peaks or optical-absorption lines. In both the UPS and UAS
spectra the lines in the solid state are shifted to lower energies
relative to the corresponding ones in the gas phase by a relaxation energy, $\sim$ 1 eV for ion states and $\sim$ 0.1 eV for exciton
states, associated with the intermolecular polarization induced
by the ion or exciton (10, 19, 20, 21, 22). In addition, these
lines also are broader in the solid state, due in part to the
interaction of the ions and excitons with intermolecular polarization fluctuations (19, 20, 21, 22, 36, 54) and to a lesser
extent to the spatial inhomogeneity of the polarization-induced
relaxation energy in the vicinity of a surface (18, 19, 37). It
is important to note in Fig's 5 and 6, however, that for hole
states in both PVP and PS, these solid-state effects are essentially fully manifested in the condensed molecular solid films.
In fact, both the intermolecular relaxation energies, $E_r$ = 1.5 ±
0.1 eV, and solid-state $\pi$-electron linewidths, $\Delta\pi \sim$ 1.0 eV, = are
the same within experimental error for the polymer and the
condensed molecular solid films.

The results derived from Fig's. 5 and 6 are that the intermolecular relaxation (polarization energy) shifts are essentially
uniform for all the molecular cation states i.e., $E_r \approx 1.5$ eV
independent of the identify of the MO involved , that the values
of these shifts are identical for PVP and PS, and that the
linewidths of the uppermost π-electron cation states are identical in PVP and PS. The equality of the intermolecular relaxation
energies for PS and PVP is surprising because of the existence of
a much larger value of the static dielectric constant in (polar)
PVP than in (nonpolar) PS, leading to the prediction of a correspondingly larger relaxation energy in the Born theory of solvation, which had been commonly used to model point charges in
dielectric media (20). Moreover, this discrepancy transcends the
adiabatic as well as the independent-electron versions of the
theory of dielectric relaxation (55). Therefore our experimental
UPS data shown in Figs. 5 and 6 have revealed a fundamental
failure of previous theories of charge-induced dielectric relaxation which have been shown to be removed by the concept of a
molecular-ion model (20).

Specifically, in the Born theory model of solvation, the
intermolecular relaxation energy is (21)

$$E_r = \frac{e^2}{R} [1 - \varepsilon(0)^{-1}], \qquad (6)$$

where e is the electronic charge, R is the radius of a sphere of
volume equivalent to that of the molecule in question, and $\varepsilon(0)$ is
the zero frequency (static) dielectric constant. Since there is
over a factor of two difference between $\varepsilon(0)$ for PVP and PS, the
Born model predicts a $E_r$ of about 4.5 eV for PVP and about 1.3 eV
for PS; in disagreement with the observed result of 1.5 ± 0.1 eV
for both PS and PVP. A molecular ion model taking into account
the interactions of the electronic (ionic) states with the longitudinal polarization fluctuations of the (dielectric) medium,
however, accounts for the observations (20). The low frequency
torsional and i.r. longitudinal polarization modes are strongly
coupled to the electronic excitations. Therefore these fluctuations homogeneously broaden the UPS lines rather than shift them
(21). The weakly coupled branches of the dielectric function then
contribute to the intermolecular relaxation in the form (56),

$$E_r = \frac{e^2}{R} [\varepsilon_\alpha^{-1}(\omega) - \varepsilon_{\alpha-1}^{-1}(\omega)], \qquad (7)$$

where $\varepsilon_\alpha(\omega)$ is the value of $\varepsilon(\omega)$ for $\hbar\omega$ just above the electronic
excitation region of the $\varepsilon(\omega)$ curve, while $\varepsilon_{\alpha-1}(\omega)$ is the value
just below the electronic excitation region of $\varepsilon(\omega)$. Since the
electronic contributions to the dielectric constant, $\varepsilon(\omega)$, for 1
eV $< \hbar\omega \lesssim 10$ eV, are essentially identical for PS and PVP, these
weakly coupled modes are expected to produce the same result in

the intermolecular relaxation shift in the UPS spectra for both PS and PVP; as observed. In addition, the associated homogeneous contributions to the linewidths are $\Delta_h \sim 0.3$ eV. Since this value is much smaller than the predicted inhomogeneous static-fluctu- ation-induced value of about 0.75 eV (21), we expect the observed UPS lines to be inhomogeneously broadened and about the same width in both PS and PVP, again in accordance with the measurements. The quantitative prediction of the experimental results by the molecular ion model concept also is very good (20).

The final point to be discussed involves the gas-to-solid shifts in the optical absorption spectra. Having established the correspondence between the electronic excitations of the con- densed monomer films and the polymer films, we can focus on a comparison of the gas phase monomer UAS spectra with the polymer film spectra. As shown in Fig's 7 and 8 the polymer UAS spectra are shifted and broadened relative to their gas phase counter- parts. As in the case of the UPS spectra, these two effects can again be quantitatively understood within the context of a molec- ular ion framework. The UAS intermolecular relaxation energy shift is twice as large in PVP (0.09 eV) as in PS (0.04 eV). This observation is consistent with the existance of a larger number of weakly coupled longitudinal polarization modes in PVP as compared with PS. In addition, the broadening of the gas phase 2-ethyl pyridine UAS spectra as compared with the gas phase ethyl benzene spectra is an indication that there are more strongly coupled intramolecular modes in 2-ethyl pyridine than in ethyl benzene; a fact consistent with the polar nature of 2-ethyl pyridine. Final- ly, the solid state (polymer) UAS spectra for PVP are more severely broadened than the UAS spectra for PS; once again consistant with the existence the larger number of i.r. active modes in PVP (56).

The purpose of this section has been to utilize a molecular ion concept to provide a satisfactory interpretation of the UPS and UVA data given in Figs. 5, 6, 7 and 8. In particular, the molecular ion concept affords an elementary prediction of the surprising result that the measured intermolecular relaxation energies are the same for both PVP and PS (20) even though the former is a polar material and while the latter is not (i.e., the two materials exhibit quite different low-frequency dielectric responses).

We summarize here the features of the UPS and UVA data which lead to the molecular ion concept for these aromatic pendant group polymers. First, the spectra of the polymers PVP and PS are essentially identical to those of condensed model molecular moieties 2-vinyl pyridine and ethyl benzene, respectively. Second, the solid-state spectra are related to the gas-phase spectra of these model moieties by an essentially constant shift to higher energy (lower binding energy) of all the ionization peaks by $E_r = 1.5 \pm 0.1$ eV. Third, the width in energy of the solid-state ionization peak is $\Delta\pi = 1.0 \pm 0.1$ for both polymers

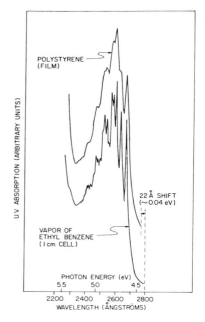

*Figure 8.   UV absorption spectra in the region of the first absorption band in benzene ($1_{A_{1g}} \rightarrow 1_{B_{2u}}$) for the ethyl benzene in the gas phase and a PS film.*

*The similarity of the molecular and polymer spectra indicates that the aromatic groups determine these lowest energy optical excitations in the polymer. The slight relaxation energy shift is discussed in the text. The vibrational progressions of both ethyl benzene and PS are clearly seen.*

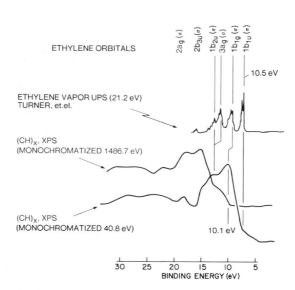

*Figure 9.   Gas-phase UPS spectra of ethylene molecules (2) compared with that of polyacetylene (57)*

at T = 300°K, although its temperature dependence was not measured.  The temperature effect was investigated, however, employing model-molecule studies, i.e. isopropyl benzene as discussed
above.

The similarity of the polymer UPS data and those of the gas-
phase model molecular moieties is incorporated into a molecular-
ion framework by envisaging the ion states in the solid to be
those of the isolated model moities modified only by their
interaction with polarization fluctuations in the surrounding
dielectric medium.  If these fluctuations cause large enough
variations in the relaxation energies, they localize the ion
states in the polymer onto individual pendant-group moieties.
That such is the case in PVP and PS was discussed above.  This
fluctuation-induced localization suffices to explain the detailed
similarity between the UPS data of the condensed model molecules
and those of the corresponding polymer (Figs. 5 and 7) for those
low-ionization-potential states which do not involve large contributions from the valence electrons in the substituted poly-
alkane backbone.  In addition,the shifts and broadening of gas-
phase spectral lines in the solid state are quantitatively predicted by a molecular ion model (20, 21, 56), the literature for
which should be referred to for a detailed mathematical discussion.

Electronic States of Polyacetylene.  In contrast to the
localized molecular ion states seen in the spectra of PS and PVP
in the previous section, we outline here very briefly the results
of a study of the fully conjugated polymer, polyacetylene, or
$(CH)_x$, using photoelectron spectroscopy (57, 59).  This work was
stimulated by the enormous change in electrical conductivity of
free standing films of $(CH)_x$ upon doping (60).  The electronic
structure of polyacetylene has been a subject of controversy for
decades (57).  For the present purposes, we focus on what would be
the delocalized nature of the lowest binding energy $\pi$-band (57) in
an isolated ideal chain of $(CH)_x$ and compare it with the $1b_{1u}(\pi)$
state in ethylene.  Although acetylene has a triple $C\equiv C$ bond,
polyacetylene contains alternating double C=C and single C-C
bonds.  The ethylene molecule, therefore, is a good model molecule
representative of the local electronic structure in $(CH)_x$.  The
lowest binding energy $\pi$-state, however, is delocalized over the
entire $(CH)_x$ chain.  Residual electrical conductivity arises from
bond-alternation-defects frozen into the polymer film during the
film growth, using a Ziegler-Natta catylist (60).  More recent
models (62) of the electrical conductivity invoke quasi-localized
states in a Fermi glass description.

The magnitude of the intermolecular relaxation energy shifts
in the solid phase UPS spectra of the polymer films relative to
the gas phase UPS spectra of the model molecules seen for the
aromatic pendant group polymers is also observed in the case of
the $(CH)_x$ versus ethylene UPS spectra, as can be seen in Fig. 9.

The facility for understanding the intermolecular relaxation energy shift is contained within the framework of the molecular ion concept and is consistent with recent models (56, 62) of the very π-band edges in (CH)$_x$. Since the optical absorption in (CH)$_x$ has been shown to be excitonic in nature (57), with large molecular exciton binding energies ($\sim$ 1 eV), the very band edge states are said to be consistent with a Fermi glass model. Thus the localization allows an interpretation of the intermolecular relaxation energy shift on the same grounds as for the aromatic pendant group polymers. Recent photovoltaic effect measurements (63) have verified the existance of an excitonic absorption edge at a photon energy corresponding to what is called the gap energy in (CH)$_x$ and the existance of a free e-h pair absorption edge about 1 1/2 eV to higher photon energy. These observations thus vindicate the observation of the intermolecular relaxation energy in the UPS spectra that are otherwise expected only for molecular insulating solids with localized wave functions.

Appendix: Sample Preparation and some other Experimental Details
Relevant to the Study of Polystyrene and Poly(2-Vinyl Pyridine)

A 40.8 eV UPS spectrum of a nominally 20Å thick film of polystyrene (PS) is also shown in Fig. 6. Rather careful solution casting procedures, coupled with immediate insertion into the spectrometer vacuum chamber for drying, was necessary to produce films of uniform thickness, relatively free from surface contamination. The rationale for the thin samples was discussed above. The 40.8 eV He II resonance line was chosen for use for two reasons: its energy, when used with a monochromator to filter out other lines, enables examination of a wider range of energy of MO's than the usual 21.2 eV He I resonance line; and there is less influence over most of the spectrum from background due to inelastic electrons. The effect of surface contamination was studied by comparing 21.2 eV and 40.8 eV UPS spectra, as well as by subsequently exposing the samples to laboratory air for short periods of time and then examining the spectra. The slight peak at about 14 eV binding energy in the PS spectrum relative to the spectrum of condensed ethyl benzene, in Fig. 6, is due to surface contamination.

The first optical absorption bands of benzene and pyridine are known to occur near 5 eV. The optical data on 1000 Å thick, solution cast, PS and PVP as well as on the model molecules ethyl benzene and 2-ethyl pyridine in the gas phase were recorded on a Cary 14 double beam instrument. The data are therefore limited to $\hbar\omega$ < 6 eV by the optics of the instrument. Condensed molecular solid spectra were not necessary since the photoemission studies had already shown that the polymer and condensed molecular spectra were equivalent. Therefore a comparison of gas-phase model-molecule spectra with that of the corresponding polymer spectra was sufficient.

## Abstract

    Intermolecular relaxation effects are a central issue in the interpretation of the ultraviolet photoelectron spectroscopy (UPS) of molecular solids.   These relaxation effects result in several significant characteristics of UPS valence spectra.   Intermolecular relaxation phenomena lead to localized electron molecular-ion states, which are responsible for rigid gas-to-solid molecular spectral energy shifts, spectral line broadening, and dynamic electronic localization effects in aromatic pendant group polymers.

## Literature Cited

1.   Price, W.C. and Turner, D.W. (Eds.), Phil. Trans,

2.   Turner, D.W., Baker, C., Baker, A.D. and Brundle, C.R., Molecular Photoelectron Spectroscopy, (Wiley Interscience, London, 1970).

3.   Siegbahn, K., Nordling, C., Fahlman, A., Nordberg, R., Hamrin, K., Hedman, J., Johannson, G., Bergmark, T., Karlsson, S.E., Lindgren, I. and Lindberg, B., ESCA-Atomic, Molecular and Solid State Structure Studied by Means of Electron Spectroscopy, Nova Acta Regiae Soc. Sci. Upsaliensis Ser. IV, Vol. 20, 1967.

4.   Siegbahn, K., Nordling, C., Johansson, G., Hedman, J., Heden, P.F., Hamrin, K., Gelius, V., Bergmark, T., Werme, L.O., Manne, R. and Baer, Y., ESCA Applied to Free Molecules, (North-Holland, Amsterdam, 1969).

5.   Cardona, M. and Ley, L., in Photoemission in Solids, I., M. Cardona and L. Ley, Eds. (Springer-Verlag, Berlin, 1978).

6.   Siegbahn, K., J. Elec. Spec., 1974, 5, 3.

7.   Kunz, C., "Synchrotron Radiation:   Overview", in Photoemission from Solids II., M. Cardona and L. Ley, Eds. (Springer-Verlag, Berlin, 1979).

8.   Clark, D.T., "Application of ESCA to Structure and Bonding in Polymers", in Characterization of Metal and Polymer Surfaces, Vol. 2, L.H. Lee, Ed. (Academic Press, New York, 1977), and the many references therein.

9.   Pireaux, J.J., Riga, J., Candano, R, Verbist, J.J., Delhalle, J., Delhalle, S., Andre, J.M. and Gobillon, Y., Physica Scripta, 1977, 16, 329, and other works by the above authors.

10. Duke, C.B., Salaneck, W.R., Paton, A., Liang, K.S., Lipari, N.O. and Zallen, R., "Equivalence of the Electronic Structure of Molecular Glasses, Gases and Crystals", in Structure and Excitations of Amorphous Solids, G. Lucovsky and F.L. Galeener, Eds, AIP Conf. Proc. No. 31 (AIP, NY, 1976).

11. Grobman, W.D. and Koch, E.E., "Photoemission from Organic Molecular Crystals", in Photoemission from Solids II, Cardona, M. and Ley, L., Eds. (Springer-Verlag, Berlin, 1979).

12. Fujilira, M. and Inokuchi, H., Chem. Phys. Lett. 1972, 17, 554.

13. Fadley, C.S., "Basic Concepts of X-ray Photoelectron Spectroscopy" in Electron Spectroscopy: Theory, Techniques and Applications, C.R. Brundle and A.D. Baker, Eds. (Academic Press, London, 1978).

14. e.g. Wittel, K. and McGlynn, S.P., Chem. Reviews, 1977, 77, 745.

15. e.g. Hohlneicher, G., Ecker, F. and Cederbaum, L., in Electron Spectroscopy, D. Shirley, Ed., (North Holland, Amsterdam, 1972), p. 647-659.

16. e.g. Cederbaum, L.S. and Domcke, W., Adv. Chem. Phys., 1977, 36, 205 and references therein.

17. Manne, R. and Åberg, T., Chem. Phys. Lett., 1970, 7, 282.

18. Salaneck, W.R., Phys. Rev. Lett., 1978, 40, 60.

19. Duke, C.B., Fabish, T. and Paton, A., Chem. Phys. Lett., 1977, 49, 133.

20. Duke, C.B., Salaneck, W.R., Fabish, T.J., Ritsko, J.J., Thomas, H.R. and Paton, A., Phys. Rev., 1978, B18, 5717.

21. Duke, C.B., Mol. Cryst. Liq. Cryst., 1979, 50, 63.

22. Duke, C.B., Surf. Sci., 1978, 70, 674.

23. Duke, C.B. and Schein, L.B., Physics Today, Feb. 1980, p. 42-48.

24. Salaneck, W.R., Duke, C.B., Eberhardt, W. and Plummer, E.W., Phys. Rev. Lett., 1980, 45, 280.

25. Citrin, P.H., Eisenberger, P. and Hamann, D.R., Phys. Rev. Lett., 1974, 33, 965.

26.   Gadzuk, J.W., Phys. Rev., 1976, B12, 5458.

27.   Gadzuk, J.W., Phys. Rev., 1978, B20, 515.

28.   Norton, P.R., Tapping, R.L., Broida, H.P., Gadzuk, J.W. and Waclowski, B.J., Chem. Phys. Lett., 1978, 53, 465.

29.   Norton, P.R., Tapping, R.L. and Goodale, J.W., Surf. Sci., 1978, 72, 33.

30.   Salaneck, W.R., Lipari, N.O., Paton, A., Zallen, R. and Liang, K.S., Phys. Rev., 1975, B12, 1493.

31.   Nielsen, P., Sandman, D.J. and Epstein, A.J., Solid State Commun., 1975, 17, 1067.

32.   Plummer, E.W., Salaneck, W.R. and Miller, J.S., Phys. Rev., 1978, B18, 1673.

33.   Salaneck, W.R. and Zallen, R., Solid State Commun., 1976, 20, 793; II Nuovo Cimento, 1977, 38B, 248.

34.   Kerns, D.R. and Calvin, M., J. Chem. Phys, 1961, 34, 2026.

35.   Lipari, N.O. and Duke, C.B., J. Chem. Phys., 1975, 63, 1768.

36.   Gosar, P. and Choi, S., Phys. Rev., 1966, 150, 529.

37.   Phillpot, M.R. and Turlet, J.M., J. Chem. Phys., 1976, 64, 3852.

38.   Fadley, C.S., Baird, R.J., Siekhaus, W., Novakov, T. and Bergstrom, L., J. Elec. Spec., 1974, 4, 93.

39.   Clark, D.T. and Thomas, H.R., J. Polym. Sci., Polym. Chem., 1977, 15, 2843.

40.   Wyckoff, R.W.G., Crystal Structures (Intersicence, New York, NY, 1971) Vol. 6, Part II, p. 454.

41.   Carnovale, F., Nagy-Felsobuki, E. and Peel, J.B., Aust. J. Chem., 1978, 31, 483.

42.   Duke, C.B., in Tunneling in Biological Systems, B. Chance, D.C. DeVault, H. Frauenfelder, R.A. Marcus, J.R. Schrieffer and N. Sutin, Eds. (Academic Press, New York, 1979), p. 31-65.

43.   Duke, C.B., Ann. N.Y. Acad. Sci., 1978, 313, 166.

44. Huang, K. and Rhys, A., Proc. Roy. Soc., 1950, A204, 406.

45. Markham, J.J., Rev. Mod. Phys., 1959, 31, 956.

46. Duke, C.B. and Mahan, G.D., Phys. Rev., 1965, 139, A1965.

47. Gadzuk, J.W., Phys. Rev., 1978, B14, 5458.

48. Gadzuk, J.W. and Sunjic, M., Phys. Rev., 1975, B12, 524.

49. Himpsel, F.J., Schwenter, N. and Koch, E.E., Phys. Stat. Sol. (b), 1975, 71, 615.

50. Herzberg, G., Molecular Spectra and Molecular Structure III: Electronic Spectra and Electronic Structure of Polyatomic Molecules, (VanNostrand, Princeton, NJ., 1966).

51. Pestemer, M. and Bruck, D., in Methoden Der Organischen Chemie, E. Muller, Ed., (Georg Thieme Verlag, Stuttgart, 1955), Vol. 3, Pt. 2, p. 702.

52. Boyer, R.F., in Encyclopedia of Polymer Science and Technology, N.M. Bikales, Ed. (Wiley, New York, 1970), Vol. 13, p. 256-257.

53. Funt, B.L. and Ogryzlo, E.A., J. Poly. Sci., 1957, 25, 279.

54. Schein, L.B., Duke, C.B. and McGhie, A.R., Phys. Rev. Lett., 1978, 40, 197.

55. Jortner, J., Mol. Phys., 1962, 5, 257.

56. Duke, C.B., to be published.

57. Duke, C.B., Paton, A., Salaneck, W.R., Thomas, H.R., Plummer, E.W., Heeger, A.J. and MacDiarmid, A.G., Chem. Phys. Lett., 1978, 59, 146.

58. Salaneck, W.R., Thomas, H.R., Duke, C.B., Paton, A., Plummer, E.W., Heeger, A.J. and MacDiarmid, A.G., J. Chem. Phys., 1979, 71, 2044.

59. Salaneck, W.R., Thomas, H.R., Bigelow, R.W., Duke, C.B., Plummer, E.W., Heeger, A.J. and MacDiarmid, A.G., J. Chem. Phys., March, 1980.

60. Chiang, C.K., Fincher, C.R., Jr., Park, Y.W., Heeger, A.J., Shirakawa, H., Louis, E.J., Gau, S.C. and MacDiarmid, A.G., Phys. Rev. Lett., 1977, 39, 1098.

61.  Boschi, R., Murrell, J.N. and Schmidt, W., <u>Discuss, Faraday Soc</u>., 1972, 54, 116.

62.  Rice, M.J. and Mele, E., Preprint.

63.  Ozaki, M., Peebles, D., Weinberger, B.R., Chaing, C.K., Gau, S.C., Heeger, A.J., MacDiarmid, A.G., <u>Appl. Phys.</u> <u>Lett</u>., 1979, 35, 83.

RECEIVED January 15, 1981.

# 12

# Band Structure Calculations and Their Relations to Photoelectron Spectroscopy

JEAN-MARIE ANDRÉ, JOSEPH DELHALLE, and JEAN JACQUES PIREAUX

Facultés Universitaires Notre-Dame de la Paix, 61, rue de Bruxelles, 5000-Namur, Belgium

## Methods For The Determination of One-Electron Bands

Since the first theoretical works of the sixties (1) on LCAO techniques in polymer quantum chemistry, the field has known a rapid development and standard SCF calculations on regular polymers are now routinely performed. In those methods, the translational symmetry is fully exploited (and consequently assumed) in order to reduce to manageable dimensions the formidable task of computing electronic states of an extended system.

In the theoretical description of regular polymers, the monoelectronic levels (orbital energies in the molecular description) are represented as a multivalued function of a reciprocal wave number defined in the inverse space dimension. The set of all those branches (energy bands) plotted versus the reciprocal wave number (k-point) in a well defined region of the reciprocal space (first Brillouin zone) is the band structure of the polymers. In the usual terminology, we note the analogy between the occupied levels and the valence bands, the unoccupied levels and the conduction band.

The one-dimensional periodic lattice of a polymer consists of a large number, $2N + 1$ ($N \to \infty$) of unit cells of length $a$ each containing $\Omega$ nuclei at positions $\vec{A}_1$, $\vec{A}_2$, ..., $\vec{A}_\Omega$ relative to the cells origins $ja\hat{z}$, $j = 0, \pm 1, \pm 2, ..., \pm N$, and $2n_o$ electrons distributed along the chain. A strict electroneutrality of the cells ($2n_o = Z_1 + Z_2 + ... + Z_\Omega$) is necessary. The electrons are assumed to doubly occupy a set of one-electron orbitals, $\phi_n(k,\vec{r})$, of Bloch type written as periodic combinations of $\omega$ basis functions, $\chi_p(\vec{r})$.

$$\phi_n(k,\vec{r}) = (2N + 1)^{-\frac{1}{2}} \sum_{j=-N}^{+N} \sum_{p=1}^{\omega} C_{np}(k) \, e^{ikja} \chi_p(\vec{r}-\vec{P}-ja\hat{z}) \quad (1)$$

where $\vec{r}$ is the position vector measured from an arbitrary but fixed origin, $\hat{z}$ is a unit vector in the direction of lattice periodicity, P the center of the basis function $\chi_p(\vec{r})$, and k is a point in the first Brillouin zone (BZ), $[-\pi/a, +\pi/a]$, of the

0097-6156/81/0162-0151$05.00/0
© 1981 American Chemical Society

polymer; $L_{RC}$, the length of this Brillouin zone, is equal to $2\pi/a$; $\omega$ is the size of the basis set.

The $\phi_n(k,\vec{r})$'s represent the wavefunction of a single electron in the periodic potential created by the nuclei and the other electrons. The optimal set of those polymer orbitals for a given atomic basis set is obtained in the usual way by solving the Hartree-Fock equations. The SCF monoelectronic operator has the explicit form

$$F(\vec{r}_i) = -\tfrac{1}{2} \nabla^2(\vec{r}_i) - \sum_{h=-N}^{+N} \sum_{u=1}^{\Omega} Z_u |\vec{r}_i - \vec{A}_u - ha\hat{z}|^{-1} +$$

$$\sum_{n'=1}^{n_o} \sum_{k'} \{2J_{n'k'}(\vec{r}_i) - K_{n'k'}(\vec{r}_i)\} \tag{2}$$

where $n_o$ is the number of doubly occupied bands. The Coulombic and exchange terms are detailed as

$$J_{n'k'}(\vec{r}_i) \quad \phi_n(k,\vec{r}_i) = \int \phi_{n'}^*(k',\vec{r}_j) \ r_{ij}^{-1} \ \phi_{n'}(k',\vec{r}_j) \ dv_j \ \phi_n(k,\vec{r}_i) \tag{3}$$

$$K_{n'k'}(\vec{r}_i) \quad \phi_n(k,\vec{r}_i) = \int \phi_{n'}^*(k',\vec{r}_j) \ r_{ij}^{-1} \ \phi_n(k,\vec{r}_j) \ dv_j \ \phi_{n'}(k',\vec{r}_i) \tag{4}$$

The terms included in equation 2, are respectively the kinetic operator, the attraction of a single electron with all nuclei centered in all cells, the averaged electrostatic potential of all electrons and the averaged exchange interaction.

As it is the case for complicated problems where expansions into known functions are required, the final computational expressions are cast in matrix form. By forming the expectation value of the monoelectronic Hartree-Fock operator and by applying the variational procedure for the LCAO coefficients, $C_{np}(k)$, the following system of equations, of size $\omega$, is obtained:

$$\sum_p C_{np}(k) \left[ \sum_j e^{ikja} (F_{pq}^j - E_n(k) S_{pq}^j) \right] = 0 \tag{5}$$

The compatibility condition of this system

$$|\sum_j e^{ikja} (F_{pq}^j - E(k) S_{pq}^j)| \equiv 00 \tag{6}$$

gives the band structure, $\{E_n(k)\}$, as a multivalued function of k in the reduced zone scheme. We note that $F_{pq}^j$ is a matrix element of the monoelectronic operator, $F(\vec{r}_i)$, between the atomic orbitals $\chi_p$ and $\chi_p$ centered in cell j. $S_{pq}^j$, an overlap integral, has the same meaning for the unit operator. Both matrix elements decrease exponentially with the distance between the orbitals giving rise to a natural convergence of the summation over cells appearing in the secular systems (eq. 5) and determinants (eq. 6). Let us note that, instead of having real symmetric matrices as in molecular calculations, we have here hemitian complex matrices. In our programs, the complex eigenvalue problem (eq. 5) is solved for selected k-points (in our work, 21 equidistant k-points are requested in half a Brillouin zone) by the very efficient and self-contained routine CBORIS (2) using the Householder similarity transformation and the so-called QR algorithm of Francis which makes use in the diagonalization process of a factorization into the product of a unitary matrix Q and an upper-triangular matrix R.

The key problem is thus the numerical evaluation of the matrix elements $F_{pq}^{j}$ of eq. 5. The most complete and correct way is to evaluate those quantities from first principles, i.e. when the geometry of the polymer chain and the set of basis functions are fixed, all necessary terms are rigorously computed. This approach is known as "ab initio". Ab initio programs are now available for polymers and currently applied in several groups (3). However, it has to be realized that sophisticated ab initio methods, already time-consuming for middle- and large-sized mole-cules, become very onerous when applied to polymers of chemical interest. Furthermore, computing times are largely dependent on the size and the quality of the basis sets used.

In some cases, the cheaper semi-empirical techniques could, in principle, provide an interesting alternative but, generally, they suffer from crude approximation and arbitrary simplications. This makes interpretative works especially ticklish; for example, standard extended Hückel calculations reproduce fairly well band structures of hydrocarbon polymers but fail to interpret photo-electron spectra of highly polar systems like fluoropolyethyl-enes. CNDO (Complete Neglect of Differential Overlap) produce generally non significant density of states, even in the refer-ence polyethylene case. However, it should be pointed out that in clusters of polyacetylenes, good agreement has been obtained by spectroscopic CNDO parametrization (4).

This is a reason why we have concentrated on techniques which offer the advantages of both semi-empirical (computationally fast) and ab initio (more reliable) schemes; our intention is not to avoid ab initio calculations which remain necessary for an accurate description of electronic states but we feel that, in most cases, excellent qualitative (or even quantitative) informa-tions are obtained from the cheaper techniques presented below.

An interesting approach for comparison with XPS experiments is a generalization of the Floating Spherical Gaussian Orbital (FSGO) technique. In this method, each electron pair is repre-sented by a single Gaussian basis function whose exponent and position are obtained by a variational procedure or by reference to selected model-molecules. This allows to calculate band structures within a simple and fast computing scheme. The order of one-electron energies of various alkanes computed with this FSGO technique relates well with the results obtained when using more extended atomic basis sets and confirms this procedure as a valuable approach for interpreting XPS spectra. Similarly, variations in bond lengths and geometrical structures are in good agreement with experimental data. Such trends hold equally well for hydrocarbon polymers; X-ray scattering factors and properties directly related to electron density distribution are well estimated. Successful comparisons of theoretical electronic density of states with XPS spectra have been recently obtained for polyethylene, fluoropolyethylenes and polypropylene (5).

However, this method looses some of its advantages when applied to unsaturated systems. Extensions have been proposed but they generate the full complexity of SCF (Self Consistent Field) calculations. In a first approach, generally known as "Christoffersen's molecular fragment approach", only 1s Gaussian-Type Orbitals (GTO) are used. Part of these orbitals are placed at points of space other than on nuclei, for example, in bonding regions. In conjugated systems, the σ-framework is adequately described by somewhat localized FSGO's while the π-density is reproduced by antisymmetrical combination of s-gaussians which are symmetrically placed above and below the central atoms involved in the π-bonding. In a second approach (Whitten's approach), we no longer make use of localized orbitals but instead we keep the notion of atomic orbitals constructed as representations of Slater-Type Orbitals (STO). This approach differs from the usual ones by the choice of 2p orbitals which, as with the previous case, are formed by non-nuclei-centered antisymmetrical combination of two s-GTO's (6).

In addition to their versatility and their applicability to a wide class of systems, both the original FSGO method and its extensions have the merit of having outlined the existence of a crucial numerical problem related to the physical long-range domain of electrostatic interactions between electrons and between electrons and nuclei; mainly, in polar situations (even a CH bond has some degree of polarity), the correct balance between attraction and repulsion is usually not properly taken into account due to the slow convergency properties of 1/r-like series. Accurate methods using Fourier-transforms or multipole expansion techniques are now available for correctly computing those effects (7).

Methods based on Simulated Ab initio Molecular Orbital technique (SAMO) or on the application of Linear Combination of Localized Orbitals have been proposed. The a priori advantages are a negligible cost (typically of the order of magnitude of an extended Hückel calculation) and the ab initio character of the approach. They suffer however from a rather tedious generation of a high number of matrix elements and it is still impossible to correct significant discrepancies in the shapes and the widths or energy-bands (probably due to the neglect of long-range effects) (8).

A last but promising procedure is based on the use of effective Hamiltonians. In such approaches, it is supposed that the valence Hamiltonian that reproduces the Fock operator is a sum of kinetic and various effective atomic potentials for atoms within their characteristic chemical environment. In practical computations, those effective potentials are, for example, chosen in a non-local form of Gaussian projectors with spherical or non-spherical symmetry. Parameterization of these potentials is performed by least square fitting of corresponding valence Fock operators for small model molecules (9).

As an indication of the computing time needed for polymer calculations, an "exact" ab initio Gaussian-lobe STO-3G band structure of polyethylene requires about 55 hours on Digital DEC 2050. In such a calculation, all electronic interaction are evaluated between 34 carbon atoms and 68 hydrogen atoms. Interactions up to infinity are handled by a multipole expansion. This important computing time is partly due to the large amount of interactions taken into account in the calculation and to the lobe Gaussians used. The use of cartesian Gaussians instead of lobe ones and the explicit inclusion of a shell atomic structure as in recent molecular programs would probably reduce this figure by a factor of 10 or more. For a sake of comparison, the time required in the "Christoffersen molecular fragment" approach, is 25 minutes while it is only of the order of 2 minutes in the model potential technique.

For polymers, the explicit consideration of the helical symmetry (10) will result in additional computing time savings since the sizes of the matrices are considerably reduced. Our experience is a factor of 3 for screw axis of order 2 and a factor as large as 10 for a screw axis of order 3.

As a practical conclusion of this section, we are allowed to say that we have now at our disposal a number of efficient methods for calculating in reasonable machine times band structures of stereoregular polymers.

## Relations Between One-Electron Bands And PS Spectra

To our knowledge, the first published valence XPS spectrum of a macromolecule goes back to 1972 with the work of Wood et al. (12) where an experimental spectrum of $C_{36}H_{74}$ was compared to a CNDO/2 calculated band structure of all-trans-polyethylene. However, it has to be realized that band structure is not measurable directly and one has to apply transformations to bring calculated data in a form readily comparable to experiment. We should like to stress at this place that, in author's opinion, band structure plots do not offer the best representation of valence bands properties particularly in those systems where a

large number of bands lies in a narrow energetic region as it is
the case for polymers. We have thus implemented and systematized
a three-step procedure which produces a "theoretical XPS spectrum" based on a strict one-electron picture (13).    In its
presently simplified form, this procedure corresponds
     a)   to the calculation of the density of states histograms,
     b)   taking into account cross-section effects, and
     c)   convoluted for simulating experimental resolution.
    The first step is to calculate over the energy bands a
function referred to as the energy distribution of the joint
density of states. For polymers it amounts to a one-dimensional
integration over the Brillouin Zone (BZ) of a rather complicated
integrand such as found in the theoretical evaluation of optical
spectra. For sufficiently high photon energies the final states
are rather unstructured and the function to be compared with
experiment may be simplified into

$$J(E) \sim \sum_{\substack{i=1 \\ BZ}}^{occ} \delta[E-E_i(k)] \; P_i^{\omega}(k) \; dk \qquad (6)$$

where the summation extends over all occupied states and $P_i^{\omega}(k)$ is
the probability for the ejection of an electron in state $|i,k>$ by
a photon of energy $h\omega$.
    Making use of Koopmans' theorem the procedure can be simpli-
fied into the evaluation of the number of electronic energy states
per unit energy range and per unit length $\mathcal{D}(E)$:

$$\mathcal{D}(E) = \frac{a}{\pi} \sum_{i=1}^{occ} \left| \frac{dk}{dE_i(k)} \right| \qquad (7)$$
$$E_i(k) = E$$

Electronic densities of states distribution are not straightforward to evaluate numerically but the problem has been well
analyzed and various approaches to compute these functions have
been proposed (14).
    One has then to take the photoionization cross section into
account. It induces an appreciable modification of the general
shape of the density of states. The effect is rather difficult to
evaluate rigorously.
    A practical and quite successful approach is the intensity
model proposed by Gelius (15) which involves Mulliken population
analysis and considers relative atomic photoionization cross-
section, $\sigma_p$'s, to approximate $P_n^{\omega}(k)$:

$$P_n^{\omega}(k) \sim \sum_p C_{np}^{*}(k) \; [\sum_q S_{pq}(k) \; C_{nq}(k)] \; \sigma_p \qquad (8)$$

The $\sigma_p$'s are the relative photoionization cross-sections of a
particular atomic subshell; they have been obtained by a fitting
procedure on reference systems.
    In practice, one computes histograms, $\bar{I}(E_n)$, rather than
$J(E)$:

$$\bar{I}(E_n) \sim \frac{1}{\Delta E} \sum_{i=1}^{occ} \int_{E_n - \frac{E}{2}}^{E_n + \frac{\Delta E}{\Delta 2}} \left[ \left| \frac{dk}{dE_i(k)} \right| \quad x \ P_i^\omega(k) \right] \quad dE$$

$$(9)$$

Lastly, it is generally assumed that 0.5eV is the best possible resolution for solid state XPS measurements and the experimental resolution function is reasonably well reproduced by a Gaussian of full width $\Gamma$ at half maximum of 0.7eV. A final "theoretical XPS spectrum" is obtained after correction of the basic density of states function by cross-section effects and convolution by the experimental resolution function (16):

$$I(E_n) \sim \int_{-\infty}^{+\infty} \bar{I}(E) \quad \exp \ [-(E-E_n)^2/2\sigma^2] \ dE \qquad (10)$$

Despite all the simplifications, the resulting curve is a reasonable starting point for preliminary investigations and we shall see in the next section that encouraging results have already been obtained.

Comparisons Between Theoretical and Experimental XPS Valence
Spectra

The previously discussed three-step procedure has been applied to the test case of an all-trans-polyethylene chain. The comparison of different theoretically determined spectra and of the observed one is excellent. The positions of both theoretical and experimental peaks agree surprisingly well and both fine structures are directly comparable (17). As an example, the FSGO band structure of all-trans-polyethylene, its density of states and both the experimental and theoretical XPS spectra are displayed in Figure 1. Due to the difficult problem of assigning an absolute energy origin to XPS spectra of insulators, we have superimposed the theoretical XPS spectrum and the experimental one by bringing into coincidence the most intense peak of the C-C band. As it can be noticed from the figure, the overall agreement is quite satisfactory. This is especially true at large binding energies, the low intensity region (top most levels) is unreliable from the experimental point of view since it comprises those one-electron states for which the C-H orbitals (very low relative cross sections) are dominant in the linear combination.

A natural development of this work was to study the effects of structural perturbations like conformational changes on the valence bands of stereoisomers. We have considered the four conformations relevant to the interpretation of the vibrational spectrum of polyethylene (18). The theoretical calculations on those four conformers of polyethylene T, TG, G and TGTG' revealed so important differences in the shape of densities of states that

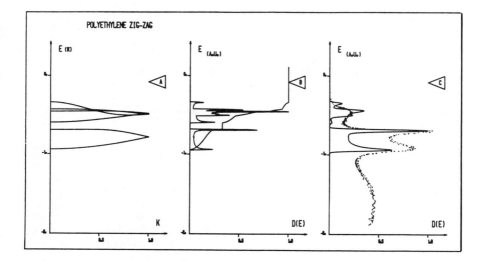

*Figure 1. Electronic structure of all-trans polyethylene: (A) valence band struc-
ture; (B) density of states histogram and its integration curve; (C) experimental
( · · · ) and theoretical (———) XPS spectra.*

they should be observable provided those conformations could be isolated which is not the case unfortunately. The theoretical XPS spectra of those four possible conformations are reproduced in Figure 2. It is seen at first sight that conformational changes do affect the general structures of the density of states. For example, the C-C bands are affected. Their bottom energy values are almost constant while a modification takes place at the top of the bands giving rise to important changes in the band widths. Another aspect is that an energy gap appears in the C-C band of polyethylene for which we have successive trans-gauche conformations (TG and TGTG'). A non-negligible fact is that trans and gauche are very easily distinguished by the fine structure of the C-C peak which is clearly doublet for all trans (T) polyethylene while being definitely triplet for all-gauche (G) compound. Those effects could probably be experimentally observed if well-characterized samples of those four polyethylenes are available (19).

This previous computer experiment on various conformations of polyethylene predicts definite observable influence of the conformation on the electronic density of states distributions. It was thus incentive to investigate a real case and prove the existence of observable conformational effects on the valence electronic density of states of iso- and syndiotactic polypropylenes. In the case of an isotactic polypropylene, the geometrical arrangement is that of an isoclined 2*3/1 helix (20). A tentative interpretation of the experimental XPS spectrum by assuming a zig-zag planar conformation has been reported (21) and is illustrated in the upper part of Figure 3. No satisfactory agreement as to the peak structure was obtained. For example, an intense and sharp peak emerging in the middle of the experimental clump does not appear in the theoretical zig-zag calculation. More-over, only four distinct peaks are observed in the calculated spectra instead of five in the experiment. The previous and inspiring computer experiment suggests the role of conformational effects. Thus, the problem has been reinvestigated by computing the properties of an isotactic polypropylene in its actual helix form. The results are shown in the lower part of Figure 3. As a consequence of the conformational effects, the right structure now appears in the C-C clump: the missing intense and sharp peak is now present and the outermost sub-bands split into two components; peaks intensities also follow the right order (22). The conformational effect could probably be nicely illustrated in the case of syndiotactic polypropylene which does exist in the zig-zag (23) and in the helical form (24). Theoretical XPS density of states of the artificial zig-zag isotactic polypropylene and of experimental conformations of helical isotactic and of both zig-zag and helical syndiotactic polypropylene are presented in Figure 4. Those results were obtained by a model potential technique. They reproduce the already discussed differences between zig-zag and helical isotactic polypropylene. They show

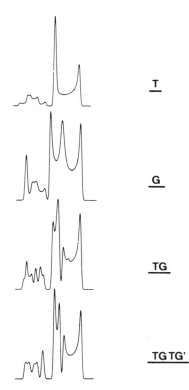

*Figure 2. Theoretical XPS spectra of four possible conformations of polyethylene (T, G, TG, TGTG′). Results were obtained using a model potential technique.*

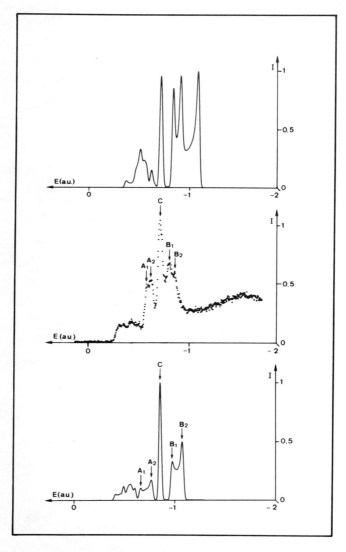

*Figure 3. XPS spectra of isotactic polypropylene: (A) theoretical (extended Huckel) for the zig-zag planar conformation; (B) experimental; (C) theoretical (extended Huckel) for the helical 2\*3/1 conformation. Intensities and energies are given respectively in arbitrary and atomic units.*

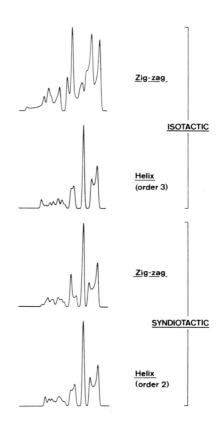

*Figure 4.   Theoretical XPS density of states obtained by a model potential technique for isotactic and syndiotactic polypropylenes*

some expected modifications in the syndiotactic conformations. More experimental investigations are now in progress to check the validity of these conclusions.

Following the way just indicated, the effects of a systematic substitution on the electronic structure of models can be examined. The series of polymers obtained via fluoro-substituted ethylene is ideally well designed for this purpose as most compounds are available to the experimentalists and fluorine is likely to induce some of the strongest electronic effects on the band structure. As observed in other cases, it is found that the valence bands constitute a real and unique fingerprint of the polymer. Indeed, they provide complementary and sometimes less ambiguous information than core levels. In the case of fluoro-polymers, adequate theoretical methods (i.e. those able to take into account the large redistributions effects due to fluorine electronegativity) like FSGO, model potential or <u>ab initio</u> ones produce theoretical spectra in close agreement with experiment. An example is given in Figure 5 where theoretical XPS density of states obtained by the FSGO technique (but not including any cross section effects) are compared to experimental results.

Valuable information for future applications to other fluoropolymers or other copolymers can be obtained from such comparisons. For example, the energy location of C-F and bottom of C-C peaks is found to be constant in each spectrum (C-F = 40eV, bottom of C-C = 20eV). A possible use of those peaks is to act directly as internal reference points for experimentalists especially as the C-F band can always be unambiguously assigned (<u>25</u>).

Those examples clearly demonstrate the interest of a combined theoretical-experimental approach to extend the field of application of photoelectron spectroscopy in polymer physical chemistry.

Another challenging project would be to put more effort in developing techniques capable of predicting electronic spectra of substitutionally disordered polymers. For example, these methods could be valuable for the investigation of copolymers of various types (<u>26</u>):

| block | random | alternating |
|-------|--------|-------------|
| $-(A)_n-(B)_m-$ | ...A-B-B-A-B-A-A- | $-(AB)_n-$ |

One may think of fluoropolymers as working examples (since they contain the highly electronegative fluorine atoms) and consider several arrangements:

  - Perfect head-to-tail polyvinylidene fluoride

$$-CH_2-CF_2-CH_2-CF_2-CH_2-CF_2-CH_2-CF_2-CH_2-CF_2-$$

    h     t     h     t     h     t     h     t     h     t

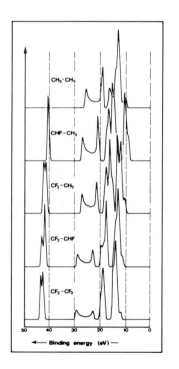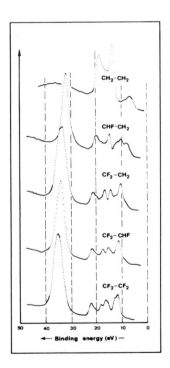

*Figure 5. Theoretical (FSGO) and experimental XPS valence spectra of various fluoro-polyethylenes (theoretical results do not include any cross-section effects)*

- perfect alternating copolymer of ethylene and tetrafluoroethylene

$$-CH_2-CH_2-CF_2-CF_2-CH_2-CH_2-CF_2-CF_2-CH_2-CH_2-CF_2-CF_2-$$

       h    h    t    t    h    h    t    t    h    h    t    t

- somewhat less perfect but more realistic distribution
  in polyvinylidene fluoride for

$$-CH_2-CF_2-CH_2-CF_2-CF_2-CH_2-CH_2-CF_2-CH_2-CF_2-$$

       h    t    h    t    t    h    h    t    h    t

## Conclusions

It is relevant to stress upon the fact that a combined
approach (experimental XPS technology and theoretical band structure calculations) is able to provide a good insight into the
electronic structure of regular polymers.   It is also to be
pointed out that simple computational techniques like FSGO, model
potential or extended Hückel sometimes provide a cheap alternative to more complete ab initio results.  From a practical point
of view, the original three-step procedure (calculation of band
structures, of density of states and of simulated density of
states) turns out to be a necessary complement to interpret fine
effects on the valence structures.   Indeed, it has the merit of
opening new inroads into the areas of stereoregular polymers but
refined investigations are certainly needed in the near future to
confirm and/or correct some conclusions.  This paper has mainly
discussed the work developed by groups at the university of Namur
(Belgium).  We would like to stress that complementary contributions have been made at the University of Durham (England) (D.
Clark, D. Kilcast,...), at IBM-San Jose (P. Grant, I. Batra,...),
at Xerox (C. B. Duke, W. Salaneck, H. R. Thomas,...) and should
compete with this work.
Some notations like T, G, TG, TGTG' or 2*3/1 helix are
perhaps not too familiar to non-chemist readers; A detailed
explanation of those symbols can be found in standard reference
books like:  "A. J. Hopfinger, Conformational Properties of
Macromolecules, Academic (1973), New York" or "H. Tadokoro,
Structure of Crystalline Polymers, Wiley (1979), New York."

## Abstract

Applications of quantum mechanical methods to polymers has
become increasingly important over the last decade.  One of the
several reasons for that is the parallel improvement of experimental techniques for investigating electronic properties of
polymers like Photoelectron Spectroscopy (PS) and especially X-

ray induced Photoelectron Spectroscopy (XPS), sometimes known as
ESCA.

The present paper will first review shortly the way of
performing Hartree-Fock (HF) calculations for ground state properties of polymers. By use of the Koopmans' theorem, the
corresponding HF density of states is of direct interest as an
interpretative tool of XPS experiments. A practical way of
correlating band structure calculations and XPS spectra is thus
presented. In the last part, we illustrate the type of mutual
enrichment which can be gained from the interplay between theory
and experiment for the understanding of valence electronic properties.

## Literature Cited

1.    Ladik, J., Acta Phys. Hung., 1965, 18, 173; Ladik, J., Acta
      Phys., Hung., 1965, 18, 185; André, J.M., Gouverneur, L.,
      Leroy, G., Internat. J. Quant. Chem., 1967, 1, 427; André,
      J.M., Gouverneur, L., Leroy, G., Internat. J. Quant. Chem.,
      1967, 1, 451; Del Re, G., Ladik, J., Biczo, G., Phys. Rev.,
      1967, 155, 977; André, J.M., J. Chem. Phys., 1969, 50, 1536.

2.    Zupan, J., Ann. Soc. Scient, Brux., 1975, 89, 337.

3.    André, J.M., Comput. Phys. Comm., 1970, 1, 391; Kertesz, M.,
      Acta Phys., Hung., 1976, 41, 127; Suhai, S., Ladik, J., Solid
      State Commun., 1977, 22, 227.

4.    André, J.M., Kapsomenos, G.S., Leroy, G., Chem. Phys.
      Letters, 1971, 8, 195; Delhalle, J., André, J.M., Delhalle,
      S., Pireaux, J.J., Caudano, R., Verbist, J.J., J. Chem.
      Phys., 1974, 60, 595; Duke, C.B., Paton, A., Salaneck, W.R.,
      Thomas, H.R., Plummer, E.W., Heeger, A.J., MacDiarmid, A.G.,
      Chem. Phys. Lett., 1978, 59, 196.

5.    Andre, J.M., Delhalle, J., Demanet, C., Lambert-Gerard,
      M.E., Internat. J. Quant. Chem., 1976, S10, 99; André, J.M.,
      Lambert-Gerard, M.E., Lamotte, C., Cull. Soc. Chim. Belges,
      1976, 85, 745; Andre, J.M., Bredas, J.L., Chem. Phys., 1977,
      10, 367.

6.    Christoffersen, R.E., Adv. Quantum Chem., 1972, 6, 333;
      Whitten, J.L., J. Chem. Phys., 1963, 39, 349; Whitten, J.L.,
      J. Chem. Phys., 1966, 44, 359; Bredas, J.L., André, J.M.,
      Fripiat, J.G., Delhalle, J., Gazz. Chim. Ital., 1978, 108,
      307.

7.    Piela, L., Delhalle, J., Ann Soc. Scient. Brux., 1978, 92,
      42; Piela, L., Delhalle, J., Internat. J. Quant. Chem.,
      1978, 13, 605; André, J.M., Fripiat, J.G., Demanet, C.,

Bredas, J.L., Delhalle, J., Internat. J. Quant. Chem., 1978,
S12, 233; Bredas, J.L., André, J.M., Delhalle, J., Chemical
Physics, 1980, 45, 109; Delhalle, J., Piela, L., Bredas,
J.L., Andre, J.M., Physical Review B., (accepted).

8.   Duke, B.J., O'Leary, B., Chem. Phys. Letters, 1973, 20, 459;
Duke, B.J., Eilers, J.E., O'Leary, B., Chem. Phys. Letters,
1975, 32, 602; O'Leary, B., Duke, B.J., Eilers, J.E., Advan.
Quant. Chem., 1975, 9, 1; Delhalle, J., André, J.M.,
Delhalle, S., Pivont-Malherbe, C., Clarisse, F., Leroy, G.,
Peeters, D., Theoret. Chim. Act Berl., 1977, 43, 215.

9.   Nicolas, G., Durand, P., J. Chem. Phys., 1979, 70, 2020;
André, J.M., Burke, L.A., Delhalle, J., Nicolas, G., Durand,
P., Internat. J. Quant. Chem., 1979, S13, 283.

10.  Blumen, A., Merkel, C., Phys. Stat. Sol., 1977, B83, 425.

11.  Most of the computer programs for polymers cited in this
paper are routinely running in our laboratory in Namur.
Informations about them can be obtained upon request to the
authors;
ab initio program producing lobe gaussian, FSGO or molecular
fragment-type results:  Program MONIKE
extended Huckel program:  Program EHCO
model potential program:  Program MODPTA

12.  Wood, M.H., Barber, M., Hillier, I.H., Thomas, J.M. J. Chem.
Phys., 1972, 56, 1788.

13.  Delhalle, J., Thelen, D., André, J.M., Comput. and Chem.,
1979, 3, 1.

14.  Delhalle, J., Internat. J. Quant. Chem., 1974, 8, 201;
Delhalle, J., Chem. Phys., 1974, 5, 306; Delhalle, J.,
Delhalle, S., Ann. Soc. Scient. Brux., 1975, 89, 403;
Delhalle, J., Delhalle, S., Internat. J. Quant. Chem., 1977,
11, 349.

15.  Gelius, U., in "Electron Spectroscopy", ed. D.A. Shirley,
North-Holland, Amsterdam, 1972, p. 311-334.

16.  André, J.M., Delhalle, J., Delhalle, S., Caudano, R.,
Pireaux, J.J., Verbist, J.J., Chem. Phys. Letters, 1973, 23,
206.

17.  Delhalle, J., André, J.M., Delhalle, S., Pireaux, J.J.,
Caudano, R., Verbist, J.J., J. Chem. Phys., 1974, 60, 595;
Bredas, J.L., André, J.M., Delhalle, J., Chemical Physics,
1980, 45, 109.

18.  Snyder, R.G., J. Chem. Phys., 1967, 47, 1316; Morosi, G.,
     Simonetta, M., Chem. Phys. Letters, 1971, 8, 358.

19.  Delhalle, S., Delhalle, J., Demanet, C., André, J.M., Bull.
     Soc. Chim. Belges, 1975, 84, 1071; Delhalle, J., André,
     J.M., Delhalle, S., Pivont-Malherbe, C., Clarisse, F.,
     Leroy, G., Peeters, D., Theoret. Chim. Acta (Berl.), 1977,
     43, 215.

20.  Natta, G., Corradini, P., Ganis, P., Makromol. Chem., 1960,
     39, 238; Natta, G., Corradini, P., Nuovo Cimento, 1960, 15,
     40.

21.  Pireaux, J.J., Riga, J., Caudano, R., Verbist, J.J.,
     Delhalle, J., Delhalle, S., André, J.M., Gobillon, Y., Phys.
     Scripta, 1977, 16, 329.

22.  Delhalle, J., Montigny, R., Demanet, C., André, J.M.,
     Theoret. Chim. Acta (Berl.), 1979, 50, 343.

23.  Natta, G., Peraldo, M., Allegra, G., Makromol. Chem., 1964,
     75, 215.

24.  Natta, G., Pasquon, I., Zambelli, A., J. Amer. Chem. Soc.,
     1962, 84, 1488.

25.  Delhalle, J., Delhalle, S., André, J.M., Bull Soc. Chim.
     Belges, 1974, 83, 107; Pireaux, J.J., Riga, J., Caudano, R.,
     Verbist, J.J., André, J.M., Delhalle, S., Delhalle, J., J.
     Electron Spectr., 1974, 5, 531; Delhalle, J., Delhalle, S.,
     André, J.M., Pireaux, J.J., Riga, J., Caudano, R., Verbist,
     J.J., J. Electron Spectr., 1977, 12, 293; André, J.M.,
     Advan. Quantum Chem., 1980, 12, XXX.

26.  Delhalle, J., "Recent Advances in the Quantum Theory of
     Polymers," André, J.M., Bredas, J.L., Delhalle, J., Ladik,
     J., Leroy, G., Moser, C., eds.,LNP 113 (Springer-Verlag,
     Berlin, 1980), p. 255.

RECEIVED February 18, 1981.

# Electronic Structure of Polymers

## X-Ray Photoelectron Valence Band Spectra

J. J. PIREAUX[1], J. RIGA, R. CAUDANO, and J. VERBIST

Facultés Universitaires Notre-Dame de la Paix, Laboratoire de Spectroscopie Electronique, 61 rue de Bruxelles, 5-5000 Namur, Belgium

The first X-ray photoelectron spectra of polymers valence bands and core levels were recorded almost ten years ago (exactly in 1971 and 1972) for polytetrafluoroethylene (1), polyethylene (2), and nitroso rubbers (3). That same period was the beginning of an exploding interest of the spectroscopists for the technique better known under the acronyms XPS (X-ray Photoelectron Spectroscopy) and ESCA (Electron Spectroscopy for Chemical Analysis). Indeed, since that time, the technique has grown into a valuable method to study the electronic structure of materials, particularly polymers. But, contrary to all the other fields of research (on metals, semiconductors, compounds,...), applications to the polymeric materials were, until recently, almost only concerned with core levels analysis; in numerous series of polymers, emphasis was put on the study of the gross and fine chemical composition, the chemical bonds, the electronic charge distributions, of (co)polymers as clean and pure materials, or after some reaction (aging, casing, adhesion, oxidation, sputtering, electrical or plasma discharge...).

Until recently, the valence bands of polymers have received the attention of very few scientists: XPS studies of valence band spectra are still scarce; although they are measuring the less bound electrons of the materials, those that are directly involved in the bonds between the atoms of the molecules, and consequently those that are containing the best potential inform-

[1] Chercheur Qualifie of the Fonds National de la Recherche Scientifique, Belgium.

**Work performed under the auspices of the IRIS program-Institute of Research in Interface Science-supported by the Belgian Ministry for Science Policy.

ation to understand the molecular structure of the polymers. In fact, the recording of polymer valence band spectra and the interpretation of these data require on one hand very careful and patient experimentation, on the other hand, the support of elaborate and accurate quantum mechanical calculations and/or the comparison with other types of experimental data,... in order to extract all the information hidden behind the experimental data.

It appeared recently that XPS valence band data of polymers, that were known to contain valuable information related to the substituents effects (<u>primary structures</u>), could also be used as powerful -and sometimes unique, compared to a core level analysis- fingerprints to characterize as subtle systems as those concerned with purely carbon containing chains, isomerism, and crystalline structure... what will be called the <u>secondary structure</u> of the polymers.

The purpose of the present paper is to review this field of research - a summary of this review appeared in (<u>4</u>). After a brief introduction (part 2) to the XPS technique itself, specially oriented to the study of the polymer valence bands, given in order to stress upon the inherent difficulties (and limitations) of the experiment, we shall present specific examples of the various types of informations that can be acquired on the polymer molecular and electronic structure; (part 3) substitution effects in the valence bands; (part 4) structural isomerism, stereoisomerism, and geometrical crystalline structure; (part 5) other types of information.

## Experiment

In view of the number of monographies, review articles,... published on the XPS technique, it is useless to describe here in length all the details of the experiment. Here, we will only recall the elementary concepts for the non-experts in the field and direct the readers to the literature for a complementary presentation if necessary (<u>5</u>). Particular topics, that are proper to the study of polymers will be described in greater length.

The XPS Measurement. In an XPS spectrometer, the studied material is exposed inside a vacuum chamber to a flux of X-rays (energy 1 keV). The kinetic energy of the photoelectrons ejected from the sample is measured by an appropriate analyzer. This energy is directly related to the binding energy of the electrons inside the sample; on a wide scan XPS spectrum, the unscattered electrons result in characteristic peaks: their energies serve to identify the elements in the material (atomic composition), to characterize the molecular environment of these atoms (chemical analysis, see inset A of Figure 1), and, by the measurement of the photoelectric lines ratios, to reach some quantitative results. Such type of measurement from the core level peaks can usually be

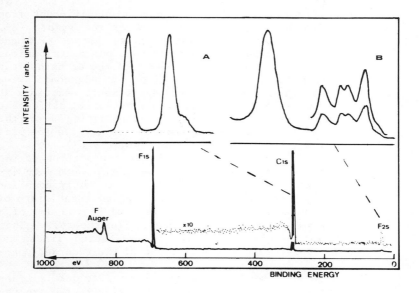

*Figure 1.   Wide survey scan of an ethylene tetrafluoroethylene copolymer. Inset A represents the G-1s core level peaks; inset B pictures the copolymer valence band.*

achieved rather accurately in a few tens of minutes. Let us here
make a few remarks of importance:

1.  The sample being studied in vacuum (usually better than
    $10^{-8}$ Torr), it must be a solid, and, if liquid at room
    temperature (like for pre-polymers with low molecular
    weight), it must be cooled by some refrigerant circula-
    tion. This limits also the possibility to study high
    temperature effects on such materials.
2.  As the photoelectrons reaching the analyzer escape from
    a 50-100 Å layer at the top of the studied solid, the
    technique is extremely surface sensitive, i.e. it
    could detect impurities (a composition different from
    the bulk) at the extreme surface of the sample (6).
    Great care must then be taken in the preparation and
    handling of the samples to avoid spurious contributions
    from contamination. This drawback of the technique can
    be advantageously turned out to an advantage when
    studying (in an angle resolved experiment) reactions at
    the surfaces, e.g. note that the substantial difference
    of escape depth for electrons originating from the
    valence levels compared to core levels may also be used
    to obtain information on a depth profile (7).
3.  It helps to the precision of the interpretation of the
    spectra, to use a monochromatized X-ray source to ir-
    radiate the sample. This can be conveniently done for
    the Al $K_{\alpha}$ line (energy 1486.6eV), the apparent width of
    the photoelectron peaks being reduced by the use of a
    monochromator from 0.9 to 0.6eV (in theory). This
    helps also to remove spurious excitation lines and
    reduce appreciably the background of the recorded spec-
    tra.
4.  For a valence band analysis, the need to use a mono-
    chromatized X-ray source is almost evident; the excita-
    tion line having satellites at 10-12eV (for Al $K_{\alpha}$, with
    a relative intensity of 10%), cannot probe without
    disturbing the spectrum a valence electron region the
    width of which ranges from 20 to 35eV (for carbon
    polymers, to substituted polymers) (11). It is also of
    prime importance to remove the Bremsstrahlung continuum
    of the radiation source, to improve considerably the
    sensitivity and reliability of the fine details con-
    tained in the valence bands.
5.  Being insulators, the polymers present a particular
    problem during the XPS analysis due to the ejection of
    the photoelectron their surface becomes (sometimes in-
    homogeneously) charged, what perturbs the measurements
    by displacing the position, and distorting the shape of
    the photoelectron peaks, and preventing any direct
    calibration of the electron binding energy scale.

6.  Radiation damage might occur during the study of poly-
    mers. In our investigations, only the chlorine-con-
    taining compounds were found to present a color change,
    but no loss of chlorine could be detected on the XPS
    core level spectra.

The advantages of using a X-ray source instead of a UV lamp
to study the polymers are many fold:

1.  With X-ray excitation, it is possible to probe deeper
    in binding energy in the molecular orbitals (M.O.) of
    the valence bands and accessorily, to let the user
    benefit of the possibility of a core level analysis.
2.  Intrinsically, a UV lamp has a much better resolution
    (a few meV), that in fact turns out to be very useful to
    study gases. But in a solid, the molecular orbitals are
    grouped into bands, broadened by vibrational excita-
    tions, and become unresolved so that it has been proven
    that a photoelectron valence band spectrum recorded
    with a UV lamp do not offer substantial advantages
    compared to a XPS experiment.
3.  The photoelectric cross section for a given electronic
    orbital varies with the energy of the excitation
    source: on a UV valence band spectrum, the carbon 2p
    orbitals appear with larger intensity than the C2s
    ones. The reverse is true when a X-ray excitation
    source is used: the cross section is greater for M.O.
    with major 2s character. This enhances the intensity
    of the 2s bands that are generally simpler to interpret
    than the 2p ones. It must be said that the comparison
    between the two (X-ray and UV) excited spectra can be of
    great help to assign the symmetries of the probed M.O.
    (compare e.g. polystyrene (X-ray) of (10), and (UV) of
    (9).
4.  When excited with a UV lamp, the photoelectrons have a
    kinetic energy range (0-40eV) where the escape depth is
    small ($\simeq$ 5Å); the contamination problem is then crucial
    in this case. On the contrary, when excited with a X-
    ray source, the electrons emitted from the valence
    levels have a larger escape depth, so that the surfaces
    features do not contribute as crucially to the valence
    band spectrum than to the core level peaks.
5.  For a valence band analysis, the need to use a mono-
    chromatized X-ray source is almost evident; the excita-
    tion line having satellites at 10-12eV (for Al $K_\alpha$, with
    a relative intensity of 10%), cannot probe without
    disturbing the spectrum a valence electron region the
    width of which ranges from 20 to 35eV (for carbon
    polymers, to substituted polymers) (11). It is also of
    prime importance to remove the Bremsstrahlung continuum
    of the radiation source, to improve considerably the

sensitivity and reliability of the fine details contained in the valence bands.

The Role of the Valence Band.  The valence band represents
only a very small part of the photoelectron spectrum (inset B of
Figure 1):  due to the very low photoelectric cross-sections for
electrons having low binding energies, the recording of one
valence band spectrum requires many hours.  In spite of this fact,
it is in principle possible to perform on valence band data a
similar (elemental and chemical) analysis than on the core
levels.  But, in a valence band, and for the same energy window,
the photoelectron lines are more numerous and generally
associated within bands.  Why spend more time for a valence band
analysis to obtain similar results than with a XPS analysis of the
core levels?  This experience is not in fact unrealistic, as the
molecular orbitals analyzed in a valence band spectrum involve
those electrons that are directly participating in the molecular
bonds: it will be shown indeed that such experiments can furnish
other types of information on the polymer molecular structure,
that are not available from the core level photoelectrons.

Practical Details.  To face the problems and requirements
announced above, the spectra presented here and chosen among the
most recent contributions from the authors' laboratory, were
obtained under the following conditions:

1.    The photoelectron spectra have been recorded with a
      Hewlett-Packard spectrometer (HP5950A) using a mono-
      chromatized Al $K_\alpha$ radiation.
2.    The charging effect on the polymers is controlled by
      the use of an electron flood-gun whose current and
      energy are adjusted to neutralize the surface electro-
      static charges:  the charging effect become stabilized,
      but not absolutely compensated.  This external electron
      source does not allow therefore to measure absolute
      energies automatically, but helps to partially cancel,
      or homogenize the charging effect on the whole insula-
      tor surface, in such a way that the recorded spectra are
      much better resolved and very stable on the energy
      scale during the long accumulation runs of the valence
      band spectra.
3.    The preparation of the polymer material as a very thin
      film (deposited from a solution onto a metallic sub-
      strate) was found to be the most satisfactory to reduce
      the charging effect.
4.    A reproducible calibration of the recorded spectra was
      obtained by mixing the polymer with a reference com-
      pound having similar electrical properties (e.g. poly-
      ethylene, or polytrifluoroethylene), and studying the
      mixture under different settings of the electron bom-

bardment source. Absolute binding energies, expressed relatively to the Fermi level of the spectrometer are obtained with a precision usually of the order of 0.1- 0.2eV (that means that several independent measurements were consistent with each other to this $\Delta E$).

5.    The recording of one valence band spectrum requires typically one overnight accumulation. But one valence band spectrum is recognized as satisfactory only when a few similar spectra have been recorded separately on different samples. Part of the spectra presented here (Figures 7, 8, 9, 10, 12) were digitally smoothed; the others are shown as recorded.

6.    The polymers were all studied, without further chemical treatment, as received from their sources, most generally universities or industrial research laboratories, where in the most cases they were characterized by the conventional techniques.

Like for the core levels, a valence band photoelectron spectrum is composed of peaks or bands having the following characteristics:

-    the binding energy (expressed relatively to a reference value);
-    the intensity;
-    the peak/band width and shape.

Each of these characteristics can be interpreted to infer the molecular and electronic structure of the polymers.

Comparison with Theoretical Calculations. It appears that the polymer valence bands are (very) difficult to interpret without the aid of a theoretical basis, or a model, or of the use of a reference spectrum obtained from a model compound. Indeed, Quantum Chemical theory is nowadays able to calculate band structure and density of states for polymers, to simulate the limited resolution of the spectrometer, and to modulate these theoretical density of states to account for the photoionization cross sections that vary considerably for valence bands of polymers containing different types of atoms, and electrons with various symmetries. Consequently, one is able now to predict theoretically the energies of the various molecular orbitals, but also their relative intensities. As it is proven elsewhere (12) and in another review at this conference (13), the support of such theoretical results of an appreciable and appreciated value is to extract all the information that are hidden behind the fine details of the valence bands of polymers. But we will not discuss further this theoretical aspect of the study of the electronic structure of the polymers. The purpose of the next sections will be to show that, in spite of the time and special care required to analyze valence band spectra, the particular and unique role played by the low energy orbitals to describe the bonding of the atoms within the molecule, makes it worthwhile to undertake their study.

## Substitution Effects in the Valence Bands

To discuss the information available from XPS spectra of polymers, it is necessary first to understand the electronic structure of the simplest of these compounds, polyethylene (PE), the spectrum of which will constitute the basis for the interpretation of all the other studied compounds.

<u>Polyethylene and Other Model Compounds</u>.  The polyethylene XPS valence band spectrum is reported on Figure 2.  Superimposed on a background generated by the scattering of the photoelectrons on their way outside the solid, two structures are resolved on this spectrum.  The first one at high binding energy has two peaks with different magnitudes at 18.8 and 13.2eV; the second structure contains several fine components (they are just emerging from the statistical background on a better spectrum (15); they are not easily discerned due to the low intensity of this band.

A satisfactory interpretation —in agreement with the theoretical calculations— of this valence band is to assign the low energy structure to a C–H band (with C2p and H1s polymer orbitals), and the high energy structures to the C–C band (with a dominant C2s character) in the polymer: the leftmost peak is the bonding combination of the C2s–C2s orbitals, the rightmost is the antibonding analogue (14, 15, 16).

In order to better understand this valence band spectrum, we found it interesting to investigate the effect of one-dimensional periodicity on the broadening of energy levels (in an atom or in a molecule), which gives rise to the energy bands (in a solid). Indeed, polymers exist with a wide possible range of molecular weight (M.W.) –or chain lengths–; could we think of distinguishing polymers with different M.W.'s?

The alkanes series is suitable to study such an effect; the successive molecules with an increasing number of carbon atoms in the almost linear chain are considered as progressive steps in the formation of an ideal one dimensional solid, polyethylene (17, 18).  On the Molecular Orbital point of view, the electronic levels of carbon increase in density, to finally form a band structure.  Selected spectra recorded for the lightest alkanes are reported on Figure 3.  All show distinct C2s and C2p regions separated by an energy gap; the number of levels in the C–C(C2s) region is equal to the number of carbon atoms in the alkane chain. As these levels are confined in a limited energy range, the spacing between each of them decreases and soon, a band will be formed.  From which minimal chain length, will the valence band spectrum approximate the band structure of the infinite solid polyethylene?  Figure 4 shows that the essential features of the polyethylene valence band are already present in the n-tridecane solid: the experimental resolution does not allow to resolve all

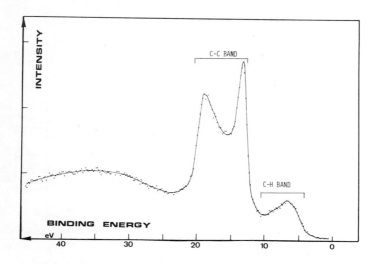

*Figure 2.    XPS valence band spectrum of polyethylene (4)*

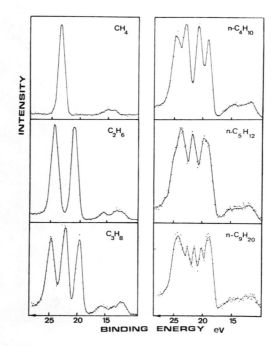

*Figure 3.    XPS valence band spectra of the lightest alkanes: (from* top *to* bottom *and* left *to* right) *methane, ethane, propane, n-butane, n-pentane, n-nonane, successively.   The measurements were performed in the gas phase.*

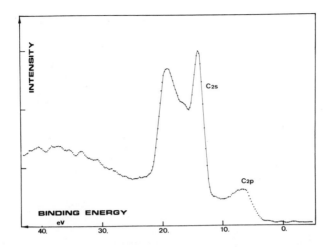

*Figure 4.    XPS valence band spectrum of* n-*tridecane (*n-$C_{13}H_{28}$*)*

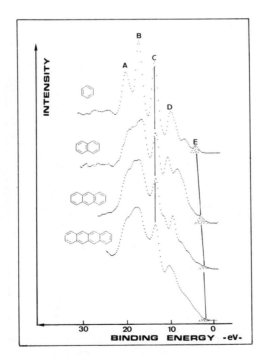

Physica Scripta

*Figure 5.    XPS valence band spectra of the acenes (from* top *to* bottom*): benzene, naphthalene, anthracene, and tetracene (20)*

the components of the C2s band, the width of which (7.6eV) approaches the value measured for polyethylene (7.7eV).

As we observed that a dozen of carbon atoms in the alkane chain is sufficient to simulate the electronic properties of polyethylene, it becomes evident that the XPS technique, by the analysis of the valence bands, will not be able to distinguish between polymers with different high molecular weights.

But, the same experiment proves on another hand that the XPS technique is not too short range sensitive: it is shown that the valence band photoelectrons leaving the polymer sample bring information on the electronic structure of one "slice" of the polymeric skeleton containing at least a dozen carbon atoms.

Another experimental evidence for this type of limitation is provided by the XPS valence band measurements of the polyacenes (benzene, naphthalene, anthracene, and tetracene) (19, 20). Passing from benzene to tetracene (see Figure 5), the number of molecular orbitals increases and very soon the C2s bands (peaks A, B, C) become less resolved. Particular in this case is the evolution observed for the π electrons (peak E) that progressively become more delocalized; this effect is important to be measured as it is directly related to the electrical properties of the polyacenes.

A third example of this effect is evidenced by a study of the valence band structures of the series of n-phenyl molecules converging to the polymer p-polyphenyl whose valence band spectrum can be seen in Figure 6 (21). This experience confirms the results obtained for the alkanes and the polyacenes.

When one (or more) hydrogen atom(s) of the polyethylene monomeric unit is (are) replaced by a heteroatom, an aliphatic or an aromatic group, modifications are induced in the valence band spectrum; new peaks, band shift and/or splitting, redistribution of the electronic population among the molecular orbitals will denote the new bonds created in the molecule. Similar effects will be observed for the insertion of heteroatom(s) between two carbon elements of the polymer skeleton.

Hydrocarbon Based Polymers. The substitution of one hydrogen atom in the $-CH_2-CH_2-$ unit by some short carbon chains induces subtle modifications in the electronic structure (molecular orbitals) of the polymers. Though these modifications cannot be easily evidenced on the XPS carbon 1s core level spectra, it appears that the XPS valence band structures are much more sensitive to these substitutions and that they become unique and readable fingerprints of the polymers (10, 22). We will not speak here of the C1s shake-up data that were revealed useful to distinguish between saturated and unsaturated bonds (this field with various applications was recently reviewed (23)).

Up to now, only the valence bands of the simplest and commercially most used polymers have been recorded, and their molecular orbitals assigned (10); these are polypropylene (PP),

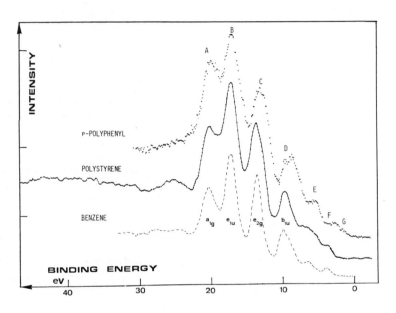

*Figure 6.   XPS valence band spectra of benzene (solid phase), polystyrene, and
p-polyphenyl (4)*

poly(1,butene) (PB), poly(vinyl cyclohexane) (PVCH) for the
saturated compounds, and poly(styrene) (PS), and polyphenyl (PPH)
for the compounds containing unsaturated bonds.   Some other
series of polymers were also studied, but the stress was put on
another subject than the complete understanding of the valence
molecular orbitals; they will be discussed in another section of
this paper.

In polypropylene, one methyl group replaces an hydrogen atom
in the polyethylene repetitive unit; this results in an intense
structure located in the middle of the C-C (C2s) band.   In
isotactic polypropylene moreover, the bonding and antibonding
structures of this band are split in accordance with theoretical
calculations performed on isotactic polypropylene in an helical
conformation (10, 44).

Substituting a hydrogen in PE by a $-CH_2-CH_3$ radical modifies
more drastically the valence band spectrum of poly(1-butene).   An
interpretation of this spectrum can be extrapolated from the data
recorded for the saturated hydrocarbon identified with the monomeric unit, n-butane.   The two spectra differ however in some
details (relative intensities of the bands, or splittings of the
peaks) due to the new bonds created in the polymer by the bonding
of the $-CH_2-CH_3$ group to the polyethylene skeleton (10).

The electronic structure of poly(vinyl cyclohexane) is in a
similar way explained with the XPS spectrum of solid cyclohexane
(10); but in this case, the similarity with the PE spectrum is
very low.   Coming from PE, to PP, PB and PVCH, the length of the
monomeric unit becomes longer through the side group that is
imposing its shape to the CC band.   This C2s structure constitutes
a real fingerprint of the electronic structure of the polymer;
almost immediately, the polymer can be identified through its C-C
band shape (22).

Polystyrene (PS) and p-polyphenyl (PPH) are two $\pi$ electronic
systems whose photoelectron valence band spectra can be compared
to the data recorded for solid benzene (Figure 6).   The assignment
of the peaks is straightforward; the bands are similar in shape,
the photoelectron peaks are located at about the same binding
energies.   Looking more carefully to the spectra, we note slight
differences; in PPH, band D is broadened and split due to an
interaction of the lowest lying carbon 2s $\sigma$ and the highest C2 p$\sigma$
M.O.'s from various units into the polymer chain; the PPH $\pi$ bands
have a different shape, disclosing an interaction of the $\pi$
electrons between different phenyl groups within different polymer molecules; the binding energy of the lowest occupied $\pi$
molecular orbital is measured at 1.8eV, compared to 2.9eV in
polystyrene.   This sole measurement explains the different electrical properties of the two polymers; the low binding energy of
these $\pi$ electrons for PS (compared to the C2p band of the purely
saturated hydrocarbon polymers) shows that these electrons remain
localized in the pendant benzene rings, and that they do not
participate in the bonding of the carbon skeleton of the polymer

(10). A much lower energy for these π electrons in PPH reveals an interaction between all the π electrons, that become more delocalized into the polymer material. In other words, the π electrons of PS are shown to remain in the pendant group on the polymer chain, whereas those of PPH are subject to intermolecular interactions, which explains a much more pronounced conductor character for PPH (21).

When the polymer backbone contains a large number of substituents, or long side chains, it is clear that its valence band photoelectron spectrum will contain a lot of peaks; each side group will bring its own fingerprint in the XPS spectrum, and the interpretation of the data will be very difficult. An example of this limitation has been found in the study of polydiacetylenes, which until now were not synthesized with sufficiently short side chains to reduce the number of bands (molecular orbitals) appearing in the valence spectrum (24, 25).

It is the reason why it was also important to study systematically the XPS valence band spectra of simple model polymers substituted by an heteroatom; it is necessary to discover also the fingerprint(s) of these substituents in order to understand in the future more complicated systems.

Fluoro-substituted Polymers. The fluoropolymers were between the first to be studied by the XPS technique because the substitution of F atom(s) in the $-CH_2-CH_2-$ unit induced very large modifications in the XPS core level spectra (shifts up to 8eV) that were easy to detect and interpret. The XPS valence band spectra of similar compounds, namely poly(vinyl fluoride) (PVF), poly(vinylidene fluoride) (PVF2), poly(trifluoroethylene) (PVF3), and poly(tetrafluoroethylene) (PTFE) (26, 27, 28) are also expected to reflect the induction of such strong electronic effects at the valence molecular level.

The first, and almost evident fingerprint of the substitution of hydrogen atom(s) by fluorine is the appearance in the valence band spectrum of two new structures related to the electronic properties of fluorine, namely its 2s and 2p orbitals whose bands are expected to appear in the binding energy region between 0 and 40eV. The spectra presented at Figure 7 show clearly the F2s peak appearing with a very high cross section between 30 and 40eV (the theoretical relative photoionization cross sections for the F2s, C2s, F2p and C2p orbitals are respectively 4.2, 1.0, 0.35, and 0.032 (29)). The F2p peak is less sharp, but gains in relative intensity when the number of fluorine atoms in the monomeric unit increases; it is located at about 10-12eV.

The C-C band of the fluoropolymer series still appears with the same shape than in PE, its low energy limit shifting to higher binding energy, while its high energy limit moves slightly less; this results in a narrowing of the band in the series. It is also noteworthy to remark that the energy separation between the F2s

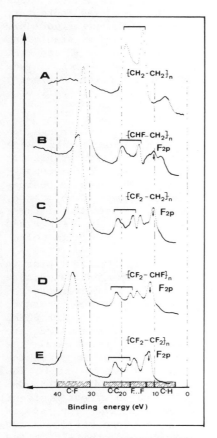

*Figure 7.    XPS valence band spectra of the fluoropolymers: PE, PVF, PVF2, PVF3, and PTFE (27)*

and the antibonding part (structure at lowest energy) in the C-C
band is constant to within 0.3eV. These peaks may then serve as
markers in each spectrum, especially as the F2s peak can always be
unambiguously assigned.

The compilation of such data constituted a firm basis that
was used to study a specific and more complicated system; the
elucidation of the electronic structure of a copolymer of ethylene (48%) and tetrafluoroethylene (52%) whose synthesis was conducted in order to maximize the alternating sequences. The
valence band spectrum of such a compound (Figure 8) was found very
similar to the one measured e.g. for poly(vinylidene fluoride).
But, by looking to the fine details of the spectrum, by simulating
the valence band of a block copolymer (by addition of PE and PTFE
spectra), and by comparison with model calculations, it was
possible to show that the C-C band width and the distance F2s-top
of the C-C band were characteristic of an ethylene-tetrafluoroethylene copolymer with dominant alternant structure (28).

Chlorine-containing Polymers. Polymers containing one
chlorine atom in various environments (other sustituents) were
studied by XPS; poly(vinyl chloride) PVC, poly(chlorotrifluoroethylene) PCTFE, an (ethylene-chlorotrifluoroethylene) copolymer, and poly(epichlorohydrine) PEPI, were chosen because besides
carbon atoms they contain chlorine in presence of hydrogen,
fluorine, and oxygen atoms. The valence band spectra of these
compounds (see Figure 9) show that features can be easily and
unambiguously assigned to a contribution from the chlorine molecular orbitals.

From the Cl3s and Cl3p levels that are expected to appear in
the studied energy range, the Cl3p band is the dominant structure
of the spectrum, as it is detected as the first (closest to the
Fermi level) and most intense peak. Its energy is not constant;
it varies in the range of 5 (for PVC and PEPI) to 8eV (for PCTFE
and the copolymer). The presence of such an intense structure
affects the rest of the spectrum as it causes a lifting of all the
other structures over an intense background, resulting from the
inelastically scattered photoelectrons through the material. The
Cl3s level appears at a varying position in the middle of the C-C
(C2s) band; it is much less resolved, but at an almost constant
energy of 12eV from the Cl3p peak. We note that the Cl3p band is
more intense than the Cl3s one, while the reverse was found for
the corresponding symmetries of the outermost F orbitals in the
fluoropolymers (compare Figures 7 and 9).

A more complete description of the features appearing in the
valence band spectra of these compounds will be presented elsewhere (30). We note here that, when the polymer contains chlorine
and fluorine atoms, its valence band spectrum becomes (very)
complicated, consisting of an almost continuous succession of
molecular orbitals with similar intensities. Their interpretation could not have been given without the aid of a relatively

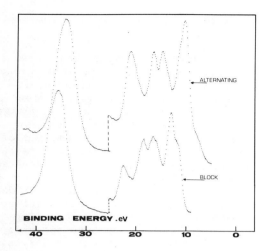

*Figure 8.   XPS valence band spectra for two ethylene–tetrafluoroethylene copolymers: (top) alternating structure; (bottom) block structure.*

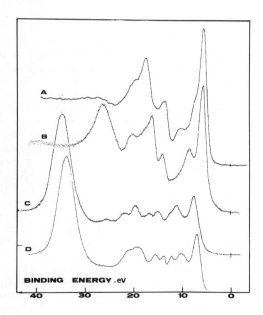

*Figure 9.   XPS valence band spectra of (A) polyvinyl chloride, (B) poly(epi chlorohydrine), (C) poly(chlorotrifluoroethylene), (D) co(ethylene-tetrafluoroethylene*

complete set of reference spectra recorded for similar polymers
(10, 30).

Oxygen Containing Polymers. Oxygen can be incorporated into
a polymer chain as part of different functional groups: alcohol,
ether (linear and branched), ester, carbonate,... Valence band
data on polyacrylates (7) and polymethylacrylates (34) have already been published, but without a complete assignment of the
molecular orbitals. Compounds belonging to other series were
studied in our laboratory (31, 32) with the aid of theoretical
calculations (34). But we chose to present here special topics
that will shed a little more light on our understanding of the
valence band spectra of polymers.
   The linear polyethers are model compounds to show the effect
of the insertion of one oxygen atom at different positions into
the polyethylene $-CH_2-CH_2-$repetitive unit. What will be the
influence on the polymer XPS valence band spectrum, compared to
the one of polyethylene (Figure 2)? Our study was conducted on
three materials; poly(methylene oxide) or PMO, poly(ethylene
oxide) or PEO, and poly(tetramethylene oxide) or PTMO, that are
conveniently pictured by the formula $-[(CH_2)_x-O]-$ , with x=1,
2, and 4 respectively. Figure 10 reports the valence band spectra
of these polymers, where from two types of informations can be
deduced:

1.  The presence of oxygen in the polymer is easily deduced
    from the appearance of a new peak located at about 27eV:
    the O2s molecular orbital, that retains probably most
    of its atomic level character (i.e. does not feel the
    influence of the other molecular orbitals defining the
    real valence band at lower binding energies). On the
    contrary, the O2p level is strongly mixed with the
    other (carbon and hydrogen) levels, as it is the medium
    that bounds the oxygen to the other atoms; it does not
    appear clearly on the XPS spectra. Noteworthy is the
    signature of the oxygen lone pairs (O2p π non bonding
    molecular orbitals); they appear as the rightmost
    structures, at the lowest binding energy. They play an
    important role in characterizing the polymer molecular
    structure, as they are intimately related to the pro-
    perties of surface wettability of such types of com-
    pounds.

2.  The interaction (bonding) of oxygen with the carboneous
    skeleton is seen e.g. in the C-C bonding band as new
    structures that are related to the overlap between the
    C2s and O2p molecular orbitals to form the new C-O bonds
    (they are best seen in the PTMO spectrum as the two new
    structures appearing in the middle of the C-C band).
    The discussion and assignment of the M.O. of the pre-
    sented spectra are given elsewhere (31, 32). We want

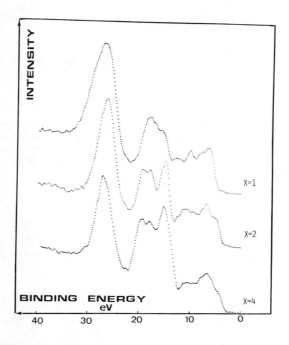

*Figure 10. XPS valence band spectra of the linear polyethers: (1) poly(methylene oxide), (2) poly(ethylene oxide), (4) poly(tetramethylene oxide) (4).*

here to stress only on the trends that are clearly seen
in the bandwidths and shapes, as a consequence of the
interactions between the carbon and oxygen orbitals.
For example, the O2s peak is measured to narrow from 5.7
to 3.8 and 3.5eV for the polymers PMO, PEO, and PTMO
respectively; this fingerprints the progressive isolation of the oxygen atom in the carbon chain (as a
reference, the O2s line width of poly(vinyl methyl
ether) PVME and poly(vinyl alcohol) PVOH are 3.7 and
3.8eV respectively). Correspondlingly, the C2s band
broadens out from 5.1 (PMO) to 7.1eV (PTMO), approaching the width measured for the infinite $-CH_2-$ sequence
in polyethylene; 7.7eV. It is noteworthy to point out
here that these trends observed in the valence band
spectra are correlated by separate theoretical calculations at the Extended Huckel level (33).

Miscellaneous. To complete the library of XPS photoelectron
valence band spectra it would be useful to add measurements on
various nitrogen, or silicon, or sulfur containing polymers.
These data are not yet available now...they would be of interest
for the XPS study of polyurethanes, polysiloxanes, resins, polymeric glues, polysulfones...

Isomerism

Introduction. Substitution effects in the valence band
spectra of polymers are the largest effects one could hope to see
with the XPS technique; indeed, like for a core level analysis, it
is when one heteroatom is included in or aside the polymer
skeleton that one expects to see the most substantial modifications in the spectra; new peaks or bands, influence (shift...) on
the existing structural characteristics. But, all the polymers
that are processed and studied nowadays do not differ exclusively
by their chemical composition. Within one series or family of
polymers, there exists compounds with various molecular weight
(we did show previously that the XPS technique is not able to
distinguish between two polymers with different high molecular
weight)., crystallinity, tacticity, isomerism (linear, branched,
ramified structures; head-to-head linkages...).
Chemical industries synthesize now more and more elaborated
materials, they exploit not only additives (for the synthesis,
the processing, the stability...), but also reach now the very
promising field of the copolymers where all the parameters cited
above could be played with for two, or three...polymers in the
same material. These new polymers are expected to become very
sophisticated and dedicated to very specific applications....
Therefore, new techniques are scanned now to widen the available
analytical tools in order to help the more classical ones (infrared absorption and nuclear magnetic resonance).

X-ray photoelectron spectroscopy, through the analysis of the polymer core levels spectra did succeed in giving partial answer to a few questions of these types. Although studies of the degree of crystallinity were negative (35, 36), some results were obtained in the study of structural isomerism, for nitroso rubbers (3), polybutyl acrylates (7), and paracyclophanes (37). The limits of the method are already noted: the core level spectra of structural isomers are virtually identical. In some particular cases, one might be able to identify isomers, using XPS with monochromatized X-ray source, studying the shake-up satellites if any, and with the help of appropriate theoretical calculations. Results obtained on copolymers are more promising to study the surface composition, topography, morphology, and the first outermost layers (e.g. co(polydimethylsiloxane and polystyrene), or co(polystyrene and poly(ethylene oxide)) (38, 39).

The XPS valence band spectra picture directly the bonds between the atoms of the molecule, and are more characteristic of the compounds (34) especially for polymers containing only carbon atoms; could the technique (with the use of complementary reference spectra, and/or theoretical density of states calculations) be sensitive enough to allow an identification of isomers?

If in the study of the polymers primary structure we did not especially take into account particular characteristic parameters of the compounds because "substitution effects" are so clearly marked, it is clear that in studying the secondary structure of the polymers, we must be very careful and critical, about the preparation and characterization of the compounds as far as the isomeric conformation, tacticity, and/or crystallinity... are concerned. It is indeed expected that the effects that must be experimentally evidenced will be (very) fine, and small in amplitude. It is the reason why the compounds, whose XPS spectra are discussed below, were synthesized and characterized by specialized laboratories in the world (practical informations will be given elsewhere (40, 41).

Isomerism is a term covering different molecular conformation effects. We will not speak here again on the type of information that can be gained on the alternating or block structure of copolymers, as was explained above for an ethylene-tetrafluoroethylene copolymer. For homopolymers, we will distinguish various categories:
- constitution isomers, i.e. polymers that have the same chemical formula, but whose bonds are differently organized, like in linear or branched structures;
- head-to-head linkages;
- stereoisomerism, i.e. polymers whose functional groups are cis- or trans- , relative to a double bond;
- or iso- and syndio-polymers, relative to the main skeleton (tacticity).

Structural Isomerism. Solely hydrocarbon based polymers (that

differ only by saturated and unsaturated bonds) can be considered
in a broadened sense as structural isomers. It has already been
shown that XPS core level spectra can easily distinguish between
the two types of compounds as shake-up satellites are related to
the existence of π electrons in the unsaturated polymers (23). We
want to mention here that similar results are achieved by an
overlook to the valence band spectra of the concerned compounds;
this was already shown for polystyrene and poly(vinyl cyclohexane), see above and (10). We present here (Figure 11) another
example of this potentiality of the technique; it concerns
poly(1-butene) and poly(butadiene 1,4)cis. These valence band
spectra are so profoundly different that they undoubtly fingerprint the compounds.

   Examples of real isomerism of constitution are very numerous;
we list here a few series of compounds which were more or less
easily identified by their XPS valence band spectra. Generally,
they are relatively simple polymers in the sense that their
monomeric unit contains few atoms; for larger systems, it is
expected that the distinction between two isomers will be less
evident. We distinguish here the two following groups of polymers, and only list the compounds already studied (the list
contains all the compounds already studied in the laboratory or
elsewhere; it is not restrictive: all the isomers studied up to now
were found different by their valence band spectra):

   1. for purely hydrocarbon polymers:
      - poly(3-methyl 1-butene)and poly(1-pentene), with a
        $C_5H_{10}$ monomeric unit.
      - poly(4-methyl 1-pentene) and a copolymer(butene-2,
        ethylene) with a $C_6H_{12}$ monomeric unit.

   2. for oxygen containing polymers:
      - poly(vinyl alcohol) and poly(ethylene oxide) with a
        $C_2H_4O$ unit; but in this case the carbon 1s core level
        spectra of the two compounds are different (24, 31,
        32);
      - poly(propylene oxide) and poly(vinyl methyl ether) with
        a $C_3H_6O$ unit;
      - poly(tetramethylene oxide) and poly(vinyl ethyl ether)
        with $C_4H_8O$ unit;
      - three isomeric poly(butyl acrylates) (7) with a $C_7H_{11}O_2$
        unit;
      - four isomeric poly(butyl methacrylates) (34) with a
        $C_8H_{14}O_2$ unit.

The interpretation of the valence band spectra of these
polymers can be found elsewhere (31, 32).

   Figure 12 is included to substantiate the ability of the XPS
technique to distinguish two constitution isomers; it compares
the core and valence spectra recorded for PPO and PVME, and shows
that, whereas the C1s levels appear very much alike, the valence
bands readily distinguish the two polymers.

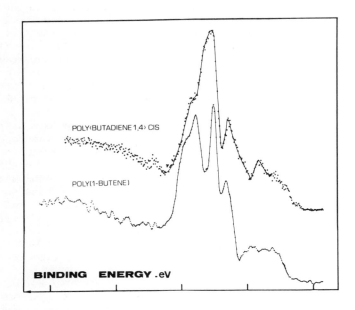

*Figure 11.* XPS valence band spectra of poly(1-butene) and poly(butadiene 1,4)cis

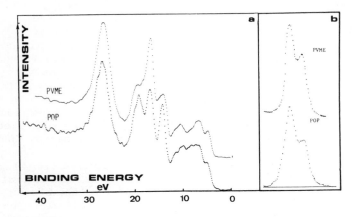

*Figure 12.* XPS valence band and core level spectra of poly(propylene oxide) and poly(vinyl methyl ether) (4)

Head-to-Head and Head-to-Tail Linkages. Errors in the link-ages between successive monomeric units of the polymers are possible (and always statistically present). The effect of head-to-head (HH) and head-to-tail (HT) bonds in the XPS core levels spectra of substituted polymers have been computed and found at the limit of the sensitivity of the technique (e.g. 42). The control of these linkages during the synthesis is difficult and the number of polymers that can be prepared with 100% of HH or HT linkages is small (43).

Two compounds of this type were placed at our disposal; isotactic polypropylene and an alternating erythro-iso-copolymer of butene-2 and ethylene. Looking to the extended chemical formula of the latter (Figure 13), it is indeed immediately evident that this copolymer can be considered as an equivalent to a HH polypropylene. The XPS analysis of the valence band spectrum of this compound reveals that its electronic structure, reflected through the C-C (C2s) molecular orbitals is entirely different from that of polypropylene (Figure 14).

In view of this favorable experience, it is now highly desirable to test the potentiality of the technique on other series of HH and HT polymers.

Stereoisomerism. The first trials to use the XPS valence band spectra to distinguish between two stereoisomers were un-successful. Cis- and trans- poly(isoprene) - with short branched chain and small substitution effects-, as well as cis- and trans-poly(1,4dichloro-2,3epoxybutene) - with longer branched chain and more intense substituent effects- did not show in our first measurements significant differences in their valence band spec-tra that could be attributed to the searched effect. Before concluding in a limitation of the technique, we note:
  - that the poly(isoprene) samples were commercially available compounds, whose (chemical and stereo-isomeric) purity was not known, and that very likely they were highly crosslinked as no care was taken to store adequately the polymers.
  - the poly(1,4 dichloro-2,3 epoxybutenes), two well char-acterized research products, are probably much too complicated systems to be distinguished by their XPS valence band spectra.

Indeed, preliminary results just acquired for cis- and trans- 1,4polybutadiene are more promising as substantial modi-fications appear in their valence band spectra (the C1s core level peaks are exactly comparable (41)). This indicates that more research in this field is worthwhile and will be carried on in the near future.

Tacticity. As the XPS core level investigation of the tacticities in the polymers revealed itself a relatively in-sensitive technique (34), it was worthwhile to test the poten-

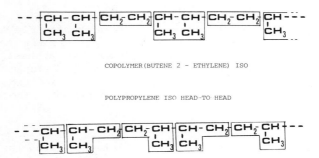

COPOLYMER (BUTENE 2 - ETHYLENE) ISO

POLYPROPYLENE ISO HEAD-TO-HEAD

*Figure 13. Extended chemical formula of a butene-2 and ethylene copolymer*

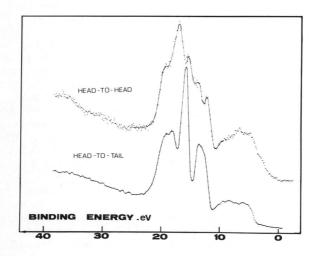

*Figure 14. XPS valence band spectra of HH and HT polypropylenes*

tiality of the XPS valence band analysis.  Studies (10, 44) of
polypropylene revealed discrepancies in the comparison of experimental and quantum chemical calculations on the valence band.  The
simplicity of the monomer structure and the actual commercial
interest of the polymer render its study more attractive.
Specially synthesized and characterized iso- (99%) and syndio-
(80-95%) tactic polypropylene were studied, and the valence band
spectra of the two stereoisomers were found profoundly different
by their C2s band shape (Figure 15); its width is identical in the
two compounds, but the bonding subband (at highest binding
energy) is markedly reduced in intensity and losses of its
substructures in syndio-tactic PP.  We note in addition that no
change in the carbon 1s level could be detected.  The data are
discussed in length with the aid of new theoretical calculations
elsewhere (13, 40).

   Polymer Conformation and Crystallinity.  Beyond the stereo-
regularity and tacticity, the geometrical conformation of the
polymer chain in the solid material could influence its electronic structure, through a modification of its valence band molecular orbitals.  Indeed, a few years ago, very characteristic band
structures were calculated for T, G, TG, and TGTG polyethylenes
(45).  More recently, Extended Huckel crystal orbital calculations showed that for isotactic polypropylene, a zig-zag planar
or a helical conformation resulted in significant changes in the
theoretical valence band spectra, supporting the idea that conformation effects could be detected experimentally by the XPS
method (44).
   Isotactic polypropylene does not exist in the two zig-zag
and helix conformations, but syndiotactic PP is a good candidate
for this search.  Indeed, according to the preparation procedure
of the polymer (46), it can exist in a helix or zig-zag planar
conformation.  Syndio-PP films were prepared following both ways,
their conformation checked by IR, and studied by XPS.  Their
valence band spectra again show distinct differences in the C2s
band (Figure 16); for zig-zag PP that is probably highly amorphous, the C-C band width increased by about 1eV, whereas the
bonding subband increased in intensity and became more structured
(40).
   That success led us to tackle another more complicated
question; is it possible with the XPS valence band spectra to
distinguish between crystalline and amorphous polymers?  It has
already been shown that core level analysis was not successful
(36).  From other measurements, mainly on semiconductor materials, it is known that, if the core level peaks of the amorphous
materials were slightly broader, the main modifications appear in
the valence band where the fine details are smeared out compared
to the crystalline material (47).
   Crystalline and amorphous polystyrene were chosen for the
XPS study.  But, the two recorded valence band spectra do not

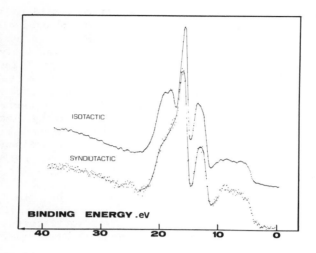

*Figure 15.* XPS valence band spectra of iso- and syndio-polypropylenes

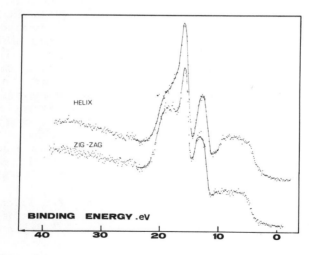

*Figure 16.* XPS valence band spectra of syndio-polypropylene in the helical and zig-zag planar conformations

appear notably different from each other. We noticed only a
slight modification of the width of the antibonding C2s band, and
a small (0.15eV, at the instrumental detection limit) shift of the
whole C-C band relative to the C-H molecular orbitals. For
completeness, we add that the two polystyrenes did not present
different C1s core peaks and C1s shake-up satellites (only the C1s
line itself is slightly broadened, from 1.1 to 1.2eV-in the
amorphous state). Is this failure due to a damage of the samples
by irradiation?)? Is this experiment a proof of one limitation in
the potentialities of the XPS technique? Further experiments are
needed to give a definitive answer, but it is already clear at the
moment - through the results presented in this review - that, like
for the studies of other types of compounds, the dominant structures appearing in the valence band spectra of the polymers are
dictated by the atomic properties (chemical composition defining
the substituent effect), and by the short range order (in the
largest sense, isomerism and conformation) in the materials. As
regards the long range ordering of the polymer chain (i.e. the
polymer crystalline form or the interchain interactions...), it
probably influences the electronic structure of the valence bands
through only (too) fine details.

## Other Informations Available from the Valence Band Spectra

For sake of completeness and only briefly (as other reviews
are specially dedicated to these subjects), we would like to
mention that XPS valence band spectra of polymers can also be used
-going beyond a sole interpretation of the molecular orbitals in
the valence band- to provide informations on some solid state
parameters and on surface structure effects of the polymers.

The low binding energy structures in the valence band of the
polymers are broad, less structured and of low intensity. Nevertheless, from these data it is possible to extract valuable
informations on the band gap (between the valence and conduction
bands) and the work function of the polymers. These properties
are relevant to understand e.g. the conductivity and adhesion of
commercial polymers. Data are available on model compounds,
alkanes (18), benzene (19) and other acenes (20), polyphenyl (21)
and other polymers (10, 48).

Being essentially a surface technique, XPS valence band
spectra also allows to monitor modifications occurring at the
polymer surface during adsorption, reactions, of degradation...
Very few contributions are, up to now, dealing with such studies
(49). The most direct use of the technique is actually a
comparison of the core and valence photoelectron line intensities
to deduce informations about the surface and the in-depth composition of the polymer, as well as about the orientation of the
macromolecular chain at the surface boundary.

## Conclusions

X-ray photoelectron spectroscopy used to study the electronic structure of polymers through the analysis of their valence band spectra was found to present very interesting potentialities, but also some limits.

If the technique was not found sensitive to distinguish polymers with different high molecular weight, we proved that it gives an instantaneous picture of the electronic structure of a rather large (10 to 20 consecutive carbon atoms) portion of the polymer skeleton.

Substitution of hydrogen atom in the $CH_2-CH_2-$ polymer skeleton results always in significant modifications in the valence bands; this is mostly interesting for the solely hydrocarbon based polymers, for which the valence band data represent a unique source of information, as they usually fingerprint the polymer (when no shake-up satellite can revalorize the core level data). General trends in the valence band spectra were also presented for polymers containing fluorine, chlorine, and oxygen atom(s).

Numerous examples of constitution isomerisms that can be solved with the aid of the valence band spectra were given. Also, for specially synthesized and characterized compounds, it was possible to show a potentiality of the technique to evidence head-to-head linkages, stereoisomers (?), tacticities and (alternant/block) structure of copolymers. If the influence of conformation in the valence band could also be evidenced, no success was obtained for differentiating crystalline and amorphous polymers.

Almost ten years after the recording of the first valence band spectra of polymers, it appears that the method has not yet reached its maturity. A fundamental, cautious, systematic, and progressive investigation of the simplest polymers is needed to discover all the informations that can be extracted from the fine details contained in the polymer valence band data. This is certainly one field of research that will progress in the near future.

## Acknowledgements

The authors wish to thank their coworkers from the Laboratoire de Spectroscopie Electronique, and the Laboratoire de Chimie Theorique Appliquee for their collaboration. Gifts of samples and fruitful discussions are aknowledged from Prof. Zambelli (Milano) and Dr. Gobillon (Solvay Co.). J.J.Pireaux is indebted to the Fonds National de La Recherche Scientifique for financial support.

Literature Cited

1.  Clark, D.T., Kilcast, D. Nature (Phys. Sc.) 1971, 233, 77.

2.  Wood, M.H., Barber, M., Hillier, I.H., Thomas, J.M., J. Chem. Phys. 1972, 56, 1788.

3.  Clark, D.T., Kilcast, D., Feast, W.J., Musgrave, W.K.R., J. Poly. Sci., 1972, AI10, 1637.

4.  Pireaux, J.J., Pol. Preprints, 1980, 21, 123.

5.  Siegbahn, K., et al. ESCA; "Atomic, Molecular and Solid State Structure Studied by means of Electron Spectroscopy," Almqvist and Wiksells, Uppsala, 1967; Verbist, J.J., "Quantum Theory of Polymers", J.M. Andre et al. (eds), D. Reidel Publ. Co., Dordrecht, 1978, p. 31; Clark, D.T., Feast, W.J., J. Macromol. Sci.-Revs. Macromol. Chem., 1975, C12, 191; Clark, D.T., "Structural Studies of Macromolecules by Spectroscopic Methods, K.J. Ivin (ed.) Wiley, N.Y., 1978, p. 3.

6.  Clark, D.T., Thomas, H.R., J. Poly Sci. Polym. Chem. Ed., 1977, 15, 2843; Cadman, P., Gossedge, G., Scott, J.D., J. Electr. Spectr., 1978, 13, 1; Cadman, P., Evans, S., Scott, J.D., Thomas, J.M., J. Poly. Sci. Polym. Lett. Ed., 1978, 16, 461; Clark, D.T., Thomas, H.R., Shuttleworth, D., J. Poly. Sci. Polym. Lett. Ed., 1978, 16, 465.

7.  Clark, D.T., Thomas, H.R., J. Poly. Sci. Polym. Chem. Ed., 1976, 14, 1671.

8.  Clark, D.T., Dilks, A., Thomas, H.R., J. Poly. Sci. Polym. Chem. Ed., 1978, 16, 1461.

9.  Duke, C.B., Salaneck, W.R., Fabish, T.J., Ritsko, J.J., Thomas, H.R., Paton, A., Phys. Rev. B., 1978, 18, 5717.

10. Pireaux, J.J., Riga, J., Caudano, R., Verbist, J., Delhalle, J., Andre, J.M., Gobillon, Y., Phys. Scripta, 1977, 16, 329.

11. Huang, T.J., Winters, H.F., Coburn, J.W., Appl. Surf. Sci., 1978, 2, 514.

12. Andre, J.M., et al. "Recent Advances in the Quantum Theory of Polymers," Springer Verlag, Berlin, 1980; Delhalle, J., therein, p. 225.

13. Andre, J.M., Delhalle, J., Pireaux, J.J., these conference proceedings.

14. Andre, J.M., Delhalle, J., Delhalle, S., Caudano, R., Pireaux, J.J., Verbist, J.J., Chem. Phys. Lett., 1973, 23, 206.

15. Delhalle, J., Andre, J.M., Delhalle, S., Pireaux, J.J., Caudano, R., Verbist, J.J., J. Chem. Phys., 1974, 60, 595.

16. Dewar, M.J.S., Suck, S.H., Weiner, P.K., Chem. Phys. Lett., 1974, 29, 220; Beveridge, D.L., Wun, W., Chem, Phys. Lett., 1973, 18, 570; Wood, M.H., Lavery, R., Hillier, I.H., Barber, M., J. Chem. Soc. Farad. Disc., 1975, 60, 173; Seki, K., Hashimoto, S., Sato, N., Harada, Y., Ishii, L., Inokuchi, H., Kanbe, J.I., J. Chem. Phys., 1977, 66, 3644; and reference 2 above.

17. Pireaux, J.J., Svensson, S., Basilier, E., Malmqvist, P.A., Gelius, U., Caudano, R., Siegbahn, K., Phys. Rev., 1976, A14, 2133.

18. Pireaux, J.J., Caudano, R., Phys. Rev., 1977, B15, 2242; Pireaux, J.J., Ph.D. Thesis Facultes Universitaires, Namur 1976, Univ. Micr. 77-70, 049.

19. Riga, J., Pireaux, J.J., Verbist, J.J., Mol. Phys., 1977, 34, 131.

20. Riga, J., Pireaux, J.J., Caudano, R., Verbist, J., Phys. Scripta, 1977, 16, 346.

21. Riga, J., Pireaux, J.J., Boutique, J.P., Caudano, R., Verbist, J., Gobillon, Y., Synth. Metals, submitted.

22. Clark, D.T., Thomas, H.R., J. Poly. Sci. Polym. Chem. Ed., 1978, 16, 791.

23. Clark, D.T., Dilks, A., J. Poly Sci. Polym. Chem. Ed., 1976, 14, 533; Clark, D.T., Peeling, J., O'Malley, J.M., J. Poly. Sci. Polym. Chem. Ed., 1976, 14, 543; Clark, D.T., Dilks, A., J. Poly. Sci. Polym. Chem. Ed., 1977, 15, 15; Clark, D.T., Adams, D.B., Dilks, A., Peeling, J., Thomas, H.R., J. Elect. Spectr., 1976, 8, 51.

24. Stevens, G.C., Bloor, D., Williams, P.M., Chem. Phys., 1978, 28, 399.

25. Knecht, J., Reimer, B., Bassler, H., Chem. Phys. Lett., 1977, 49, 327; Knecht, J., Bassler, H., Chem. Phys., 1978, 33, 179.

26. Pireaux, J.J., Riga, J., Caudano, R., Verbist, J.J., Andre,

J.M., Delhalle, J., Delhalle, S., J. Electr. Spectr., 1974, 5, 531.

27. Delhalle, J., Delhalle, S., Andre, J.M., Pireaux, J.J., Riga, J., Caudano, R., Verbist, J.J., J. Electr. Spectr., 1977, 12, 293.

28. Pireaux, J.J., Riga, J., Caudano, R., Verbist, J., Gobillon, Y., Delhalle, J., Delhalle, S., Andre, J.M., J. Poly. Sci. Polym. Chem. Ed., 1979, 17, 1175.

29. Gelius, U., "Electron Spectroscopy", D.A. Shirley (Ed.), North Holland, Amsterdam, 1972, p. 319.

30. Pireaux, J.J., Riga, J., Caudano, R., Verbist, J., Gobillon, Y., LSE 78-2-46, to be published.

31. Sevrin, C., Memoire licence 1979, Facultes Universitaires, Namur.

32. Pireaux, J.J., Sevrin, C., Riga, J., Caudano, R., Verbist, J., Gobillon, Y., to be published.

33. Puissant, C., Memoire licence 1978, Facultes Universitaires, Namur.

34. Clark, D.T., Thomas, H.R., J. Poly. Sci. Polym. Chem. Ed., 1976, 14, 1701.

35. Chujo, R., Maeda, K., Okuda, K., Murayama, N., Hoshino, K., Makromol. Chem., 1975, 176, 213.

36. Adams, D., Clark, D.T., Dilks, A., Peeling, J., Thomas, H.R., Makromol. Chem., 1976, 177, 2139.

37. Thomas, H.R., Clark, D.T., Acc. Chem. Res., in press.

38. Clark, D.T., Peeling, J., J. Poly. Sci. Polym. Chem. Ed., 1976, 14, 543.

39. Thomas, H.R., O'Malley, J.J., Macromol., 1979, 12, 323; O'Malley, J.J., Thomas, H.R., Contemp. Topics in Polym. Sci., Vol. 3, p. 215.

40. Pireaux, J.J., et al., to be published.

41. Pireaux, J.J., et al., to be published.

42. Clark, D.T., Feast, W.J., Kilcast, D., Musgrave, W.K.R., J. Poly. Sci. Polym. Chem. Ed., 1973, 11, 389.

43. Chem and Eng. News, April 9, 1979, p. 61.

44. Delhalle, J., Momtigny, R., Demanet, C., Andre, J.M., Theor. Chim. Acta., 1979, 51, 343.

45. Delhalle, S., Delhalle, J., Demanet, C., Andre, J.M., Bull. Soc. Chim. Belges, 1975, 84, 107.

46. Natta, G., Peraldo, M., Allegra, G., Makromol. Chem., 1964, 75, 215; Peraldo, M., Cambini, M., Spectrochim. Acta., 1965, 21, 1509.

47. Ley, M., Cardona, M., Pollak, R.A., "Photoemission in Solids", II., Topics in Applied Physics, Vol. 27, Springer, 1979.

48. Duke, C.B., Paton, A., Salaneck, W.R., Thomas, H.R., Plummer, E.W., Heeger, A.J., MacDiarmid, A.G., Chem Phys. Lett., 1978, 59, 146; Salaneck, W.R., Thomas, H.R., Duke, C.B., Paton, A., Plummer, E.W., Heeger, A.J., MacDiarmid, A.G., J. Chem. Phys., 1979, 71, 2044.

49. Burkstrand, J.M., J. Vac. Sci. Technol., 1978, 15, 223; Briggs, D., Brewis, D.M., Konieczko, M.B., J. Mat. Sci., 1976, 11, 1270; Brewis, D.M., Comyn, J., Fowler, J.R., Briggs, D., Gibson, V.A., Fib. Sci. Technol., 1979, 12, 41; Briggs, D., Brewis, D.M., Konieczko, M.B., J. Mat. Sci., 1979, 14, 1344; Clark, D.T., "Adv. Polymer Friction and Wear", Vol. 5A (L.H. Lee ed.), Plenum 1975, p. 241; Clark, D.T., Characterization of Metals and Polymer Surfaces", Vol. 2, (L.H. Lee, Ed.), Academic Press, 1977, p. 5; reference 11 above.

RECEIVED March 9, 1981.

# Identification of Chemical States by Spectral Features in X-Ray Photoelectron Spectroscopy

C. D. WAGNER

Surfex Company, 29 Starview Drive, Oakland, CA 94618

While the principal interest in this symposium is the application of surface analytical techniques to polymeric systems, polymers often contain, besides C, H, and O, elements of Group VII (halogens), Group VI (S, Se, Te), Group V (N, P, As, Sb), and sometimes Group IV (Si, Ge, Sn). In polymer technology the study of polymers in heterophase systems is increasingly important, with processing additives, fillers, pigments, and interfaces in composites involving many other elements. For these reasons it seems appropriate to discuss this subject in quite a general way.

From the earliest presentations of this technique (1) it was clear that a principal value of the method is the information it furnishes on the environment of the emitting atom, obtained by observing the exact energy of the photoelectrons. Chemical shifts in this energy are substantial (2) for carbon, sulfur, and nitrogen. The range in photoelectron energy for these elements in organic systems is 7–8 eV. Since line positions after charge referencing are ordinarily determinable with an error less than 0.5 eV, the technique has been of great value in analysis of polymer surfaces.

The principal limitation in identification of chemical states has been the fact that photoelectron line energy is a one parameter system. Moreover, chemical shifts in some elements are very small; for example those of Zn, Ag, Cd, In, and the alkali metals and alkaline earths encompass ranges of less than two electron volts, so that for these elements this technique has little use. Clearly there is much incentive to find other spectral features that provide information on chemical state.

Shakeup Satellites. An unusual feature in many spectra is the shakeup satellite, a line or lines usually several eV lower in kinetic energy than the parent photoelectron line. It arises when the photoelectric transition has a significant probability of generating a final ion in an excited state. The extra energy in the excited state is reflected in the energy separation of the satellite from the main line. A common example in organic systems

is that due to the $\pi^* \leftarrow \pi$ transition. In this instance the
excitation energy is in the range 5.5-8.5 eV (3); a well-known
example is that of polystyrene (Fig. 1). The satellite is
observed when the carbon atom is part of a pi-bonded system, e.g.
compounds with benzene rings, (4), pyrrole, thiophene (5), pyridine, furan, or even diolefinic or mono-olefinic systems (6). The
intensity of the satellite can be as much as 15% of the parent
peak. Similar satellites have been observed with a number of
other compounds, such as formaldehyde, carbon suboxide (7),
tetracyanoquinodimethane (8) and metal carbonyls. The satellites
have been observed with the lines of the heteroatoms as well as
with Cls, with similar, but not identical, energy separations and
intensities. The energy separation and intensity are unique for
lines from each atom in each chemical state.

Shakeup satellites are often more intense with paramagnetic
inorganic ions, such as those of the first transition series, the
rare earths, and the actinides. Those of the first transition
series can be quite complex (Fig. 2), with multiple satellites
(multiple excited states) possible. Frost, Ishitani, and
McDowell (9) for copper compounds and Matienzo et al (10) and
Tolman et al (11) for nickel compounds have furnished rather
comprehensive data on satellite energies and intensities. The
paramagnetic octahedral and tetrahedral nickel complexes exhibit
shakeup patterns, while the diamagnetic square planar ones do
not. With copper the paramagnetic cupric states have shakeup
satellites while the diamagnetic cuprous states do not. The rule
is not invariable, however, and satellites are sometimes observed
with diamagnetic species, but usually in much lower intensity.
With the transition metals the excitation mechanism usually involves charge transfer with the ligand. Brisk and Baker (12)
summarize the discussions on mechanism that have appeared.

Rare earth ions also exhibit complex shakeup satellite patterns. That of Ce3d in $CeO_2$ is especially noteworthy in the fact
that two satellites for each of $Ce3d_{3/2}$ and $Ce3d_{5/2}$ appear, one of
them more intense than the primary line and separated from it by
as much as 16 eV (Fig. 3). Actinide compounds also exhibit many
satellites in their spectra. Wide use of these features in
analytical work awaits collection of these spectra in a comprehensive review.

Multiplet Splitting. Interaction of a core vacancy resulting from the photoelectric process with unpaired electrons in a
valence shell induces multiplet splitting in the lines corresponding to the emitted electron. Thus, the 3s level in the
transition metals exhibit relatively simple splittings, often of
several eV, specific for each chemical state (13). The 2p levels
are split into multiple lines, and the effect of their convolution
is to widen the apparent split of the doublet (14). Frost, et al
(15) tabulate the 2p splitting for cobalt compounds, which varies
in the range 14.6 to 16.1 eV. There should be similar variability

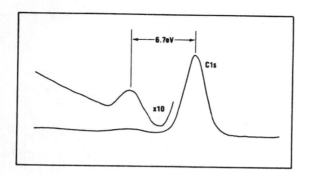

*Figure 1.    Shakeup line for the C-1s line in polystyrene (2)*

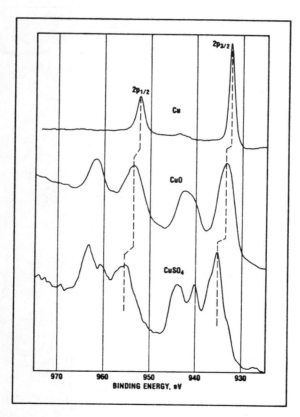

*Figure 2.    Examples of shakeup lines in the Cu-2p spectrum (2). Energy separation of the 2p lines is 19.8 eV.*

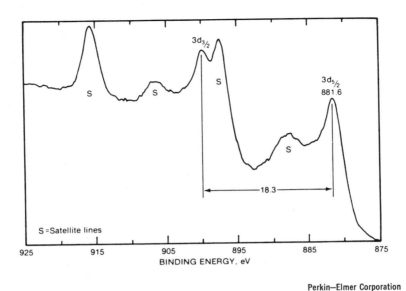

*Figure 3.    Shakeup spectra in the Ce-3*d *lines of ceric oxide (2)*

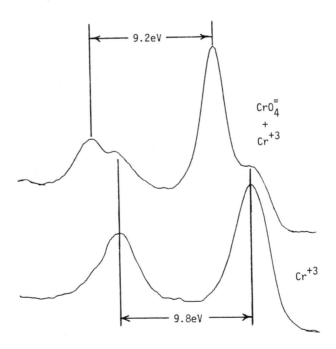

*Figure 4.    Multiplet splitting effects in the Cr-2*p *doublet*

for Ni, Fe, Mn, and Cr.   Chromate ion and chromic oxide are
readily distinguished by this effect (Fig. 4) as well as by the
binding energy of the $2p_{3/2}$ line.   All of these features, the
shakeup satellites and the multiplet splitting phenomena, have
the advantage that correction for static charge is unnecessary.

Core-Type Auger Lines, and Chemical State Plots.   The
utility of Auger lines in identification of chemical states is
being recognized.   About half of the naturally-occurring elements
present in solid state materials exhibit Auger lines in XPS
spectra with Al or Mg X-radiation (Fig. 5).   Of these, twenty two
exhibit core-type Auger transitions (final vacancies in core
levels) and sixteen have valence-type Auger transitions.   Core-
type Auger groups in the spectra of neighboring elements are very
similar, differing only in magnitude of line spacing, and have at
least one line that is narrow and intense.   Chemical shifts in
these lines are different from and ordinarily larger than those of
photoelectron lines of the same elements (16).   When line energies
of the most intense photoelectron line are plotted against those
of the most intense or sharp Auger line for compounds of the same
element, the resulting two-dimensional array becomes much more
useful for identifying chemical states.   An example is shown in
Fig. 6 for copper.   Cupric compounds are all located at higher
photoelectron binding energies and have identifying shakeup
lines.   Cuprous forms, on the other hand, have virtually the same
photoelectron binding energy as elemental copper, and no shakeup
satellite lines.   They are distinguishable, however, by the Auger
energy.   Six unknown samples containing Cu, Cl, S, and O, as well
as other metals, gave line energies denoted by the circles, and
indicate copper is probably present as $Cu_2S$, or possibly $Cu_2O$, but
clearly not CuCl or Cu metal.   Many of the chemical state plots
embodying present knowledge of line energies have been assembled
and published (17), including F, Na, Mg, Cu, Zn, Ge, As, Se, Ag,
Cd, In, Sn, Sb, Te, and I.

One feature of these plots is of additional interest.   The
diagonal grid represents the sum of the Auger kinetic energy and
the photoelectron binding energy, which we call the modified
Auger Parameter (17), $\alpha'$:

$$\alpha' = KE(Auger) + BE(Photoelectron) \qquad (1)$$

$$BE(Photoelectron) = h\nu (X\text{-ray}) - KE(Photoelectron) \qquad (2)$$

$$\alpha' = h\nu (X\text{-ray}) + KE(Auger) - KE(Photoelectron) \qquad (3)$$

$$\alpha' = h\nu (X\text{-ray}) + \alpha \qquad (4)$$

where $\alpha$ is the original Auger Parameter, as defined (18), the
difference in kinetic energy between the Auger and photoelectrons.   Since the chemical shifts of Auger and photoelectrons

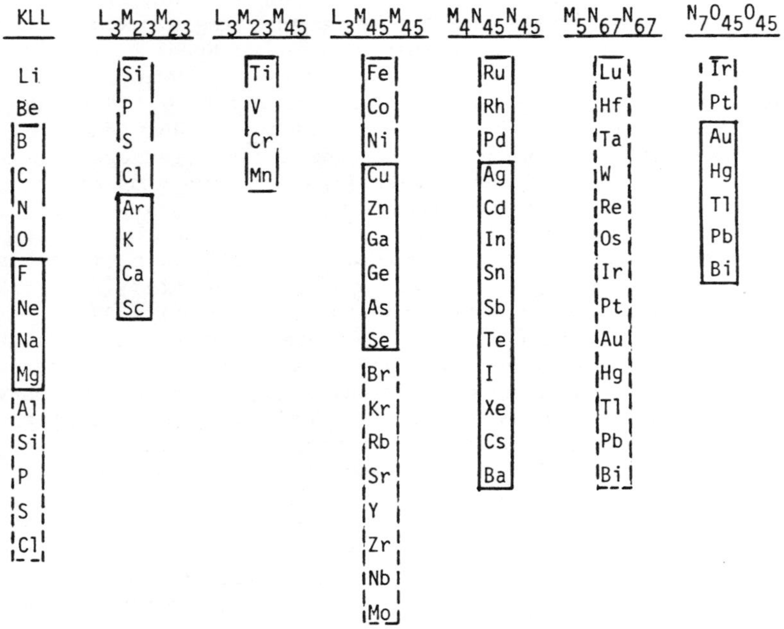

| KLL | $L_3M_{23}M_{23}$ | $L_3M_{23}M_{45}$ | $L_3M_{45}M_{45}$ | $M_4N_{45}N_{45}$ | $M_5N_{67}N_{67}$ | $N_7O_{45}O_{45}$ |
|---|---|---|---|---|---|---|
| Li | Si | Ti | Fe | Ru | Lu | Ir |
| Be | P | V | Co | Rh | Hf | Pt |
| B | S | Cr | Ni | Pd | Ta | Au |
| C | Cl | Mn | Cu | Ag | W | Hg |
| N | Ar |  | Zn | Cd | Re | Tl |
| O | K |  | Ga | In | Os | Pb |
| F | Ca |  | Ge | Sn | Ir | Bi |
| Ne | Sc |  | As | Sb | Pt |  |
| Na |  |  | Se | Te | Au |  |
| Mg |  |  | Br | I | Hg |  |
| Al |  |  | Kr | Xe | Tl |  |
| Si |  |  | Rb | Cs | Pb |  |
| P |  |  | Sr | Ba | Bi |  |
| S |  |  | Y |  |  |  |
| Cl |  |  | Zr |  |  |  |
|  |  |  | Nb |  |  |  |
|  |  |  | Mo |  |  |  |

*Figure 5.    Analytically useful x-ray–generated Auger lines of the elements ((— —) valence type; (——) core type, conventional x-rays; (– – –) core type, higher energy)*

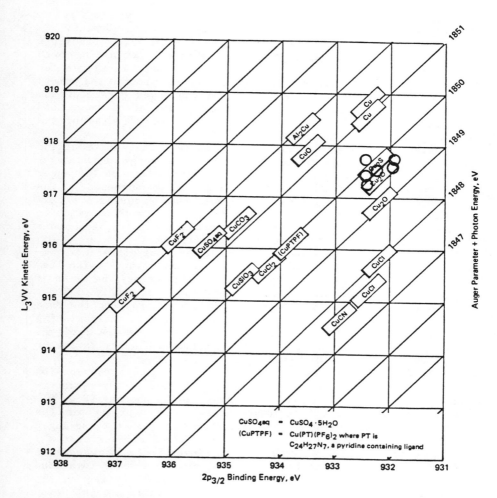

*Figure 6. Applications of the chemical state plot to unknown copper samples. Duplicate points for certain chemical states represent data from different sources (17).*

differ, the Auger Parameter is also a function of chemical state, and is very accurately determinable, since charge corrections cancel. For this reason the points representing chemical state on the two-dimensional grid are represented as rectangles with the narrow dimension perpendicular to the Auger Parameter grid lines. The modified Auger Parameter was adopted (17) because it is independent of photon energy and is always a positive value.

Castel, Hazell, and West (19) extended the Auger Parameter concept further when applying it to minerals containing both aluminum and silicon. They determined the Auger Parameter for aluminum and for silicon for each material, and plotted the points on a two-dimensional plot with the silicon Auger Parameter the ordinate and the aluminum Auger Parameter the abscissa. Good dispersion was achieved and the points are not affected by the accuracy of charge referencing.

Screening energy for the vacancy in the final state of the photoelectron and Auger processes affects the energy of the Auger or photoelectron, with a considerably larger effect on the Auger electron because of the larger charge being screened. The energies on the plots correspond to bulk homogeneous states. A chemical state present in atomic or molecular dimension in or on another state will have a different screening energy, and modified energies for the Auger and photoelectrons.

The differences in the Auger Parameter among different chemical states represent, to a first approximation, twice the differences in screening energy (18). Cognizance must be taken of the possibility that screening energies in a given sample will not be typical of the bulk for that compound, when applying the data in the plots. Thus, a metal in the form of atoms in an insulating polymer will exhibit a reduced Auger Parameter, and a rare gas atom trapped in a metal will exhibit an enhanced Auger Parameter over that from the bulk solid state.

Use of Al and Mg radiation limits the elements that can be treated in this way, because many initial states of Auger transitions require energies larger than those in the photons (Fig. 5). Castle and West (20) have demonstrated that utility of the bremsstrahlung component of the radiation to generate these Auger lines. An evaluation of this technique (21) discloses that Al, Si, P, S, and Cl especially provide KLL Auger lines in reasonable intensity from the bremsstrahlung component of unmonochromatized Al and Mg radiation. With a beryllium window substituting for the usual aluminum one, the KLL line is about one-third the intensity of the 2p line for Al and Si. Moreover, the signal to noise is more favorable at kinetic energies above that corresponding to the Fermi level.

A more general approach is to employ X-ray anodes of gold (22), silver (22), zirconium, (19), silicon (23) or titanium to generate X-ray photons of higher energy. Two-dimensional chemical state plots for Si (2p, KLL), Br (3d, LMM) and W ($3d_{5/2}$ $M_5N_{67}N_{67}$) have been developed with the use of Au M X-rays and published (17). A more recent version for silicon, utilizing mostly the bremsstrahlung from a magnesium anode, is shown in Fig. 7. Auger Parameter values from reference (20) for a number of minerals are included. Of additional interest are data points for the chlorosilanes (24), obtained in gas phase. For them the kinetic energies observed have been modified by adding 5.0 eV to the vacuum level referenced data, to make them comparable to the other Fermi level referenced data. The lower Auger Parameter indicates the screening energy for the gas phase forms to be less than that of any of the solid state materials. Moreover, differences within the group indicate that substitution of Cl for $CH_3$ bonded to silicon increases the screening energy.

An interesting observation has been that of intense shakeup satellites on high energy KLL Auger lines for such diamagnetic species as K, Ca, and Ti compounds (25, 26). When anodes generating higher energy photons come into wider use we can expect these shakeup lines to be useful in analysis.

Valence-Type Auger Lines. Valence-type Auger lines originate from Auger transitions with final vacancies in valence levels. They are all accessible with Al or Mg radiation, and their kinetic energy ranges up to several hundred eV. Since the kinetic energy of the Auger electron is determined by the difference in energy between the initial ion and the final ion in the transition, the distribution of Auger lines is uniquely dependent upon the molecular orbital pattern and the selection rules of the transition. The Auger group thus has a unique pattern for each compound (27, 28). Some observations have been made in solid state on C(KVV) (29, 30, 31), N(KVV) (32), O(KVV) (33, 34) and F(KVV) (29). With fluorine the Auger line is borderline between valence and core-type, and marked changes occur as one goes from fluoride ion in NaF to the covalent complex ions in $NaBF_4$ and $KSbF_6$ and then to the covalent polymer $(CF_2)_n$. Work is also being reported with the LVV Auger lines in Al, Si, P, S, and Cl (35). Observations of line shape rather than line energy permits generation of these lines by use of electron beams as well as X-radiation. Of course studies using fast electrons require care to use minimum exposure to avoid radiation damage. Some of the above studies were conducted with ionization by electron beam.

A particularly great variety of line intensity distributions has been observed in the O(KLL) lines in the solid state (33). The energy interval between the $KL_1V$ and the most intense KVV line is variable from 20 eV, observed with metal oxides and metallic anions, to 24 eV, observed with carboxylic acids, carbonates, chlorates, and nitrate ions. Carboxylate polymers and nitrate polymers are similar. With those species that show the wider spacing, the KVV line is split into two major components, with the second component at about 4 eV higher energy. This second component, appearing as a shoulder with many compounds, actually is the more intense in chlorate and nitrate. The entire oxygen

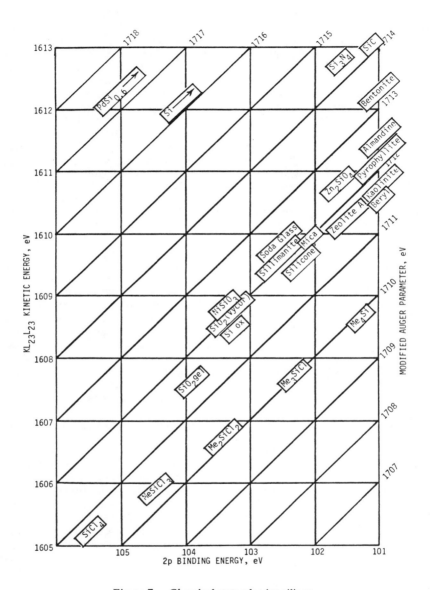

*Figure 7.    Chemical state plot for silicon.*

*Si and PdSi$_{0.6}$ are off the plot at 99.2, 1616.9 and 99.6, 1617.6 eV, respectively (with Au4$_{7/2}$ set at 83.8 eV). Materials along right border are from Ref. 20 (Auger parameter only); chlorosilanes are from Ref. 24; 5 eV were added to kinetic energies to approximate Fermi level referencing.*

Auger group of lines is reproduced in Fig. 8 for several types of compounds.

Auger transitions of the valence type are recordable for B, C,N,O, and F(KVV); Al, Si, P, S, and Cl (LVV); Ti, V, Cr, Mn, Fe, Co, and Ni (LM$_{23}$V and LVV); Ru, Rh, and Pd (MVV); and Ir and Pt (NVV). The fluorine KVV, copper LVV, silver MVV, and gold NVV are borderline between valence and core-type Auger lines (36). Some of the valence-type Auger groups in the transition metals are particularly informative, as, for example, in nickel (Fig. 9).

While the valence-type Auger lines are sometimes broad and irregular in outline, they often include a dominant and relatively sharp peak. This can be plotted in a two-dimensional chemical state plot similar to those of core-type Auger lines. Data from compounds of F, Ti, Mn, Fe, Co, Ni, Rh, and Pd have already been assembled in such plots (17). Similar plots for oxygen as inorganic oxides and in organic compounds are shown in Figs. 10 and 11 (33). Classes of compounds are shown to have oxygen line energies that place them in unique positions on the plots.

## Acknowledgements

The assistance of R. H. Raymond of Shell Development Company and J. F. Moulder of Physical Electronics Division of Perkin-Elmer Corporation in obtaining many of the spectra of pure compounds is acknowledged. The author is grateful to the latter company for permission to reproduce Figures 1, 2, and 3 herein from reference (2).

## Abstract

Chemical shifts in the line energies of the Cls, Nls, Ols, Fls, and the strong photoelectron lines of other elements have been useful for the identification of chemical states. Application is limited, however, because only one property is variable and with some elements the range in chemical shifts is very small.

Other spectral properties can aid in the characterization of the surface species: (1) Shakeup satellite lines are observed with certain organic structures, as well as with many paramagnetic inorganic ions. Their intensities and separations from the main line are specific for each chemical state, (2) Multiplet splitting in paramagnetic atoms varies with chemical state, (3) Valence-type Auger line groups, such as those for C, N, O, Si, P, and S have intensity distributions that depend upon molecular orbitals involved in the final state in the Auger transition, and (4) Core-type Auger lines, such as those of F, Si, P, 'S, Cl, As, Se, I, and most of the metals, have at least one component that is narrow and intense. Chemical shifts involving these are different and usually larger than those of the photoelectron lines. Use of both chemcial shifts is a more powerful tool in identifying chemical states than that of either alone.

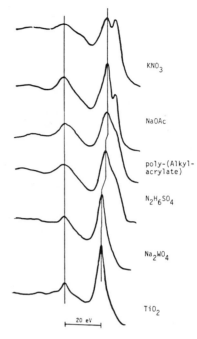

*Figure 8.    Oxygen KLL group for a variety of compounds (33)*

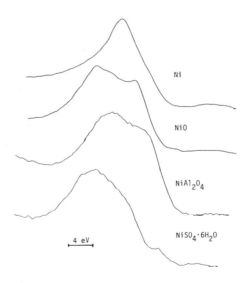

*Figure 9.    Variations in the nickel LVV Auger lines in different chemical states*

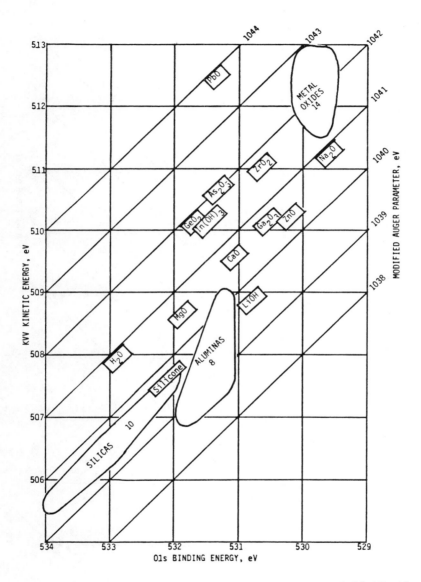

*Figure 10.    Chemical state plot for oxygen in oxides and hydroxides (33). Numbers represent the number of compounds included in the area.*

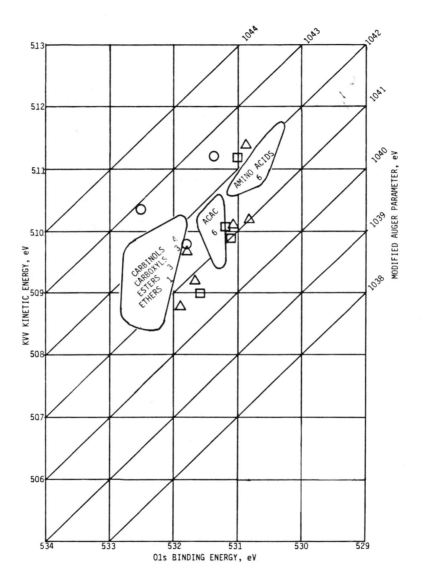

*Figure 11.   Chemical state plot for oxygen in carbon–oxygen compounds (33)*
*((○) carbonyl; (△) carboxylate; (□) carbonate)*

## Literature Cited

1.  Siegbahn, K., Nordling, C., Fahlman, A., Nordberg, R., Hamrin, K., Hedman, J., Johansson, G., Bergmark, T., Karlsson, S., Lindgren, I. and Lindberg, B., "Atomic Molecular, and Solid State Structure Studied by Means of Electron Spectroscopy, ESCA," Almqvist and Wiksells, Uppsala, 1967.

2.  Wagner, C.D., Riggs, W.M., Davis. L.E., Moulder, J.F. and Muilenberg, G.E., "Handbook for X-Ray Photoelectron Spectroscopy," Physical Electronics Division, Perkin-Elmer Corporation, Minneapolis, Minnesota, 1979.

3.  Clark. D.T., Adams, D.B., Dilks, A., Peeling, J. and Thomas, H.R., J. Electron Spectros. Relat. Phenom., 1976, 8, 51.

4.  Ohta, T., Fujikawa, T. and Kuroda, H., Chem. Phys. Lett., 1975, 32, 369.

5.  Pignataro, S., DiMarino, R., Distefano, G. and Mangini, A., Chem. Phys. Lett., 1973, 22, 352.

6.  Carlson, T.A., Dress, W.B., Grimm, F.A., and Haggerty, J.S., J. Electron Spectros. Relat. Phenom., 1977, 10, 147.

7.  Gelius, U., Allen, C.J., Allison, D.A., Siegbahn, H. and Siegbahn, K., Chem. Phys. Lett., 1971, 11, 224.

8.  Monroe, B.M. and Swingle, R.S., II, J. Electron Spectros. Relat. Phenom., 1976, 9, 479.

9.  Frost, D.C., Ishitani, A., McDowell, C.A., Molec. Phys., 1972, 24, 861.

10. Matienzo, L.J., Yin, Y.O., Grim, S.O. and Swartz, W.E., Inorg. Chem., 1973, 12, 2764.

11. Tolman, C.A., Riggs, W.M., Linn, W.J., King, C.M. and Wendt, R.C., Inorg. Chem., 1973, 12, 2772.

12. Brisk, M.A. and Baker, A.D., J. Electron Spectros. Relat. Phenom., 1975, 7, 197.

13. Carver, J.C., Schweitzer, G.K.and Carlson, T.A., J. Chem. Phys., 1972, 57, 973.

14. Gupta, R.P. and Sen, S.K., Phys. Rev., 1974, B10, 71.

15. Frost, D.C., McDowell, C.A. and Ishitani, A., Molec. Phys., 1974, 27, 1473.

16.  Wagner, C.D. and Biloen, P., Surf. Sci., 1973, 35, 82.

17.  Wagner, C.D., Gale, L.H. and Raymond, R.H., Anal. Chem.,
     1979, 51, 466.

18.  Wagner, C.D., Far. Disc. Chem. Soc., 1975, 60, 291.

19.  Castle, J.E., Hazell, L.B. and West, R.H., J. Electron
     Spectros. Relat. Phenom., 1979, 16, 97.

20.  Castle, J.E. and West, R.H., J. Electron Spectros. Relat.
     Phenom., 1979, 16, 195; ibid. 1980, 18, 355.

21.  Wagner, C.D. and Taylor, J. Ashley, J. Electron Spectros.
     Relat. Phenom., 1980, 20, 83.

22.  Wagner, C.D., J. Vac. Sci. Tech., 1978, 15, 518.

23.  Castle, J.E., Hazell, L.B. and Whitehead, R.D., J. Electron
     Spectros. Relat. Phenom., 1976, 9, 247.

24.  Kelfve, P., Blomster, B., Siegbahn, H., Siegbhan, K.,
     Sanhueza, E. and Goscinski, O., Phys. Sci., 1980, 21, 75.

25.  Fahlman, A., Hamrin, K., Axelson, G., Nordling, C. and
     Siegbahn, K., Z. Phys., 1966, 192, 484.

26.  Nishikida, S. and Ikeda, S., J. Electron Spectros. Relat.
     Phenom., 1978, 13, 49.

27.  Siegbahn, K., Nordling, C., Johansson, G., Hedman, J., Heden,
     P.F., Hamrin, K., Gelius, U., Bergmark, T., Werme, L.O.,
     Manne, R. and Baer, Y., "ESCA Applied to Free Molecules,"
     Elsevier, New York, 1969.

28.  Moddeman, W.E., Carlson, T.A., Krause, M.O., Pullen, B.P.,
     Bull, W.E. and Schweitzer, G.K., J. Chem. Phys., 1971, 55,
     2317.

29.  Wagner, C.D., Anal. Chem., 1972, 44, 967.

30.  Rye, R.R., Madey, T.E., Houston, J.E. and Holloway, P.H., J.
     Chem. Phys., 1978, 69, 1504.

31.  Kleefled, J. and Levenson, L.L., Thin Solid Films, 1979, 64,
     389.

32.  Larkins, F.P. and Lubenfeld, A., J. Electron Spectros.
     Relat. Phenom., 1979, 15, 137.

33.   Wagner, C.D., Zatko, D.A., and Raymond, R.H., <u>Anal. Chem.</u>, 1980, 52, 1445.

34.   Fiermans, L., Hoogewijs, R. and Vennik, J., <u>Surf. Sci.</u>, 1975, 47, 1.

35.   Bernett, M.K., Murday, J.S. and Turner, N.H., <u>J. Electron Spectros. Relat. Phenom.</u>, 1977, 12, 375.

36.   Powell, C.J., <u>Solid State Comm.</u>, 1978, 26, 557.

RECEIVED February 18, 1981.

# Chemical Labels to Distinguish Surface Functional Groups Using X-Ray Photoelectron Spectroscopy (ESCA)

C. D. BATICH

Central Research and Development Department, E. I. du Pont de Nemours and Co., Experimental Station, Wilmington DE 19898

R. C. WENDT

Polymer Products Department, E. I. du Pont de Nemours and Co., Experimental Station, Wilmington, DE 19898

Polymer surface chemistry is frequently dominated by the number and types of functional groups present on the surface. Because different groups can have the same elemental composition, a knowledge of atomic ratios is generally inadequate. Much of the ESCA work with polymers attempts to distinguish functional groups by looking at small chemical shifts and deconvoluting overlapping peaks. This approach works best for systems containing fluorine, where chemical shifts are large. Some researchers have tried to avoid this problem by specifically tagging or derivatizing certain functional groups and analyzing for a novel element which has been introduced. This paper will review what has been done using the second approach and will outline what is desirable for such a study. Our results for carboxylic acid group analysis will also be discussed.

Some common functional groups which are difficult to distinguish are shown in Figure 1. Carboxylic acids, esters, ketones, alcohols and aldehydes have overlapping oxygen 1s spectra. While it is generally possible to distinguish between singly and doubly-bonded oxygen, e.g. between alcohols and ketones, combinations of these moieties may require meticulous curve resolution and an expert eye. Many organic nitrogen species are overlapping, including amide, amine and nitrogen in certain heterocyclic structures. Carbon-carbon unsaturation, as well as phenyl groups, are difficult to distinguish from saturated species. Clark ([1]) has made progress using shake-up structures as distinguishing features, but the amount of shake-up structure present seems to be a variable, depending on the matrix in certain cases, according to Thomas et al.([2]).

An early attempt to identify a functional group by derivtization and ESCA study, was published by Riggs and Dwight in 1974 (Fig. 2) ([3]). Treatment of polytetrafluoroethylene (PTFE) with sodium in ammonia produced a surface depleted in fluorine. To

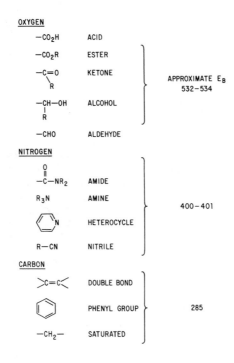

*Figure 1. Poorly resolved functional groups*

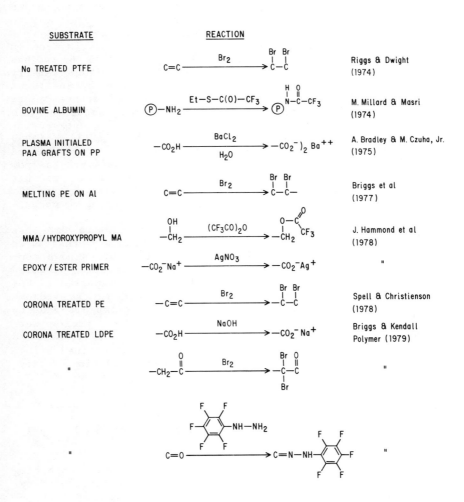

*Figure 2.   XPS studies using polymer surface derivatization*

determine whether unsaturation was present, the surface was exposed to bromine and then analyzed for the presence of this
element. The Na-treated PTFE surface did show increased bromine,
whereas the untreated PTFE surface did not. A study of bovine-
serum-albumin labeling with ethylthioltrifluoroacetate at about
the same time showed a rough relationship to known amino-group
concentration (4). Polyacrylic acid grafts on polypropylene were
studied using an ion-exchange interaction with a Ba$^{++}$ salt (5).
Briggs initially studied the effects of melting polyethylene on
aluminum metal using the bromine test for unsaturation (6) and has
since been examining corona-treated polyethylene with a variety
of labels (7). This system has also been studied by Spell and
Christensen (8). Hammond et al. studied a methyl methyacrylate
and hydroxypropyl methacrylate copolymer series reacted with
trifluoroacetic anhydride (9). They also treated an epoxy-ester
primer with silver salts and found reasonable labeling (10).
These first experiments appear to have been done on well-defined
substrates and a fuller description of the hydroxyl work has now
been presented. Briggs has used two methods to characterize his
system: (1) an ion exchange method using careful washing, and (2)
reaction of the carbonyl function obtained, forming a phenyl-
hydrazone.
        Almost all of the work which has been done to date has
involved a substrate of loosely defined stoichiometry, that is to
say, an experimentally treated polymer surface of unknown composition. In several cases indicated, chemical reaction produced
addition of an oxygen species, which was not expected. For
instance, no oxygen-containing reagent was used in the sodium-
treated PTFE or in some of $N_2$ or Ar corona-treated low-density
polyethylene, yet copious amounts of oxygenated species were
formed. What is needed for a method that can be said to
quantitatively label the surface of a polymer is outlined below.
        Method Requirements for quantitative derivatization of poly-
mers:

Primary

1.  Quantitation of major peaks on the clean, pre-labeled, poly-
    mer.

2.  Quantitative label of standard polymer surfaces containing
    several levels of independently determined functionality,
    including controls with zero levels.

3.  Selectivity when other known functional groups are present.

4.  Well defined experimental conditions reproducible between
    laboratories.

5.  Survey scan run and all major (10% of label) peaks quantified
    - especially the common contaminants C, O, Pb and Si.

Secondary

6.   Sensitivity (High cross-section).

7.   Stability.

8.   Consistency with other observations on "real-world" samples;
     e.g., adhesion, contact angle, etc.
     These conditions, as far as we can determine, have so far not
all been fulfilled for any derivatization system.
     In addition to the reactions which have been examined, a
number have also been suggested or tried under non-optimum conditions.   In 1977, at a Northeast ESCA Users Group (NEUG) Meeting,
Wendt discussed possible labeling reactions using Na, Ag or Pb as
labels for carboxylic acids and also the bromine in carbon
tetrachloride method or bromine gas for unsaturation.   Hydrochlo-
ric acid was proposed as a good label for free amino groups,
formation of dinitrophenylhydrazone for carbonyls and also chloride label for alcohols.   Dwight Williams, at a subsequent NEUG
meeting in May of 1979 also presented a number of possible
reactions:   mercuric acetate, bromine, iodine and ICl reaction
with unsaturation, HCl labeling of expoxides, HI for ethers,
carbon disulfide with amines and Ag with thio-groups.
     In most studies, few experimental variables are specified.
Some of the more important complicating factors are indicated in
Figure 3.   The definition of a surface is also a variable (Figure
4).   A zone of reaction between aqueous solutions and hydrocarbon
(e.g. PE) substrate has been suggested by Whitesides (11), but
this would probably encompass several microns depth for relatively non-polar solutes.   For simplicity, we define a surface in
terms of the depth being measured and will attempt to maintain a
homogeneous layer to that depth.   For any material, the fraction
of signal attained for a given depth is assumed to be an exponential function.   To see 90% of the functional groups present, one
must quantitatively derivatize to a depth about 2 $\lambda$, and for 99%
label, one must react to about 5 $\lambda$, in other words about 100Å
(Figure 5).   Adequate diffusion and reaction may be a significant
problem with aqueous solutions of certain ions.   For flat samples,
this effect can be mitigated to some extent by doing a variable
angle experiment.   Figure 6 shows the depth function for 99 and
90% label at different take-off angles.

Carboxylic Acid Group Model System

     As a carboxylate standard system, we initially used a co-
polymer consisting of 10% methacrylic acid (MAA) randomly distributed in ethylene.   As extruded, variable amounts of oxygen
relative to carbon were found, and it was decided that contaminants or predominantly hydrocarbon segments were being segregated
to the surface.   As shown in Figure 7, the theoretical oxygen-to-
carbon ratio is 0.034, while we observed samples containing less
than and also more than this amount of oxygen.   By scraping the

XPS RELATED
   1. ROUGHNESS
   2. SURFACE SEGREGATION
   3. VOLATILES LOST IN VACUUM

REACTION
   1. CONTAMINATION
   2. DIFFUSION OF REACTANTS

POLYMER
   1. STRESS
   2. RADICALS GENERATED DURING PREPARATION, i.e. CUTTING
   3. THERMAL HISTORY
   4. UNSUSPECTED ADDITIVES, e.g. La−Pb IN POLYESTERS
   5. MOLECULAR WEIGHT DISTRIBUTION (END GROUP CONCENTRATION)
   6. CONFORMATION

*Figure 3.    Variables in a polymer labeling study*

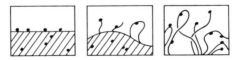

*Figure 4.    Three idealized views of a polymer surface (($\bullet$) $CO_2H$; ($\curlyvee$) $(CH_2)_n$)*

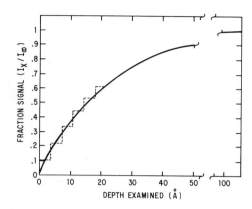

**Figure 5.** *The fraction of signal obtained for a given depth below the surface (relative to the signal obtained from a homogeneous infinitely thick solid) calculated from $I_x/I^\infty = 1 - exp(-X/\lambda)$ and assuming the mean free path of electrons $\lambda = 22$ Å. Note that 63% of the intensity is obtained from the top 22 Å (at $x = \lambda$), 90% from the top 50 Å, and 99% from the top 100 Å. Steps are introduced to show scale of atomic dimensions, i.e., jumps every 3 Å for a graphite single crystal.*

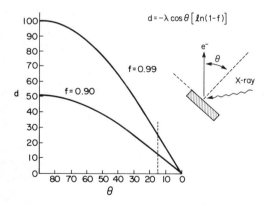

**Figure 6.** *Effective escape depth d vs. angle of ejected electrons $\theta$ for the case of 99% recovery and 90% recovery (f = fraction recovered)*

| | O/C |
|---|---|
| THEORY | 0.034 |
| EXTRUDED SURFACE | |
| SAMPLE A | 0.026 |
| SAMPLE B | 0.112 |
| SAMPLE C | 0.131 |
| SCRAPED SURFACE | |
| SAMPLE A | 0.038 |
| SAMPLE B | 0.039 |

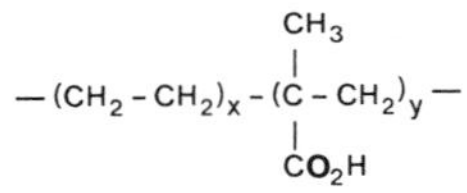

$$- (CH_2 - CH_2)_x - (\underset{\underset{CO_2H}{|}}{\overset{\overset{CH_3}{|}}{C}} - CH_2)_y -$$

*Figure 7.   Oxygen:carbon ratios measured before and after scraping the surface of a 10% MAA/ethylene copolymer*

| | MEASURED | | THEORY | |
|---|---|---|---|---|
| % MAA | INITIAL | 20 MIN WAIT | DRY | .5% $H_2O$ |
| 15% | | | | |
| AS IS | 447 | 353 | 515 | 556 |
| SCRAPED | 541 | 450 | | |
| 10% | | | | |
| AS IS | 386 | 310 | 337 | 378 |
| SCRAPED | 381 | 327 | | |
| 8.7% | | | | |
| AS IS | 361 | 327 | 292 | 332 |
| SCRAPED | 321 | 294 | | |

*Figure 8.   Oxygen:carbon ratios for E/MMA copolymers ($\times 10^4$)*

polymer, which had been extruded as a thick film, we were able to approach theoretical oxygen levels. This introduced roughness features up to 60 $\mu$m in height. ESCA quantitation was carried out by simply dividing the measured peak areas by Scofield's calculated photoionization cross sections when using instruments (12) with fixed analyzer pass energy. For fixed retarding ratio runs, an additional correction factor for kinetic energy was used (E$^{1.75}$). E/MAA copolymer was available with three different levels of methacrylic acid, and each sample was measured "as is" and then also after being scraped (Figure 8). It appears that about 0.5% water is present initially in the sample. After exposure to vacuum for about 20 minutes, most of this water is removed.

Labeling with aqueous silver nitrate is a method which has been discussed at Du Pont and elsewhere, and which has been shown to be qualitatively useful. It was known that samples of hydrolyzed polyesters reacted with silver nitrate more than ones which had not been exposed to hydrolysis for any length of time. However, a large number of standard polymer samples were analyzed with aqueous silver nitrate and gave results above and below the amount of silver predicted, depending on the rinsing method. Less rinsing lets excess silver ion remain and also high-binding-energy nitrogen (nitrate). In general, a significant amount of water was absorbed (increased oxygen) in each experiment, and this remained on the polymer for several hours of measurement time. Most of the results show that even brief rinsing with deionized water lowered the silver-ion concentrations to levels below a stoichiometric labeling of acid functionality.

A grazing angle experiment was done with one of the AgNO$_3$-treated polymers, which had a surface level of oxygen slightly less than theoretical. More of the silver ion was detected near the surface in contrast to the oxygen although total silver was less than expected. We take this to indicate that aqueous AgNO$_3$ does not penetrate far into the polymer on short reaction times and some of what is present is easily washed off. As we conducted the test, it was not a quantitative procedure (Figure 9). The three variable-angle results indicated were obtained on the same sample and collected about 15 minutes apart to evaluate any time-dependent behavior.

A number of other methods were also tried on the carboxylic acid standards. Thallium ethoxide (thallium is very poisonous and has caused human fatalities, being about as toxic as methyl mercury. It is used in mercury vapor lamps, tungsten filaments, catalysis, and organic synthesis. The biological half-life is about 8 days with removal primarily in urine. Hexacyanoferrates (II) (e.g., Prussian blue) are recommended as antidotes (13) was suggested to us by B. Trost; however, the first experiments with this material showed Pb rather than Tl being detected. Solutions of TlOEt in anhydrous ethanol would occasionally deposit excess amounts of thallium but would generally be very close to antici-

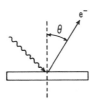

|           |   | Ag/C X 10⁴ | O/C X 10³ |
|-----------|---|------------|-----------|
| THEORY    |   | 168        | 34        |
| SAMPLE    | A | 219        | 101       |
|           | B | 60         | 51        |
|           | C | 35         | 59        |
|           | D | 73         | 48        |
|           | E | 38         | 54        |
| θ         | 20° | 49       | 59        |
|           | 80° | 59       | 44        |
|           | 20° (repeat) | 46 | 49     |

*Figure 9.    Reaction of 10% MAA/E copolymer with 0.1N aqueous AgNO₃*

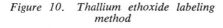

1.   IMMERSION IN TlOEt          0.5 MIN UNDER N₂

2.   WASH 3 X FOR 10 SEC WITH EtOH

*Figure 10.    Thallium ethoxide labeling method*

3.   PRE-CHAMBER 30 MIN

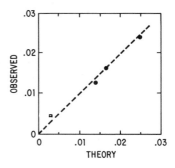

*Figure 11. Thallium: carbon atomic ratio from zero adjustable parameter theory vs. observed values*

pated levels.   Phosphorus tri-chloride generally gave sub-stoichiometric amounts.   Trying to form the mixed anhydride with trifluoroacetic anhydride also gave sub-stoichiometric amounts which
were not stable during analysis.

The final conditions which we found consistently to give a
quantitative analysis for carboxylic acids are shown in the
Figure 10.   Neat thallium ethoxide is a clear, not very viscous or
volatile liquid at room temperature and pressure.   Simple immersion of the polymer in this tetramer causes rapid penetration and
reaction.   Excess reagent can be washed out easily with ethanol to
leave the thallium label behind.   To remove excess ethanol, the
sample was left in the vacuum prechamber for 30 minutes before
measurement.   Some excess oxygen is still seen.   The samples are
stable and have been found to keep the same thallium level even
after being in the spectrometer under X-rays for more than an
hour.   Much less time is needed for analysis since the $Tl(4f_{5/2})$
cross section is about 12 times that of $C(1s)$.   Carboxylic acid
salts of thallium are known to be very stable to air, moisture and
light (14).   Preliminary experiments on an analogous mixed
acid/ester system show that the ester functionality is not label-
ed, while the acid is.

The labeling results are shown in Figure 11 for the model
systems, and show about 9.5% relative standard deviation.   A
scraped PE sample was used as a zero level guide but picked up a
small amount of Tl.   If one assumes that the residual oxygen
present on the PE after scraping may be carboxylic, the square
point on the graph can be included.   This method makes use of no
empirically derived factors or calibration methods.   It is expected to be valid for low concentrations (less than $\sim 30\%$) of
randomly incorporated acid groups.   Presumably, the presence of
one ionized acid group would hinder the ionization of its nearest
neighbor.

In principle, another standard surface can be prepared by
reacting a polyimide surface with NaOH to create the soluble salt
of the polyamic acid (Figure 12).   The reaction of NaOH is rapid
for the imide bonds, but the second reaction, which would cleave
the polymer chain, is slow.   The Na salt is soluble in water.
Since the polymer chain length is of the order of 1000Å, over ten
times the ESCA sampling depth, one should be able to make a partly
soluble surface during the initial stages of reaction.   The outer
surface, seen by ESCA, should be the soluble salt, while the inner
tail which is still part of the bulk polymer is unreacted,
insoluble polyimide.   A kinetic study of such a system is
summarized in Figure 13 where the atom ratio, Na/N, is plotted vs.
the minutes of exposure to 10 N NaOH.   All films were rinsed in
deionized water for five minutes to leach out imbibed NaOH and
dried overnight before analysis.   Complete conversion to the
polyamic acid corresponds to Na/N being 1.

The initial reaction attains 70% conversion to the polyamic
acid salt in five seconds, 97% conversion after five minutes.   The

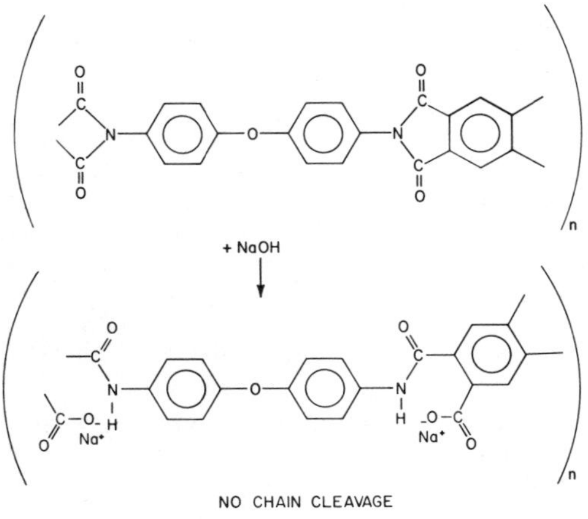

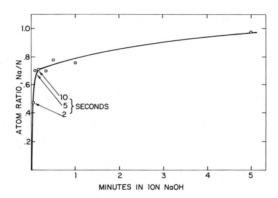

Figure 12.    Base hydrolysis reaction of polyimide

Figure 13.    Base hydrolysis reaction of polyimide

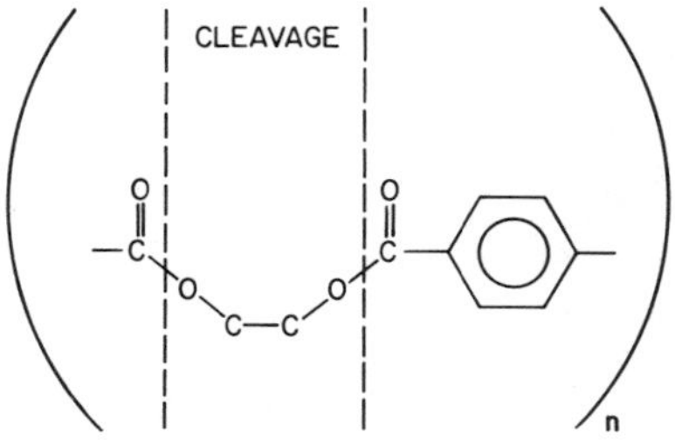

Figure 14.    Base hydrolysis of polyester
              showing cleavage

sharp break in the rate of Na uptake is not due to a sudden change
in reaction kinetics but to a complex composite of diffusion and
reaction during the reaction step, and diffusion and solution
during the rinse step.   The initial rapid uptake of Na is consistent with the observed bulk dissolution rate, 30 to 60 Å/sec,
and probably indicates the true reaction rate with the polymer.

Exchange reactions of this Na salt with Ag and Pb were not
reproducible and were high relative to Na, probably because of
copious precipitation of the hydroxides on the strongly basic
surface.   For this surface, Na ion itself is an excellent tag for
acid groups.

Using a similar method, we examined the reaction of polyethylene terephthalate with NaOH.   The reaction is shown in Figure
14.   In contrast to the polyimide case, this hydrolysis causes
chain cleavage.   Because a water rinse completely removes the Na,
we used a more complex scheme to study the kinetics.

After reaction with caustic, the films were dried with a
paper towel, dried for a day in nitrogen and analyzed.   The
resulting Na is called "total Na."   It includes the Na salt,
imbibed Na, and Na from the solution on the surface not removed by
the towel-drying step.   The sample was then rinsed in deionized
water for five minutes and dried as before.   No Na was ever
detected after this rinse.   Finally, the sample was immersed in
0.01 N NaOH for one minute to exchange Na for H, dried as before,
and analyzed.   The measured Na level is called the insoluble Na.

Results are shown in the Figures 15 and 16.   The first is a
plot of the atomic ratio of Na to carboxyl carbon vs. the minutes
of reactions.   Total Na increases rapidly for five minutes;
insoluble Na levels off quickly at the 0.02 level.   The plotted
values for the 1 N NaOH are the net after substracting the 0.036
Na per carboxyl, corresponding to zero reaction time (presumably
the Na left by the towel-drying operation).   A few seconds in 1 N
NaOH and up to five hours in 0.01 N NaOH gives a constant level of
9 Na atoms per 1000 carboxyls (0.009 Na/C).   Bulk end group
analysis of the polymer shows the average acid level is about 3
per thousand carboxyls, about one-third of the observed surface
value.

Figure 16 shows the longer-time data.   Total Na increases to
0.30 in twenty-four hours and presumably increases beyond this;
the insoluble Na remains at the 0.02 level as expected.   Exposure
of the surface to Ag ions gave reasonable values in this case.

Clearly, neat thallium ethoxide and dilute sodium hydroxide
can be used to tag carboxylic groups on polymer surfaces so they
can be meausred by ESCA.   Correct interpretation of the results is
aided by understanding the kinetics of each surface reaction.

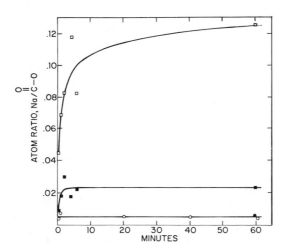

*Figure 15.   Base hydrolysis of polyester (total Na: (□, 1N; ○, .01N); insoluble Na: (■, 1N; ●, .01N))*

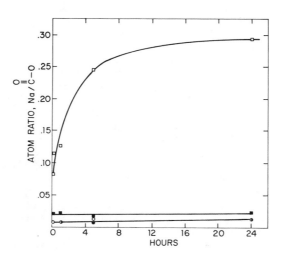

*Figure 16.   Base hydrolysis of polyester (total Na: (□, 1N; ○, .01N); insoluble Na: (■, 1N; ●, .01N)*

## Literature Cited

1.  Clark, D.T. and Thomas, H.R., J. Polym. Sci.: Polym. Chem. Ed., 1978, 16, 791-820 and other papers in this series.

2.  O'Malley, J., Thomas, H.R. and Lee, G., Macromolecules, 1979, 12, 496.

3.  Riggs, W.M. and Dwight, D., J. Electron Spectrosc., 1974, 5, 447-460; Dwight, D. and Riggs, W., J. Colloid and Interface Sci., 1974, 47, 650-660.

4.  Millard, M. and Marsi, M., Anal. Chem., 1974, 46, 1820-1822.

5.  Bradley, A. and Czuha, M. Jr., Anal. Chem., 1975, 47, 1838-1840; Czuha, M. and Riggs, W., Anal. Chem., 1975, 47, 1836-1838.

6.  Briggs, D., et al, J. Mater. Sci., 1977, 12, 429.

7.  Briggs, D. and Kendal, C., Polymer Communications, 1979, 20, 1053-1054.

8.  Spell, H. and Christenson, C., "Paper Synthetics Conference (Proceedings), Atlanta (TAPPI), 1978, p. 283-291.

9.  Hammond, J., Holubka, J., Durisin, A. and Dickie, R., "Abstract from Colloid and Interfacial Science Section ACS Miami Meeting, September, 1978.

10. Hammond, J., Holubka, J. and Dickie, R., "Preprint from Organic Coatings and Plastic Chemistry, Miami ACS Meeting, September, 1978.

11. Whitesides, G., et al, J. Amer. Chem. Soc., 1977, 99, 4736-4745.

12. Scofield, J., J. Elect. Spectrosc., 1976, 8, 457.

13. Zitko, V., Tech. Rep. Fish. Mar. Ser. (Can.), 1975, 518; "Documentation of the Threshold Limit Values, 3rd ed., (Amer. Conf. of Govt. Industrial Hygienists), 1971, p. 256.

14. McKillop, A., Bromley, D. and Taylor, E.C., J. Org. Chem., 1969, 34, 1172, "Thallium (I) carboxylates in contrast to silver carboxylates, are easily prepared and purified by recrystallization, and are stable indefinitely."

RECEIVED February 18, 1981.

# Interactions of Ion Beams with Organic Surfaces Studied by XPS, UPS, and SIMS

J. WAYNE RABALAIS

Department of Chemistry, University of Houston, Houston, TX 77004

The interactions of ions with organic surfaces can be broadly classified as "physical" or "chemical" depending on the nature of the ion-surface interactions potentials. Physical interactions (1) include processes such as ion implantation, sputtering, scattering, surface reconstruction, surface heating and collisional disassociation of impinging molecular ions. Such processes are typified by energetic rare gas ions and molecular ions impinging on surfaces. Chemical interactions (2) include chemical reactions of the impinging ions with the target species, chemisorption, and adsorption, i.e., processes in which the attractive potential forces for bond formation are larger than the forces governing the classical collision dynamics. Such processes are typified by active atomic or molecular ions impinging on active surfaces. For such cases the ion energy may be only slightly above thermal energies. The purpose of this paper is to provide examples of both types of interactions and to illustrate the applications of these methods for characterization and alteration of organic surfaces. The techniques of X-ray and UV photoelectron spectroscopy (XPS and UPS) and secondary ion mass spectrometry (SIMS) under ultra high vacuum conditions are employed in studying these effects.

As an example of physical interaction, the secondary ion mass spectra (SIMS) of some polymer surfaces are presented in Figure 1. In SIMS a solid sample is bombarded by primary ions of typically a few kiloelectron volts kinetic energy. The energy transferred to the surface results in classical sputtering of secondary atoms, molecules, and clusters in the form of positive, negative, and neutral species. Using electrostatic lenses the ions can be drawn into a mass spectrometer for analysis under UHV conditions yielding an extremely sensitive elemental surface analysis. SIMS can be performed in a "static" mode in which the ion flux is low enough to avoid destruction of the outermost surface layer during measurement (3, 4). Portions of the SIMS of Teflon, polypropylene, and polystyrene are identified in Figure 1. These spectra show that SIMS can be used to identify polymer

0097-6156/81/0162-0237$05.00/0

surfaces and that clusters characteristic to the specific polymer groups can be detected. For example, the Teflon spectrum is rich in $C_mF_n^+$ clusters, the polypropylene spectrum exhibits many $C_mH_p$ clusters. The positive spectrum of Teflon exhibits Na and K impurities. These ions are ubiquitous in SIMS (due to their extraordinary stability) unless extreme care and cleanliness are used in sample preparation. Although polymer studies by SIMS are still in their infant stage (5), applications to determination of the degree of crosslinking and crystallinity of polymers are forthcoming.

The ionic cluster compositions observed in SIMS should be readily related to the surface bonding features because the existence of strong, well-defined bonds between specific atoms in organic molecules determine, to a large extent, the dominant cluster types that will be sputtered. However, as will be shown below, the moieties that tend to sputter as a unit due to strong intramolecular bonds can be altered due to reactions, collisions, and fragmentation that occur as the moiety leaves the surface. The large quantity of energy deposited by the impinging primary ion results in a transient volume of high-energy content at the surface (4) in which ions, atoms, molecules, and radicals can escape into the vacuum. The observed SIMS clusters therefore consist of those secondary ions which (a) are stable enough to survive this high-energy content region, (b) are formed as a result of low-energy ion-molecule reactions occurring in the selvedge (the high-pressure plasma-like region which constitutes the solid-vacuum interface), and (c) the fragment ions produced by unimolecular decompositions of the ions formed in processes (a) and (b) during their flight to the detector. The final observed clusters are thermodynamically the most stable entities that can survive the selvedge plasma.

The origin of most ionic clusters in SIMS can be traced to one of at least three different ionization mechanisms. These are outlined as follows:

1.  cationization and anionization - In cationization and anionization, cations and anions, respectively, are attached to molecules. Typical examples are the attachment of metal ions, protons, or hydride ions to molecules (6). The ion attachment is believed to occur in the selvedge as a result of concurrent sputtering of the above mentioned ions and liberation of organic molecules by sputtering and/or thermal evaporation.

2.  electron transfer - This process involves transfer of an electron either to or from a molecule as it is sputtered and/or thermally evaporated and travels through the selvedge. The stabilizing collisions and electrical forces which operate in the selvedge are known to provide large charge-exchange cross sections in the form of resonance or Auger ionization processes (7).

3.  direct ion ejection - This process involves direct sputter-

ing and/or thermal evaporation of cations and anions that already exist as such in the solid (6). Classical dynamics can be used to describe the momentum transfer from the primary projectile to the lattice species.
Most of these ionization processes can be illustrated by reference to the positive and negative SIMS of formic acid frozen at $77^{o}K$, Figs. 2 and 3. Cationization and anionization processes are clearly operative as indicated by the clusters $H(HCOOH)^{+}$ and $HCOO(HCOOH)^{-}$. The cluster series $H(HCOOH)^{+}$ and $HCOO(HCOOH)^{-}$ are most likely a result of ion-molecule reactions in the selvedge region. Unimolecular fragmentation is also occurring, as evidenced by the large number of fragment peaks between the cationized and anionized parent clusters. The small ions such as $C^{-}$, $CH^{-}$, and $HCO^{+}$ are fragment ions while species such as $C_2^{-}$, $C_2H^{-}$, $O_2^{-}$, $H_3O^{+}$, and $H(H_2O)_2^{+}$ result from rearrangement reactions in the selvedge.

As an example of <u>chemical interaction</u>, the results of low energy $N_2^{+}$ bombardment of graphite, diamond, graphite monofluoride (CF), and teflon are shown in Figure 4. The $N_2$ ions are charge-exchange neutralized by resonance or Auger neutralization while they are still several Angstroms from the surface (6, 8). If the energy of the impinging $N_2$ ions is considerably above the molecular dissociation limit, collision dissociation occurs at the surface to produce hot N atoms. These hot atoms penetrate into the lattice, losing energy in collision cascades until they are eventually thermalized. The penetrating atoms can react on initial or subsequent collisions or after becoming thermalized if there is chemical affinity between the atoms and target. If the appropriate chemical affinity does not exist, the majority of the foreign atoms diffuse through the lattice until they reach a surface and escape. Such is the case for inert gas ions impingent on organic surfaces, where the inert atoms, lacking the requisite chemical affinity for reaction, diffuse through the lattice and are eventually evolved. As a result, high ion doses are required in order to observe implanted inert gas ions in organic solids. In contrast, low ion doses ($\sim 10^{14}$ ions/$cm^2$) of $N_2^{+}$ are sufficient to observe nitrogen in organic solids (9, 10, 11) because the hot N atoms which are produced have very short diffusion lengths due to their high reactivity. The X-ray photoelectron spectra (XPS) of the carbon 1s and nitrogen 1s lines of pyrolytic graphite, CF, and teflon after 500 eV $N_2$ bombardment indicating the presence of C in different chemical environments. This is expected due to the chemical shift in C from bonding to the N. The N 1s line in graphite consists of two overlapping peaks corresponding to the binding energies of partially negative cyanide-type (398.5 eV) and neutral interstitial (400.0 eV) nitrogen. The corresponding bombardment of diamond produces only a single N 1s peak at 398.5 eV corresponding to a cyanide-type nitrogen. The absence of the 400 eV peak in diamond strongly suggests that this high binding energy N 1s peak in graphite corresponds to nitrogen trapped as a

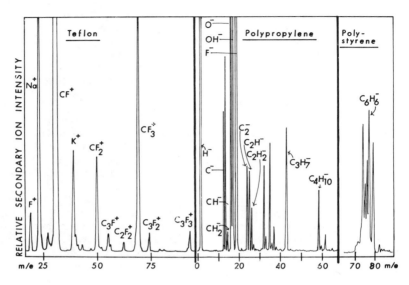

Figure 1.    Portions of the SIMS Teflon, polypropylene, and polystyrene obtained
by bombardment with 2 keV $Ar^+$ ions

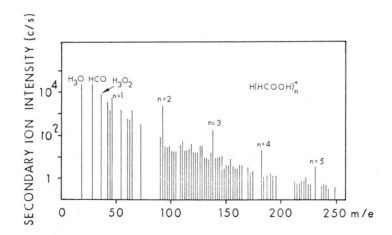

Figure 2.    Positive SIMS of formic acid frozen at 77 K

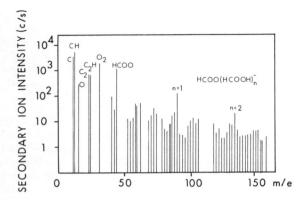

*Figure 3.    Negative SIMS of formic acid frozen at 77 K*

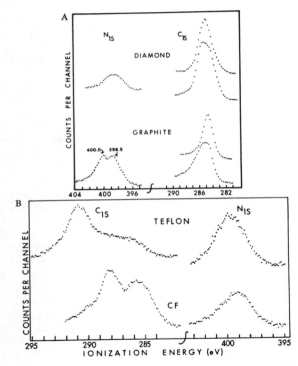

Journal of the American Chemical Society

*Figure 4.    (A) C-1s and N-1s XPS spectra of diamond and pyrolytic graphite before (1) and after (2) $N_2^+$ bombardment. (B) The C-1s and N-1s XPS spectra of Teflon and CF after $N_2$ bombardment. The C-1s spectra of Teflon and CF before $N_2^+$ bombardment consist of a single line at the position of the high binding energy component of this figure (10).*

neutral species at lattice imperfections, twinning planes, and between the layers of graphite rings. In both Teflon and CF the single C 1s line of the pure samples is split into two distinct components in the bombarded samples, i.e., a new broad C 1s line is observed at lower binding energy. This new peak can be attributed to the reduction of C by F evolution and the replacement of F atoms by less electronegative N atoms, e.g., a replacement of the type $-CF_2-$ to $-NCF-$ in Teflon or CF to $CF_{1-x}N_x$ in graphite monofluoride. The position of the N 1s XPS line is higher than those of pure cyanide-types owing to the highly electronegative fluorine atoms in the environment and, in some cases, attached to the same carbon atoms. The nitrogen does not form NF species within the lattice, for such species contain a partially positive nitrogen whose N 1s XPS energy is in the range $\sim$ 403–407 eV. The ionization energies, linewidths (in parentheses), and ion/target XPS intensity ratios are listed in Table I. The weak N 1s XPS signal in Teflon and CF as compared to graphite indicates that only a small amount of nitrogen is reacting with the carbon in those samples. This is not surprising considering the "tightness" of the Teflon and CF lattices compared to graphite. Thus the low-energy C 1s XPS peak in Teflon and CF is evidently due to both (i) reduction of carbon with subsequent nitrogen and fluorine evolution and (ii) replacement of the more electronegative fluorine by nitrogen.

In compounds of graphite it is found that the conductivity is decreased or increased according to whether the interstitial species is highly electronegative or electropositive, respectively. The UPS spectrum of graphite/$N_2^+$ shows a decreased density of states near the Fermi level, indicating that the conductivity of the reacted layer is altered. The density of states near the Fermi level is due to C $2p\pi$ electrons in orbitals perpendicular to the graphite layers. Introduction of the more electronegative nitrogen stabilizes these C 2p electrons and draws them from the Fermi level as a result of the formation of cyanide-like carbon where the electron cloud is polarized towards the nitrogen end of the CN moiety.

After reacting low energy $N_2^+$ ions with frozen benzyne and cyclohexane, the reaction products can be characterized by SIMS as shown in Figure 5. After bombardment a new peak at 26 amu corresponding to $CN^-$ is observed which is totally absent before $N_2^+$ bombardment. This observation provides direct evidence that chemical bonds are formed between the target atoms and the impinging nitrogen. It should be noted that when $N_2$ gas was allowed to physisorb on the frozen benzyne or cyclohexane surface at 77°K, no $CN^-$ was observed in SIMS; this indicates that the bombardment is responsible for the formation of carbon–nitrogen bonds. Also, there is a large increase in the $C^-/CH^-$ ratio, Figure 5, after bombardment suggesting that chemical sputtering may be occurring, i.e., the hot N atoms are reacting with the hydrogen atoms of benzyne and cyclohexane forming gaseous products which

Table I

XPS Binding Energies, Peak Wdiths and Intensity Ratios from Graphite, Diamond, Teflon and Carbon Monofluoride, Before and After $N_2^+$ Ion Bombardment *(10)*

| | C1s | N1s | O1s | F1s |
|---|---|---|---|---|
| Graphite | 284.6(1.5) | -- | -- | -- |
| Graphite/$N_2^+$ | 284.6(2.9) | 398.5($\sim$2.0), | -- | -- |
| | | 400.0($\sim$2.1) | | |
| | | N 1s/C 1s=0.18 | | |
| Graphite/$NO^+$ | 284.6(2.6) | 398.9($\sim$2.0), | 531.4(2.2), | -- |
| | | 400.2($\sim$2.1) | 533.1(2.3) | |
| | | [N 1s/C 1s=0.12] | [O 1s/C 1s=0.096] | |
| Diamond | 285.0(2.2) | -- | -- | -- |
| Diamond/$N_2^+$ | 285.0(2.9) | 398.5(2.9) | -- | -- |
| | | [N 1s/C 1s=0.072] | | |
| Teflon | 292.6(2.0) | -- | -- | 690.1(2.3) |
| Teflon/$N_2^+$ | 287.6(3.2), | 401.6(3.0) | -- | 689.6(2.8) |
| | 292.6(2.6) | | | |
| CF | 288.1(2.3) | -- | -- | 687.0(2.3) |
| CF/$N_2^+$ | 285.8(3.0), | 399.4(3.1) | -- | 687.2(2.5) |
| | 288.1(2.7) | | | |

Journal of the American Chemical Society

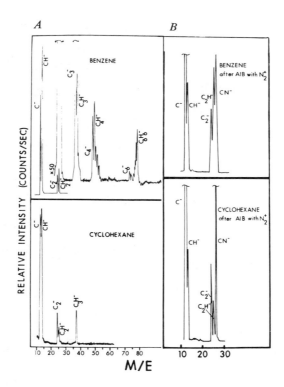

*Figure 5.* Negative SIMS spectra of benzyne and cyclohexane frozen at 77 K (A) before and (B) after bombardment with a dose of $2 \times 10^{16}$ ions/cm$^2$ of 100 eV N$_2^+$
(4)

are evolved from the surface.  This results in a depletion of the surface H and consequently an enhanced $C^-/CH^-$ ratio.

The reaction between $N_2^+$ and graphite discussed above can also be observed (12) by SIMS.  After $N_2^+$ bombardment, the SIMS of graphite exhibits new peaks at 15, 26, 27, 30, 40, 41 and 42 amu corresponding to $HN^-$, $CN^-$, $HCN^-$, $H_2N_2^-$, and $H_nC_2N_2^-$ ($n$ = 2, 3, 4), respectively.  The observation of clusters containing hydrogen atoms indicates that the nitrogen is reacting with adsorbed hydrogen on the surface as well as the graphite itself.  The partial pressure of $H_2$ in the chamber is $\sim 6 \times 10^{-10}$ torr due to the inefficiency of the turbomolecular pump for such light gases. H atoms are produced by decomposition of $H_2$ at tungsten filaments and the atoms adsorb on graphite under these experimental conditions.  The observed nitrogen-hydrogen clusters must be formed by reaction of the impinging nitrogen with adsorbed hydrogen.

These results provide direct evidence that chemical reactions occur and that new chemical bonds are formed during active ion bombardment of organic surfaces.  Also, the use of active ions for chemical synthesis via reaction of beams with surface species or by selective "chemical sputtering" is suggested.

## Acknowledgement

Acknowledgement is made to the R. A. Welch Foundation and the National Science Foundation (Grant No. CHE-7915177) for support of this research.

## Abstract

The interactions of low energy ions with organic surfaces are classified into two groups, "physical" and "chemical", depending on the nature of the ion-surface potentials.  Examples of physical interactions are provided by $Ar^+$ bombardment of teflon, polypropylene, and polystyrene via the technique of secondary ion mass spectrometry (SIMS).  The results show that SIMS provide a fingerprint of polymer surfaces and that clusters characteristic to specific polymer groups can be identified.  The dominant mechanisms for production and ionization of secondary clusters are discussed and examples of these processes are provided by the SIMS of frozen formic acid.  Examples of chemical interactions are provided by $N_2^+$ (active ions) bombardment of graphite, diamond, graphite monofluoride (CF), "Teflon", and frozen benzyne and cyclohexane via the technique of X-ray and UV photoelectron spectroscopy (XPS and UPS) and SIMS.  The results show that chemical reactions can be induced between the active ion beam and the surface yielding thin surface films of the reaction products. The reaction with diamond, CF, "Teflon", and the frozen hydrocarbons produces a cyanide-type nitrogen at the surface as characterized by XPS and UPS chemical shifts and a SIMS peak at 26 amu ($CN^-$).  The reaction with graphite yields a cyanide-type nitrogen

($N_{1s}$ = 398.5 eV) and a neutral interstitial nitrogen ($N_{1s}$ = 400.0 eV). The latter is most likely a neutral N trapped between the layers of graphite rings. The use of active ions for chemical synthesis via reaction of beams with surface species or by selective "chemical sputtering" is suggested.

## Literature Cited

1.  Carter, G. and Colligan, J.S., "Ion Bombardment of Solids," American Elsevier, NY, 1968; Townsend, P.D., Kelly, J.C. and Hartley, N.E.W., "Ion Implantation, Sputtering, and Their Applications," Academic Press, New York, NY, 1976.

2.  Lemmon, R.M., Accts. Chem. Res., 1973, 6, 65; LeRoy, R.L., Yencha, A.J., Menzinger, M., and Wolfgang, R., J. Chem. Phys., 1973, 58, 1741.

3.  Benninghoven, A., Surface Sci., 1975, 53, 596.

4.  Lancaster, G.M., Honda, F., Fukuda, Y., and Rabalais, J.W., J. Amer. Chem. Soc., 1979, 101, 1951; Honda, F., Lancaster, G.M., Fukuda, Y., and Rabalais, J.W., J. Chem. Phys., 1978, 69, 4931.

5.  Gardella, Jr., J.A. and Hercules, D.M., Anal. Chem., 1980, 52, 226.

6.  Grade, H., Winograd, N., and Cooks, R.G., J. Amer. Chem. Soc., 1977, 99, 7725; Day, R.J., Unger, S.E., and Cooks, R.G., J. Amer. Chem. Soc., 1979, 101, 501; Grade, H. and Cooks, R.G., J. Amer. Chem. Soc., 1978, 100, 5615.

7.  Hagstrum, H.D., Surface Sci., 1976, 54, 197; Phys. Rev., 1961, 122, 83.

8.  Durana, S., LeRoy, R.L., Menzinger, M., and Yencha, A.J., J. Chem. Phys., 1974, 60, 2568.

9.  Taylor, J.A., Lancaster, G.M., and Rabalais, J.W., Appl. Surf. Sci., 1978, 1, 503.

10. Taylor, J.A., Lancaster, G.M., and Rabalais, J.W., J. Amer. Chem. Soc., 1978, 100, 4441.

11. Taylor, J.A., Lancaster, G.M., and Rabalais, J.W., J. Chem. Phys., 1978, 68, 1776.

12. Lancaster, G.M., Honda, F., Fukuda, Y., and Rabalais, J.W., Chem. Phys. Lett., 1978, 59, 356.

RECEIVED December 22, 1980.

# The Modification, Degradation, and Synthesis of Polymer Surfaces Studied by ESCA

D. T. CLARK

Department of Chemistry, University of Durham, South Road, Durham City, England

Organic based polymers represent extremes in terms of both complexity of structure and sensitivity to interrogation and the single most powerful tool which has emerged for these investigations in terms of surface structure has proved to be ESCA (1). It is now over a decade since the first experiments on polymers were documented (2) and the intervening period has seen an explosive growth in the literature such that most of the requisite experiments for delineating the information levels and derived areas of applicability of the technique have broadly been accomplished. It is the wide ranging nature of these information levels which imbues the technique with capabilities far higher than might, naively have been thought from a consideration of the sum of each of the single information levels available from the ESCA experiment. For the sake of completeness we should perhaps at this stage list the available information levels in each case indicating how recent developments have expanded the scope for studies of the surface structure and its modification and for studies of the synthesis and reactions at polymer surfaces.

Figure 1 shows schematically the typical data levels and variable experimental parameters which can be employed in the application of ESCA to polymeric systems. Figure 2 indicates how, many of the information levels available from the ESCA experiment may be used to study the time and temperature dependent behavior of polymer surface phenomena. Many of the contributions to this book deal with specific applications with an emphasis on one or other of the information levels available from the ESCA technique. Before considering specific applications of our own we may profitably highlight recent developments which point the way forward for applications in the next decade.

Absolute Binding Energies. A substantial library of both experimental and theoretical data is now available and it is normally a matter of routine to identify given structural fea-

0097-6156/81/0162-0247$11.25/0

Data level

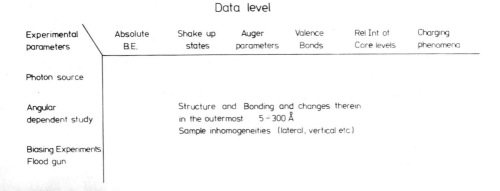

Figure 1.   Data levels in the ESCA investigation of polymers

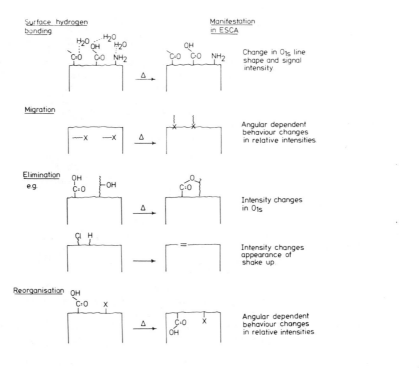

Figure 2.   Changes in surface chemistry as a function of temperature and time studied by means of ESCA

tures particularly from detailed analyses of the theoretically
and experimentally well understood Cls levels.

With good sample handling capabilities (3) it is no problem
to study low molecular weight model systems in the solid state and
for such systems it is often possible to carry out detailed non-
empirical LCAOMO SCF computations to provide a fundamental interpretation of substitutent effects (4). Theoretical studies are
also of prime importance in defining data for systems which may be
difficult to study experimentally or which may suffer significant
radiation damage. A recent compilation of theoretical data
referenced for both gas and solid state studies is depicted in
Figure 3 (5). The data refer to computations in the $\Delta$SCF
formalism in STO 4.31G basis sets and particularly noteworthy are
the data for oxygen and nitrogen functionalities. Taken in
conjunction with that previously published (6) a rather complete
set of background data is now available for the interpretation of
the oxidation and nitration of polymers and we will return to this
aspect in a later section.

Photon Energy Dependent Studies. To date most applications
in the ESCA field have been restructed to the commonly employed
$MgK_{\alpha 1,2}$ X-ray sources where the escape and sampling depths for the
core levels of usual interest typically range from 8 - 30Å and 25
- 90Å respectively. Most instrumentation now has "push button"
facility for changing photon energy and the virtues of using
harder X-ray sources is becoming apparent. Of considerable
interest is the $TiK_{\alpha}$ X-ray source where the photon energy of ∿4510
eV enables depths ∿ 300Å to be sampled and this is indicated
schematically in Figure 4. This provides a convenient step in
depth profiling to MATR (Multiple Attenuated Total Reflection
Infrared Spectroscopy) where under normal conditions the sampling
depth might be ∿0.5μ.

As an example of the use of a $TiK_{\alpha}$ X-ray source Figure 5
shows the core level spectra for a nylon film. The Ols and Nls
levels studied with the harder X-ray source clearly show that the
inherent width of the $TiK_{\alpha}$ radiation is ∿ 3 eV and consists of a
doublet of intensity ratio 2:1 and splitting 5.8 eV. The large
inherent width of the X-ray source is advantageous in this line
shape analysis since the analyzer contribution and any charging
inhomogeneity contribution in linewidth are negligible. The
width of the components of the Cls, Ols and Nls levels are
therefore identical and it is a straightforward manner to analyze
the Cls spectrum in terms of a super position of a 3 component
lineshape as indicated in the figure. The spectra, recorded with
the fixed retardation ratio (FRR) mode, show that the differential sensitivities (a convolution of instrumental and cross section factors) change by a significant but not drastic amount on
going from $MgK_{\alpha}$ to $TiK_{\alpha}$. The data are also interesting in that
they reveal the presence of a small extent of hydrocarbon contamination on the very surface since the $MgK_{\alpha 12}$ intensity ratios

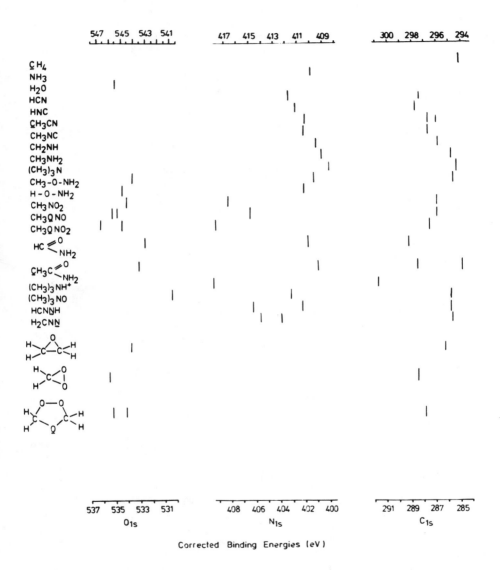

Figure 3.    Theoretically computed (Δ SCF) calculations of C-1s, N-1s, O-1s core
level binding energies

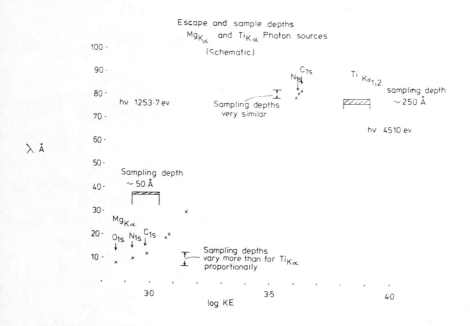

*Figure 4.    Schematic of mean-free paths as a function of KE (MgKα1,2 and TiKα1,2 sources)*

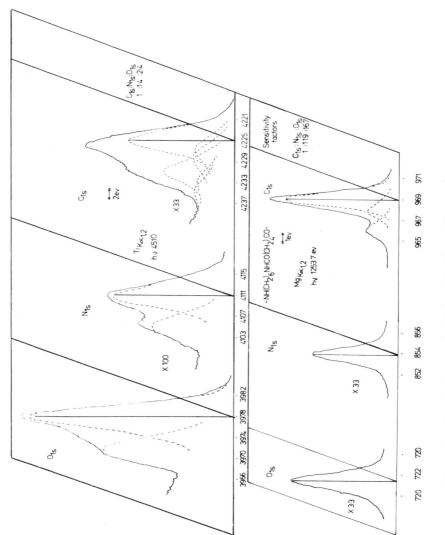

*Figure 5.    Core-level spectra for Nylon 6,6 with two different photon sources*

indicate a component at 285 eV in excess of that for the repeat unit. This is not evident for the $TiK_\alpha$ data since the contribution to the total signal from the very surface is very small. Such low level surface contamination for a film can of course more readily to be detected by angular dependent studies however for powder samples where such studies are not applicable the capability of interrogating vertical inhomogeneities by means of photon energy dependent studies is of great importance.

Angular Dependent Studies. We have recently pointed out the considerable care which must be exercised in interpreting angular (take off angle) dependent data in delineating unambiguously lateral and vertical sample inhomogeneities (7). The unrivalled capability of ESCA in providing data on vertical inhomogeneities has already produced new insights in the surface science of polymers on both an academic and technological front.

Before the advent of ESCA the question of the initial reaction of a solid surface with a given environment could not be addressed. The complexities of structures for polymers which are partially at least dictated by processing can clearly lead to correspondingly complex surface structures for systems which have undergone reaction at a gas/solid or liquid/solid interface. A typical situation might be that depicted in Figure 6. The surface, sub-surface and bulk may have crystalline and amorphous regions. For a reactive species M, for which the activation energy is small, reaction may occur for every collision and the reaction may then occur rather uniformly both laterally and vertically. If the activation energy is high, however, then both lateral and vertical inhomogeneities may occur. The situation is also complicated by the fact that the crystallinity of a given polymer system may be strongly influenced by processing conditions. Thus, uniaxial and biaxial orientation of a given polymer system may well alter quite drastically the crystallinity of a given system and this is indicated schematically in Figure 7. As we have noted with careful control of all the variables a combination of angular and photon energy dependent studies provides ESCA with unrivalled capability for depth profiling. In appropriate cases it is also possible to study lateral inhomogeneities and with improved sensitivity small area analysis becomes a distinct possibility. This can be accomplished either by means of the spectrometer lens system or by means of appropriate slit systems. At this stage however it would seem that in the next ten years we are unlikely to see spectrometers with a spot resolution size of less than $10^{-6}$ sq. cm. which is still rather gross compared with SAM.

Sample Charging. We have shown that sample charging is an important information level in its own right and with the possibility of "tuning" sample charging by means of a UV flood gun in the particular case of studies involving a monochromatised source

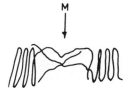

Reactive species M

Diffusion in amorphous region ≫ crystalline region

(i) Low activation energy :- uniform reaction at surface

(ii) High      "         "    :- diffusion controlled, non uniform reactivity of various surface and subsurface regions.

*Figure 6.   Surface reactions of polymers*

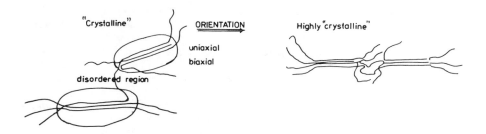

*Figure 7.   Fringed micelle structure of polymers and orientation effects on bulk morphology*

there will undoubtedly be an increasing awareness of this important information level which is traditionally thrown away (8).

In the three sections below we consider important areas of
applications of ESCA in; migration phenomena, in the synthesis of
polymer surfaces by means of plasma techniques and in reactions
initiated at the surface of cotton fibrils in the nitration of
cellulose.

## Migration Phenomena (9)

As we have indicated schematically in Figure 2 the migration
and segregation of small molecules at polymer surfaces can be
monitored by intensity changes of appropriate core levels as a
function of take off angle. A particularly intriguing application of this genre involves the migration of the low molecular
weight silicone material used as a release agent in double sided
Scotch tape commonly used for mounting purposes in ESCA experiments.

A *priori* migration of low molecular weight material could
conceivably occur by two distinct mechanisms and this is indicated schematically in Figure 8. On the left hand side migration
from one surface of a polymer to the other is by bulk migration
involving permeation through the bulk. The alternative involves
migration along surfaces as indicated schematically on the right
hand side of the figure.

An ESCA examination of a freshly exposed surface of double
sided Scotch tape (3M's company electrical insulating tape) reveals a fairly intense signal from $Si_{2p}$ and Ols levels appropriate
to silicone type material. To demonstrate the migration of this
silicone material a piece of tape was attached to a strip of
polymer and the results for high and low density polyethylene
samples (HDPE LDPE) are shown in Figure 9. By taking samples from
different sections of the polymer strip after a period of 10 days
at ambient temperature two features are evident. Firstly LDPE has
a different behavior from HDPE. Secondly both polymer films show
evidence that the silicone will migrate along the surface. This
is not entirely unexpected on the basis of the surface energetics
and the likely degree of crystallinity etc. of the surfaces of
high and low density polyethylene.

In order to differentiate between migration along a surface
versus permeation through the bulk as mechanisms for silicone to
move from one surface to another we have constructed the apparatus
shown in Figure 10. By mounting polymer films in a clamping ring
and by monitoring different portions of the film it becomes
possible to differentiate between the main alternatives for migration mechanisms and this is indicated schematically in the
figure.

The core level data for two of the positions of interest for
a LDPE film of ∿100µ thickness are shown in Figure 11. Considering firstly the room temperature data the intensities for the $Si_{2p}$

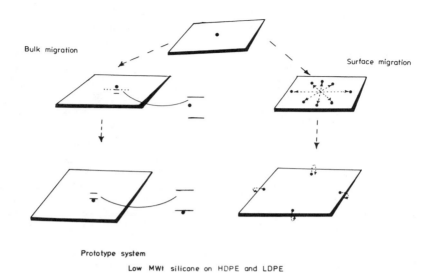

*Figure 8.    Migration phenomena: possible extremes for migration of low-molecular weight materials.*

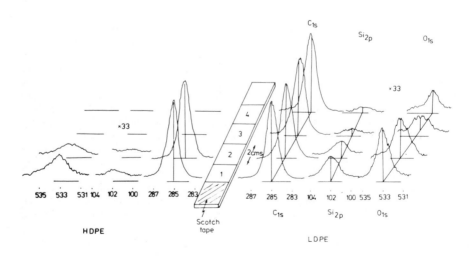

*Figure 9.    The C-1s, O-1s, and Si-2p core levels for strips of LDPE film with Scotch tape attached (diffusion studies, 10 days RT)*

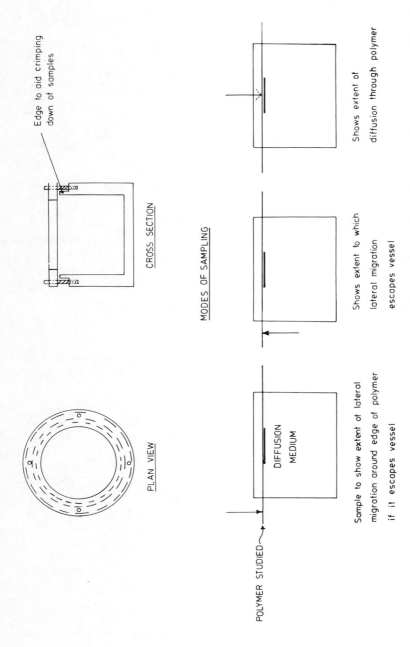

*Figure 10.   Diffusion studies apparatus and its application (1:1 scale diagram)*

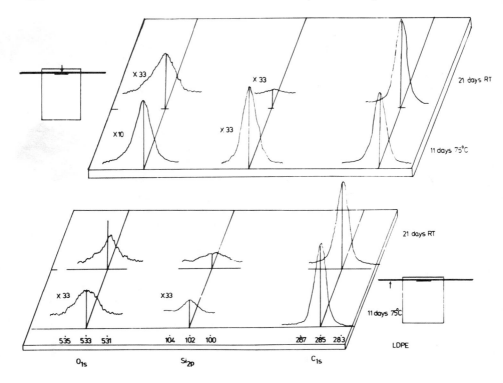

*Figure 11.   The C-1s, O-1s, Si-2p core levels for LDPE samples mounted in diffusion study apparatus*

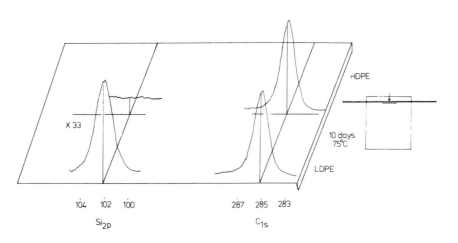

*Figure 12.   Comparison of C-1s and Si-2p levels for HDPE and LDPE samples mounted in diffusion study apparatus*

and O1s signals are somewhat similar for both positions and indicate that migration along the surface is drastically curtailed by the clamping ring arrangement. After 11 days at 75°C however the situation is substantially different the migration through the bulk being substantially faster than along the surface. The difference between HDPE and LDPE is still evident however for the bulk migration and this is clear from the data given in Figure 12.

These data show the potential for studying the mode of migration of small molecules both along and through surfaces and also point to the danger of long term exposure of samples to Scotch tape as might be used for example in securing samples for transporting purposes.

## The Plasma Synthesis Of Ultra Thin Fluoropolymer Films

The synthesis of ultra thin polymer films by plasma techniques is an area of considerable current interest from several viewpoints (10, 11). Thus the capability of taking bulk polymers of desirable bulk properties and specifically modifying the surface to alter the chemical, physical, mechanical or electrical properties has much to commend it and fields of application which have already received attention include grafting polyacrylic acid onto polyester fibre to improve the comfort factor and deposition of low friction coating fluoropolymer films onto elastomer seals for use in the oil drilling industry. Plasma deposition of ultra thin polymer films onto metals is of interest for biomonitoring electrodes and plasma coatings are also under investigation for laser fusion targets. Plasma polymers prepared from hexamethyldisiloxane have been shown to be amongst the most biocompatible of polymeric films whilst the deposition of very thin uniform pinhole free films of unusual electrical properties makes plasma polymer films of interest in device applications. The specific modification of polymer surfaces by "cool" plasmas is also an area of considerable current interest. Thus detailed kinetic studies have been made of direct and radiative energy transfer to polymer surfaces by means of inert gas plasmas, a topic of interest in the surface cross-linking of polymers and removing weak boundary layers for adhesive bonding. The surface functionalisation of polymers can also be accomplished in a cost effective manner by means of plasma techniques. The motivation for studying the fundamentals of plasma polymerizations and modifications of surfaces is therefore not difficult to understand and we consider here one aspect of the topic, the plasma polymerization of simple fluoropolymers.

The excitation of a plasma in a typical organic molecule gives rise to a number of radical cations, and excited states and anion radicals arising from electron attachment. It is not surprising therefore that transformations involving such species and reactions with the original molecule either in the gas phase

or at the gas-solid interface gives rise to a reaction pathway
which is typically quite complex. For vinyl monomers it may be
shown that under appropriate conditions plasma induced polymerizations may accompany the more general process of plasma polymerization, the former giving rise to linear systems, the latter
cross-linked products. In the glow region of the plasma where the
ion density is highest polymer deposition arises from the competitive process of polymerization and ablation, the so-called CAP
scheme (10) and the great versatility of the technique in being
able to produce thin films from virtually any organic precursor
makes the study of the process of deposition and the structure of
the accompanying films one of great interest. The fact that the
films of interest are thin (generally in the range 5 - 5000Å) and
therefore need to be studied in situ and are insoluble renders
ESCA the technique of choice for studying details of structure and
bonding. Indeed it is true to say that the mechanistic study of
the formation of such films has only made significant progress
since the advent of ESCA.

In recent papers we have shown how ESCA may be used to
establish relationship between the structure of the initial
molecular system in which the discharge is excited and that of the
polymer deposited in either glow or non-glow reations (12). The
objectives of the present programme of work on plasma polymers may
be summarized as follows:

(i)     To establish the overall stoichiometry and structural
        features present in a polymer film.
(ii)    To study the variation in (i) as a function of
        geometric factors, power, pressure etc.
(iii)   To study rates of deposition as a function of the gas
        and the site of deposition.

For a variety of reasons the work reported here pertains to
inductively coupled RF plasmas in the pressure range 50 - 200μ.
Such experiments are particularly easy to carry out and close
control can be maintained on the important variables such as
pressure flow rate, power input etc. Yasuda (10) has pointed out
the importance of W/FM (where W is the power input, F is the flow
rate and M the molecular weight) in determining deposition rate
and nature of the material (viz. powder, powdery film, film, oil).
Under the conditions of work described here polymers were found to
deposit as uniform films both on gold, pyrex and silica substrates. Three distinct apparatus's have been employed in this
work. A schematic of a typical free standing reactor for the
deposition of films which are subsequently transferred to the
spectrometer is shown in Figure 13, whilst the reactor used for in
situ deposition of polymer films which may then be directly
inserted into the spectrometer has been described in some detail
previously (11). In order to study the dependence of polymer
structure and rate of deposition as a function of deposition site
a special long reactor has been constructed and this is shown
schematically in Figure 14.

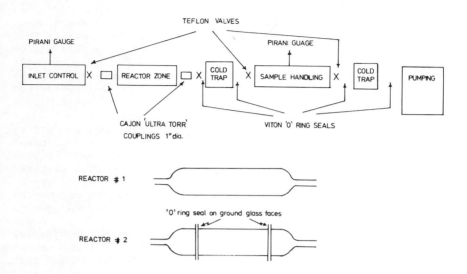

*Figure 13. Apparatus for studying RF plasma polymers*

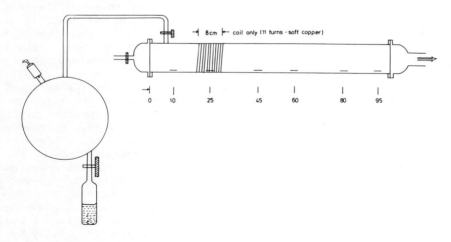

*Figure 14. Apparatus for studying site dependence of structure of plasma polymers*

It is known that plasma polymers often contain unpaired spins and as such may react on exposure to atmosphere oxygen to produce surface oxidative functionalization. In general this has not proved to be of any major significance as far as the fluoropolymers studied in this work are concerned and to emphasize this point Figure 15 shows a wide scan ESCA spectrum of a thin plasma polymer film produced from perfluorobenzene. The stoichiometry of this film is close to that of the initial "monomer" and by going to a high take off angle we can detect a very low level of oxygen signal for this film produced in an apparatus as in Figure 13.

As well as having interesting chemical, physical, electrical and mechanical properties fluoropolymers produced by plasma means are particularly suitable for study by ESCA not the least because of the very distinctive nature of the Cls levels in such systems. Figure 16 depicts typical binding energy ranges for a variety of structural features which might arise in fluoropolymer systems and with a background of such data we may straightforwardly assign spectra for plasma polymer systems.

A particularly striking example of this genre is provided by the data given in Figure 17. The stoichiometry of films can be determined from the integrated Cls:Fls intensity ratios and from the components of the Cls levels. From this data it is clear that plasma polymerization involves substantial rearrangement since trifluoromethyl groups form a significant feature of the spectra. For perfluorocyclohexane, normally considered an extremely inert fluorocarbon, the fluoropolymer exhibits a large percentage of perfluoroethyl groups and has a stoichiometry lower in fluorine than the starting material. With the advent of ESCA as a structural tool it has been shown for the first time that the deposition in the glow region of a plasma can produce well defined polymeric products over a range of pressure and power input ranges (12).

The nature of the competitive process of polymerization and deposition is such that we might anticipate that at the edges and outside of the glow region longer lived reactive species might well lead to the formation of different polymer in regions remote from the coil region in an inductively coupled plasma. Indeed in the case of conventionally polymerizable "monomers" such as the fluorinated alkenes the polymerization mechanism in the non-glow region may well be dominated by conventional addition polymerization: the so called plasma initiated polymerization.

There are two important questions which may readily be addressed by ESCA. Firstly, what is the rate of deposition of the polymer film at a given site in a plasma reactor and secondly how does the structure depend on the site of deposition? To illustrate the great power of the technique in answering these questions we consider here a recent detailed investigation of the inductively coupled plasma polymerization of pentafluorobenzene (13).

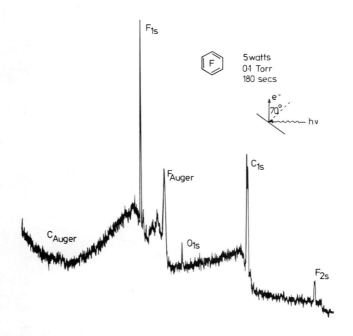

*Figure 15.   Wide-scan ESCA spectrum (MgK$_{\alpha 1,2}$) of plasma polymer produced from perfluorobenzene*

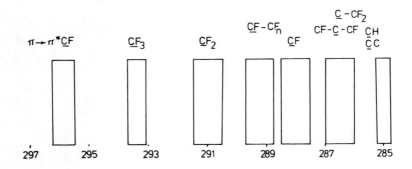

*Figure 16.   The C-1s core binding energies for fluorocarbon structural features*

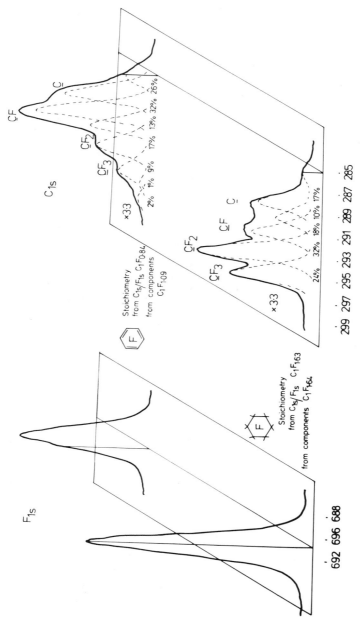

*Figure 17. Comparison of C-1s and F-1s core levels for plasma polymers produced from perfluorobenzene and cyclohexane (180 s, 10 W, 30° take-off angle)*

Cls core level spectra as a function of two take off angles for the plasma polymerization in the glow region of a free standing reactor are shown in Figure 18, at a pressure of 200μ and power loading of 10 watts. The spectra show considerable fine structure and comparison with data such as that in Figure 16 allows a straightforward assignment. The major components of the line profile consists of two peaks centered ∿ 286.7 eV and 288.7 eV corresponding to $\underline{C}$-CF and $\underline{CF}$ structural features. the components at ∿289.9 eV, 291.5 eV and 293.5 eV correspond to $\underline{CF}$-$CF_n$, $\underline{CF}_2$ and $\underline{CF}_3$ structural features. The component at ∿296 eV corresponds to the π → π* shake up satellite of the component centered ∿288.7 eV thus indicating the presence of a conjugated system. The small component at 285 eV almost certainly corresponds to a low level of hydrocarbon contamination since the relative intensity increases slightly on going to a higher take off angle.

Within experimental error the stoichiometries for the two take off angles are the same and slightly lower in fluorine content than the starting material. The presence of $\underline{CF}_3$ and $\underline{CF}_2$ structural features provides strong evidence for the extensive molecular rearrangement accompanying the polymerization reaction.

The structure and composition of the polymer deposited in the glow region remains essentially constant as a function of power and pressure over a considerable range of typical operating parameters. This Figure 19 shows the component analysis for films deposited on gold substrates in the glow region at a fixed pressure of 200μ and power inputs of 10, 20 and 30 watts.

With a knowledge of mean free paths as a function of kinetic energy it becomes possible to study the rate of film deposition as a function of the operating parameters. Typical data for the in situ deposition of polymer in a small reactor attached to the spectrometer are given in Figure 20. By monitoring the substrate $Au_{4f7/2}$ levels ($MgK_{\alpha1,2}$ radiation) the rates of deposition may be studied and the uniform nature of the deposition is evident from this data. For a five fold increase in power the deposition rate changes from ∿ 2Å per sec. to 3Å per sec. indicating that the rate of deposition is approaching the maximum at the pressure and flow rate employed for these experiments.

We now turn to the important question of the structure of the polymer as a function of site deposition. Employing the reactor configuration depicted in Figure 14, the polymer deposited at the reactor wall and at the center of the reactor has been investigated as a function of position as indicated. The glow region for a plasma excited at 200μ and 10 watts input power extends for a region ∿ 15 cms. in from of and ∿10 cm. to the rear of the coil region. Figure 21 shows the Fls, Cls and F2s levels for the polymer deposited in the glow region in front of the coil system. The spectra are essentially identical to those in Figure 18. The core level spectra for the film deposited at the edge of the glow region 45 cms. from the front of the reactor are distinctively

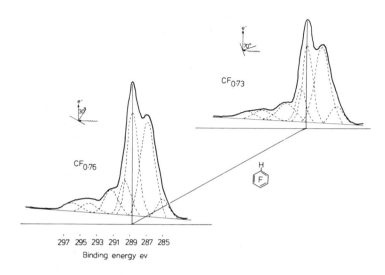

297  295  293  291  289  287  285
Binding energy ev

*Figure 18.   The C-1s core levels as a function of take-off angle for RF plasma polymer from pentafluorobenzene (10 W, 200 μ)*

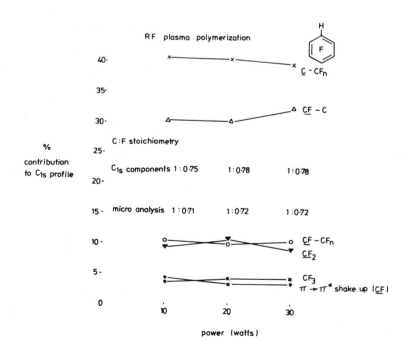

*Figure 19.   Component analysis of structural features in fluoropolymer films from RF plasmas excited in pentafluorobenzene (free-standing reactor (200μ))*

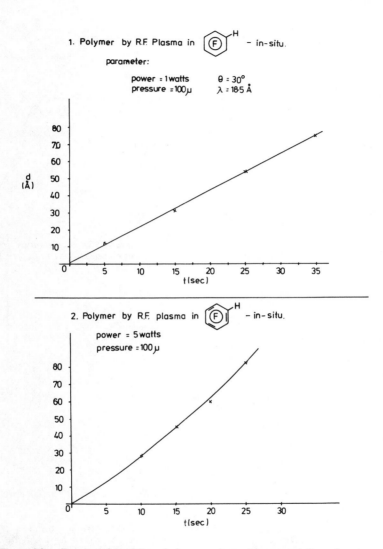

*Figure 20.   Rates of deposition of plasma polymers from pentafluorobenzene*

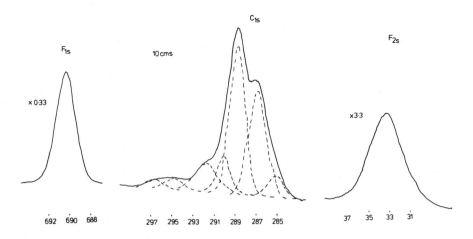

*Figure 21.* *The F-1s, C-1s, and F-2s levels for fluoropolymer produced from pentafluorobenzene (site of deposition 10 cm from input end of reactor in Figure 14)*

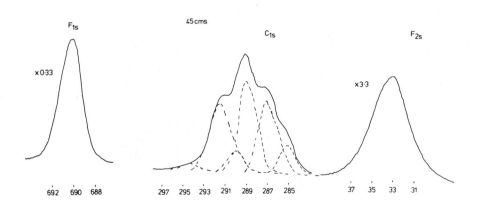

*Figure 22.* *The F-1s, C-1s, and F-2s levels for fluoropolymer produced from pentafluorobenzene (site of deposition 45 cm from input end of reactor in Figure 14)*

different as is apparent from a comparison of Figure 22 and Figure 21. Clearly the contribution from $CF_2$ structural features has significantly increased and this becomes more pronounced for films deposited in the non glow region 60 cms. from the front of the reactor (Figure 23). $CF_2$ structural features now dominate the spectra and Figure 24 shows how the derived stoichiometry of the polymer film depends on site of deposition. It may also be shown that the rate of deposition in the glow region is extremely rapid compared with the non glow regions. With the type of information provided by these ESCA studies the stoichiometry and main structural features for a given polymer film derived from a given starting material may be selected for a given application.

## An ESCA Investigation Of The Surface Chemistry Of The Nitration And Denitration Of Cellulose Materials

Introduction. The study of the nitration of cellulose particularly in the form of cotton linters has a long and chequered history (14). Indeed the discovery in 1833 by Bracconnot (15) that the reaction of nitric acid with cotton produced inflammable materials and the subsequent patent granted to Schonbein (16) in 1846 specifying the production of nitrocelluloses from cotton and mixed acids can be said to have materially changed the course of history. After the first (unscheduled) large scale explosion in 1847, work on the technical production of nitrocelluloses was somewhat restrained, after this mishap however the possibility of modifying burn rate for propulsion purposes was recognized and by 1867 the basis of the production of smokeless propellants for guns, rifles and shells was established and materially changed the course of history not only in the military sense but also in the pioneering days of the wild west.

The innovative use of camphor as a plasticizer for nitrocellulose by Parkes in 1862 (17) in the production of celluloid represents an important landmark in the emergence of polymer science and the observation of the solubility of nitrocelluloses and the production of Collodion and thin films undoubtedly accelerated the development of photography as we now know it. As an amusing side light the production of celluloid allowed the wider development of snooker and billiards. The plasticizing effect of camphor on nitrocellulose undoubtedly provided considerable impetus to the development of double and triple based propellants based on nitroglycerine/nitrocellulose formulations. The foundations of the production methods used for nitrocelluloses (Figure 25) were largely laid down in the early twentieth century and despite the technological, military and academic importance of the whole field and despite a voluminous literature much of it classified there are a number of unanswered questions. This is particularly so when the important question of surface chemistry is addressed. Thus the initial interaction of cellulose fibrils with nitrating media, the burn initiation, the stabilization,

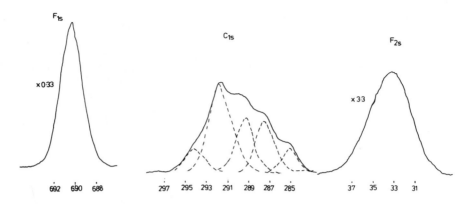

*Figure 23. The F-1s, C-1s, and F-2s levels for fluoropolymer produced from pentafluorobenzene (site of deposition 60 cm from input end of reactor in Figure 14)*

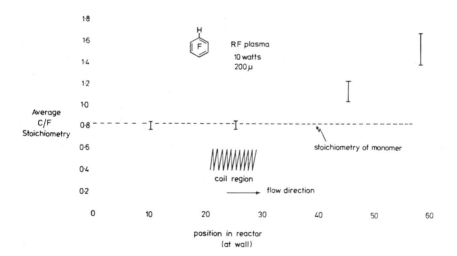

*Figure 24. Average carbon to fluorine stoichiometry as a function of site of deposition for fluoropolymers produced from pentafluorobenzene*

represent a few amongst many of the important aspects of the nitrocellulose field which depend on surface phenomena.

The wide ranging nature of both the technological and academic investigations of the cellulose and nitrocellulose field over the past hundred years or so poses certain difficulties since relevant data is widely dispersed over a number of disciplines. The most complete published survey (14) is now some 30 years old and pre-dates most of the relevant developments in both surface chemistry and mechanistic organic chemistry which are essential to the detailed understanding of heterogeneous processes at the surfaces of cellulose based materials.

The nitration of cellulose in the form of cotton linters is usually in mixed acid and on a three component diagram the zone of technical nitration is delineated by compositions in the range $HNO_3$ 17-27%, $H_2SO_4$ 66-50% and $H_2O$ 17-23% (14). Although nitrations can be conveniently effected on the laboratory scale with nitric phosphoric mixes recovery expense and corrosion problems normally preclude the use of such acid mixes on a commercial scale. Nitric acid – water mixes are not used since nitration is uneven and degradation and gelatinization of product arise at the high percentage nitric mixes required for reaction. Technical nitrations as previously noted are carried out in nitricsulphuric mixes since this affords a cheap method for production of high nitrogen content material with convenient recovery and with a lack of corrosion problems.

The main point of interest in technical nitration in mixed acid is the question of the degree of substitution and how this relates to the acid mix composition. Thus although Figure 26 shows a maximum degree of substitution of 3, in practice this is never attained and the maximum degree of substitution is ∿ 2.8. A definitive answer as to why this should be so has not thusfar been given since in a heterogeneous process the delineation between the possible explanations requires a technique which can clearly distinguish surface from bulk phenomena (cf. Figure 27).

The question of the limiting degree of substitution (usually established by a bulk nitrogen determination using a micro Kjeldahl technique) in nitrocelluloses could a priori be rationalized in terms of two extreme models.

The first can be attributed to the micro and macroscopic structure of the cellulose. Thus inhomogeneities in the bulk structure could conceivably give rise to accessible and inaccessible regions. Since nitration must depend on the diffusion of reagent throughout the bulk structure a further consideration is that there may well be a concentration profile throughout the structure. On this basis the less than maximum degree of substitution is attributable to an inhomogeneous bulk structure corresponding to regions of completely nitrated material and unreacted inaccessible regions.

An alternative and somewhat more plausible alternative is that since nitration is a reversible esterification process then

| Nitric acid | Nitric - Sulphuric | Nitric- Phosphoric |
|---|---|---|
| Uneven nitration degradation gelatinisation of product. | Rel. high nitrogen content. Cheap, ease of recovery and lack of corrosion problems. | High nitrogen content with small amount of degradation. Corrosion and recovery problems |

*Figure 25.   Nitration of cellulose: possible reaction pathways*

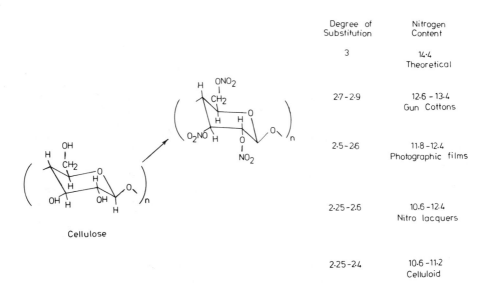

| | Degree of Substitution | Nitrogen Content |
|---|---|---|
| | 3 | 14·4 Theoretical |
| | 2·7 - 2·9 | 12·6 – 13·4 Gun Cottons |
| | 2·5 – 2·6 | 11·8 –12·4 Photographic films |
| | 2·25 – 2·6 | 10·6 –12·4 Nitro lacquers |
| | 2·25 –2·4 | 10·6 –11·2 Celluloid |

Cellulose

*Figure 26.   Nitrocelluloses DOS as a function of application*

1.  Maximum  Degree  of  Substitution  is  $\sim 2\cdot8$  why ?

    (a)   Inhomogeneities ?
        (i)  In  bulk  structure
        (ii)  In  nitrating  mix
        (iii)  Combination  of  (i)  and  (ii)

    (b)   Kinetic  and  Thermodynamic  reasons.

*Figure 27.   Points of interest in the technical nitration in mixed acid of cellulose*

the ∿ 2.8 degree of substitution typically observed represents
the equilibrium situation (cf. Figure 28).

In order to shed light on this situation we present here a
comparison of the surface and bulk nitration of cellulose in the
form of linters paper (17). Before discussing the data however it
is worthwhile briefly considering the background concerning the
bulk structure of cellulose and nitrocellulose.

The structure of cellulose or nitrocellulose may be discussed on three distinct levels (14, 18).

(i)   The structure of the macroscopic fibre generally
      about 20µ in diameter. Structural Botany has shown
      that the fibres are characterized by a number of
      distinct layers the result of the growth mechanism
      within the living plant.

(ii)  The nature of these layers which in the case of
      cellulose appear to be composed of fibrils within
      fibrils, ranging from about 1µ in diameter down to
      the so called ultimate microfibril of about 35Å dia-
      meter and thought to consist of chains of D glucose
      molecules 12 x 8 units in cross section.

(iii) The molecular level including the unit cell of the
      crystalline regions, molecular weight distribution
      etc.

For both cellulose and nitrocellulose a considerable body of
evidence has been generated on aspects (ii) and (iii) however
considerable controversy surrounds the information pertaining to
(ii). Lewis (18) has succinctly described the current state of
knowledge in which the three models depicted in Figure 29 have
been discussed.

Most of the older literature discusses the chemistry of
cellulose in terms of the well known fringed micelle model (a).
The main feature of this model is an extended crystalline structure interspersed with amorphous regions. X-ray studies and more
recent chemical evidence provide little support for such a model
and an alternative, microfibril model (b) has been developed. The
fundamental aspect of this model is that virtually 100% of the
cotton is in the crystalline form and consists of ultimate fibrils
3-5 µ in diameter. The amorphous or non crystalline regions in
this model are associated with the surface regions of these
microfibrils. The varying degree of accessibility to chemical
reagents is explained by postulating that the number of active
centers on any particular cellulose molecule will depend on
whether it lies along the corner, edge, the face or the interior
of the microfibril. The larger the fibril diameter the smaller
the proportion of non crystalline material present. In the case
of nitrocellulose experimental evidence has been presented which
is basically incompatible with such a model. A compromise between
the fringed micelle and totally crystalline microfibril theory
has recently been suggested, the main feature of the model (c)
being that amorphous regions between crystallites run in continu-

Older Theory

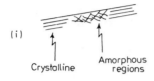

(i)

Crystalline          Amorphous
                     regions

Accessible and
Inaccessible regions.

(ii)  Nitrating Mix
      as Nitration proceeds
      $HNO_3 \downarrow$   $H_2O \uparrow$

      $\therefore$ tendency to lower nitration
      level and denitration of
      first formed nitrate.

*Figure 28.    Inhomogeneities leading to reduced DOS for bulk materials*

(b)

(i)  $\sim\!\!\!\sim O\text{-}H + 2HNO_3 \overset{K_1}{\rightleftharpoons} \sim\!\!\!\sim O\text{-}NO_2 + HNO_3.H_2O$

$HNO_3.H_2O \overset{K_2}{\rightleftharpoons} HNO_3 + H_2O$

$\sim\!\!\!\sim O\text{-}H + HNO_3 \overset{K}{\rightleftharpoons} \sim\!\!\!\sim O\text{-}NO_2 + H_2O$

$K = \dfrac{K_1}{K_2}$   Nitration - denitration equilibrium
                        rapidly established ?

(ii)  Competitive sulphonation

$\sim O\text{-}H + H_2SO_4 \rightleftharpoons \sim O\text{-}SO_3H + H_2O$

(iii)  Work up **procedure**

Hydrolysis

$\sim O\text{-}NO_2 + H_2O \rightleftharpoons \sim O\text{-}H + HNO_3$

*Figure 29.    Kinetics and equilibria involved in nitrations in mixed acids*

ous veins through the polymer and are not randomly distributed as
in the fringed micelle theory. The available data particularly
that relating to processing of nitrocellulose is compatible with
such a model with the microfibrils as in (b) being the crystallites with the intervening regions being filled with amorphous
material which differs little in density from that of the crystalline regions (see Figure 30).

In any heterogeneous process it is the surface which provides a "window" on the reaction however as we have previously
noted the studies to date have focussed on the bulk chemistry.
Miles (14) has summarized much of the important literature in the
period to circa 1950 and evidence has been presented that under
certain conditions equilibrium between nitration and denitration
is established on the typical time scale involved for nitration.
Very recent (13) C nmr studies have confirmed (19) an earlier
chemically based inference (20) that in a nitrated cellulose of a
given degree of substitution (DOS) the partial esterification
ratios (primary vs secondary $ONO_2$) are in the order 6 > 3 > 2. It
is important to recognize that this almost certainly corresponds
to an equilibrium order of reactivity since it is known that in
homogeneous reactions the rate constants for denitration of primary esters is substantially smaller than for nitration under a
given set of conditions (21). The chemical reaction sequence
involving treatment of the nitrated material with NaI in
acetonylacetone provides information on the total level of primary nitrate ester groups in the sample (20). By contrast the
high resolution (13) C nmr studies relate only to the soluble
fraction of nitrated material. The information derived from
these chemical and spectroscopic studies are valuable but still
leave important gaps in the inforamtion required to fully understand the nitration of cellulose in mixed acids. In order to shed
new light on this technologically and academically important area
we present here a detailed comparison of the surface and bulk
nitration and denitration of cellulose materials in terms of
degree of substitution and relative rates of reactions.

Experimental. Since the nitration of cellulose in the form
of linters paper may well be unfamiliar to many people we present
here a brief synopsis of the procedures involved. The general
procedures are indicated schematically in Figure 31. Commercially produced linters paper (Hercules Powder Co.) of Shirley fluid-
ity 8.8 and approximate degree of polymerization 1100 was vacuum
dried in an oven at $60°$ for 2 hrs. and stored over $P_2O_5$. This
provides a starting cellulose sample with < 2% water. Nitrations
and denitrations were accomplished by immersion in the appropriate acid mix (of sample strips (4 cm x 4 cm) for a given period
of time using an apparatus such as that depicted in Figure 32.
Bulk nitrogen determinations have been carried out using a micro
Kjeldahl.

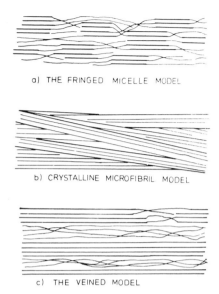

a) THE FRINGED MICELLE MODEL

b) CRYSTALLINE MICROFIBRIL MODEL

*Figure 30. Possible models for the structure of cellulose and nitrocellulose (after Ref. 18)*

c) THE VEINED MODEL

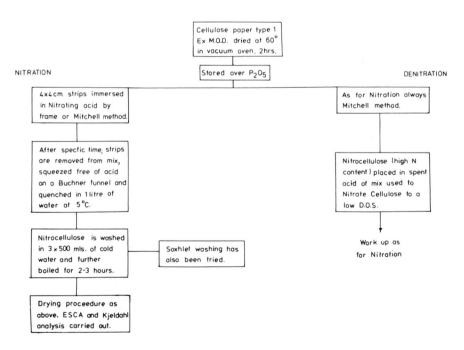

*Figure 31.   Nitration and denitration procedures (nitrations according to Mitchell procedure) (22)*

Preliminary Analysis. The core level (C1s, O1s, N1s) spectra for a typical nitrocellulose sample which has been
commercially produced is shown in Figure 33.  The C1s spectrum
shows 3 distinctive components (once a standard line shape analysis has been carried out).  Thus the central component at $\sim$287.4
eV arises predominantly from the carbons C2, C3 and C6 bearing the
nitrate ester functionality with contributions from C4 and C5.  C1
uniquely attached to two oxygens in the cyclic hemiacetal formulation of the β-Dglucopyranose ring is at highest binding energy.
The component at 285 eV arises from extraneous hydrocarbon which
we will show later is confined to the very surface of the sample.
In cellulose samples irrespective of source or type (e.g. linters
or staple fibres) the hydrocarbon signal observed is variable but
inevitably present.  The N1s signal consists of an intense high
binding energy component and (cf. Fig. 3) comparison with model
systems unambiguously identifies this as originating from the
nitrate ester groups $-O-NO_2$.  In industrial scale nitrations a low
binding energy component is often observed at $\sim$406 eV and thus is
associated with nitrite ester groups.  In the laboratory scale
nitrations where conditions can perhaps be more precisely controlled such structural features are at a much lower level.
Conventional nitrite traps (urea, ascorbic acid etc.) dissolved
in the nitrating mix to obviate the possibility of formation of
nitrite esters does not seem to remove the very low levels of such
structural features and it could be that they arise as a result of
reactions occurring during stabilization and storage.  The O1s
levels are an unresolved broad peak.

With a knowledge of sensitivity factors for the various core
levels it is possible to straightforwardly work out the degree of
substitution DOS (average number of nitro-ester functionalities
per glucose residue).  Thus the integrated C1s/N1s area ratios
(excluding the extraneous hydrocarbon component) yields a DOS of
2.3 for the phosphoric/nitric nitrating mix identical with that
determined from micro Kjeldahl bulk analysis.  The total C1s/O1s
ratio indicates that there is little residual water in the
nitrated sample.

With fibrillar samples such as linter papers any information
on vertical inhomogeneities into the sample may only be inferred
by looking at different levels corresponding to different escape
depths.  Clearly the good agreement between bulk and surface
analyses suggests that the sample is uniformally nitrated and the
extraneous hydrocarbon must therefore be localized at the surface
perhaps in the form of a patched overlayer since in most cases
there may well be significantly less than a monolayer present.
One way of establishing the purely surface nature of this hydrocarbon (since the usual angular dependent studies are not feasi-
ible) is to compare DOS averaged over different sampling depths.
Whereas for MgK$_\alpha$ the typical sampling depth will be say $\sim$50Å for
TiK$_\alpha$ with photon energy 4510 eV a figure of $\sim$300Å would be more
appropriate.

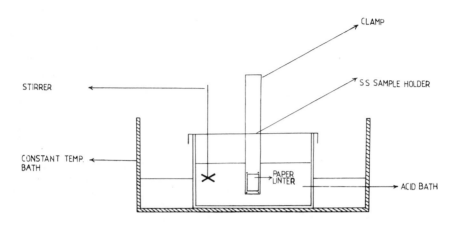

*Figure 32.    Nitrating apparatus used in ESCA studies*

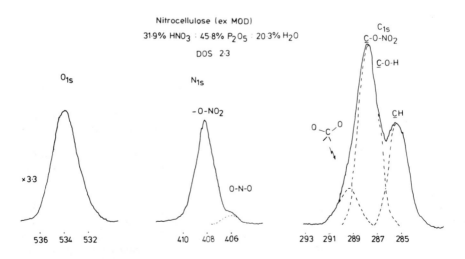

*Figure 33.    Core-level spectra for commercially produced sample of nitrocellulose*

Figure 34 shows N1s and C1s spectra for cellulose linters paper nitrated for 300 secs. in a low nitrating mix. Whereas the $MgK_\alpha$ spectra for the C1s levels show a significant contribution from the extraneous hydrocarbon the corresponding lineshape analysis for the $TiK_\alpha$ spectra is described with no contribution from hydrocarbon since even monolayer coverage ($\sim 5\text{\AA}$) would contribute a negligible contribution to the C1s levels for electrons having a mean free path at least an order of magnitude greater ($\sim 90\text{\AA}$). This shows the great value of having a variable photon source routinely available.

The relative sensitivity of nitrocelluloses to photochemical decompositions is known and it is therefore necessary to investigate the sensitivity to interrogation by means of ESCA of nitrated and denitrated cellulose samples.  It may readily be shown that on the typical time scale for the ESCA investigations photochemical degradation is negligible.  At the typical dose rates involved (typical X-ray power $\sim 150$ watts) significant signs of decomposition require irradiation periods of $> 2$ hours. The main reaction appears to be photo-reduction.  Thus the high binding energy component in the N1s spectrum appropriate to $-O-NO_2$ structural features is accompanied in the case of material subjected to irradiation for extended periods by a small peak at low binding energy attributable to $-\overset{O}{\underset{}{C}}-NH$ functionality.  Even after 5 hours irradiation however this component still only represents a small fraction of the total N1s spectrum.  We may conclude from this that X-ray degradation is negligible during the time scale of a typical ESCA investigation.

Detailed Studies Of Nitration And Denitration.  For a nitrating mix of a given composition one of the most important features of interest is the question as to how rapidly an equilibrium DOS is established in the surface regions accessible to ESCA.  As an example Figure 35 shows core level spectra for samples of the same batch of linters papers nitrated for differing periods in a nitrating mix consisting of 75% $H_2SO_4$: 22.2% $HNO_3$: 2.8% $H_2O$. The N1s levels, which provide a ready means of following the nitration, remain constant in intensity from reaction times of 1 sec. to 1 hour.  This indicates that on the ESCA depth scale, equilibrium is very rapidly established.  This is not entirely unexpected since the diffusion of nitrating mix into the outermost $50\text{\AA}$ or so of the cellulose fibrils is expected to occur rapidly.  A similar comparison is shown in Figure 36 for samples studied by means of the harder $TiK_\alpha$ X-ray source.  Here the sampling depth will be several hundred Angstroms yet the spectra recorded after reaction times of 1 sec. and 300 sec. are closely similar.  It is clear therefore that the equilibrium DOS is rapidly established in the surface regions.

The gross features of the equilibria which are involved in determining the overall DOS in the surface regions of cotton fibrils is outlined schematically in Figure 37.  For mixed acid

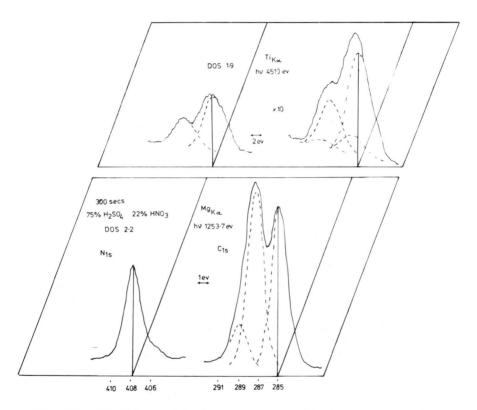

*Figure 34.   The N-1s and C-1s spectra for nitrocellulose produced in mixed acid*
*(MgK$_{\alpha 1,2}$ and TiK$_{\alpha 1,2}$ spectra)*

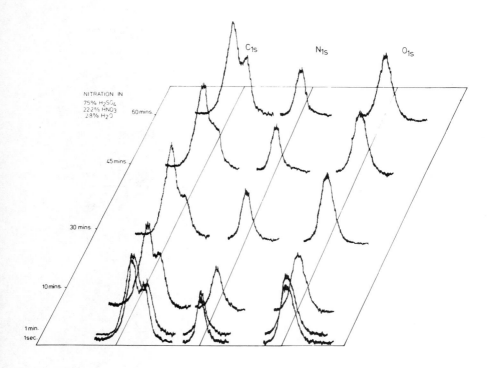

*Figure 35.  Core-level spectra as a function of time for nitration in mixed acid*

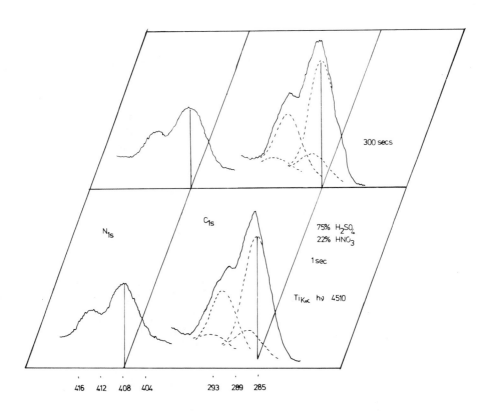

*Figure 36. The TiK$_\alpha$ spectra for nitration of cellulose for 1 s and 300 s in mixed acid*

nitrating mixes it has been proposed that the sulphuric acid
component does not diffuse into the bulk (14) and this could
obviously lead to differences in DOS for the surface and bulk
regions of samples since only in the surface regions is sulphonation competitive with nitration.   An important question which
ESCA is potentially capable of answering is whether sulphate
esters may be detected in the surface regions of nitrated material.

     $MgK_{\alpha1,2}$ spectra for the Nls, S2p and Cls regions of cotton
linters for nitrated material are shown in Figure 38.   A low level
S2p signal is detected and the high binding energy of $\sim$ 169 eV
identifies this as arising from sulphate ester groups.   It is
clear therefore that sulphate esters are formed in the very
surface regions.   Comparable studies with the $TiK_\alpha$ X-ray source
with a larger sampling depth shows virtually no evidence for
sulphate esters and ESCA therefore uniquely demonstrates the
surface nature of such groups.

     The final DOS for a nitrocellulose, at least as far as the
surface regions is concerned, depends on nitration, denitration,
sulphonation equilibria.   The fact that the DOS is rapidly
established suggests that denitration is competitive with nitration whilst the fact that even in high sulphuric mixes the DOS is
still appreciable illustrates that nitrate ester formation is
more facile than sulphate ester formation.   The rapidity with
which denitration-nitration equilibria are established in the
surface regions is nicely illustrated by the data displayed in
Figure 39.   Thus for starting material, DOS 2.7 denitration of the
outermost few tens of angstroms is rapid and dependent on acid
mix.   In 79.1% $HNO_3$ denitration in 1 sec. is to DOS 2.3 since this
mix is close to the minimum necessary for nitration to be effected
(77.8% $HNO_3$ corresponds to the monohydrate) whilst for 84.4% $HNO_3$
denitration is to a DOS of 2.5 on a 1 sec. time scale.   Corresponding spectra with a $TiK_\alpha$ X-ray source indicates that the degree
of denitration is slightly lower than for the $MgK_\alpha$ X-ray source.

     The extreme sensitivity of ESCA in the detection of the
initial stages of reactions is nicely displayed by the comparative data given in Figure 40.   This shows the nitration and
denitration of cellulosic samples.   The substantial secondary
shift of the nitrate ester group is shown by the shift to high
binding energy compared with the extraneous hydrocarbon peak ($\Delta E$
2.3 eV for the $\underline{C}$-O-$NO_2$ component compared with $\Delta E$ 1.7 eV for
cellulose itself.   The ESCA data presented, thusfar therefore
establishes that nitration-denitration-sulphonation equilibria
are rapidly established in the surface regions.

     Systematic studies have been made of nitrations in both
nitric-phosphoric and nitric-sulphuric mixes, both from a surface
and bulk point of view.   For nitric phosphoric mixes the DOS
depends on the composition and there is little evidence for
formation of phosphate esters.   The DOS in the surface region
therefore represents the equilibrium between nitration and de-

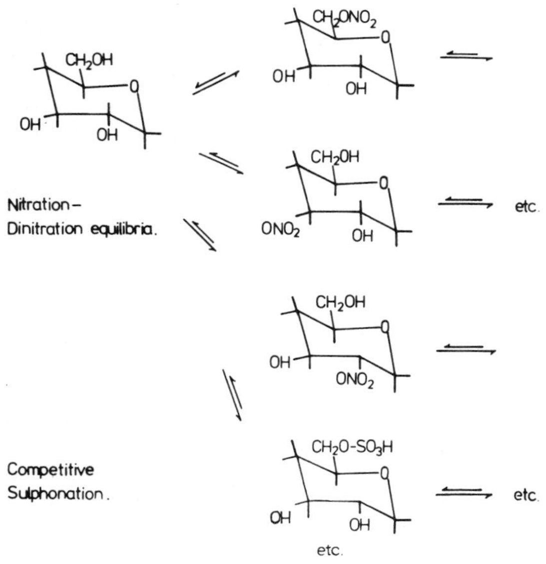

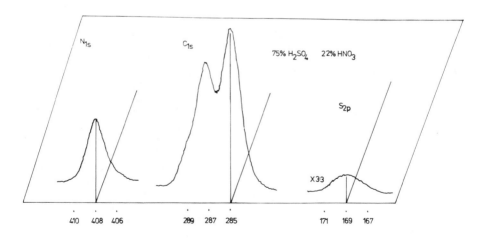

Figure 37.   *Competitive equilibria in nitrations in mixed acids*

Figure 38.   *The N-1s, C-1s, and S-2p core-level spectra for nitrated material (detection of sulfate esters)*

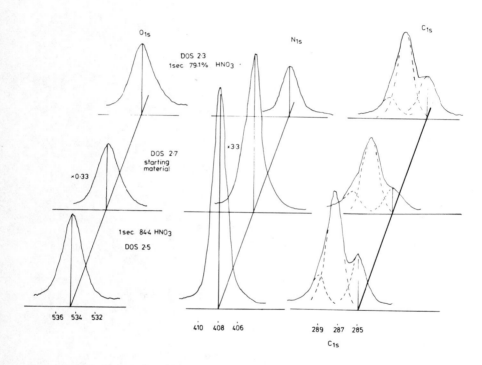

*Figure 39.   Denitration of nitrocellulose DOS 2.7 as a function of denitrating mix* ($MgK_{\alpha_{1,2}}$)

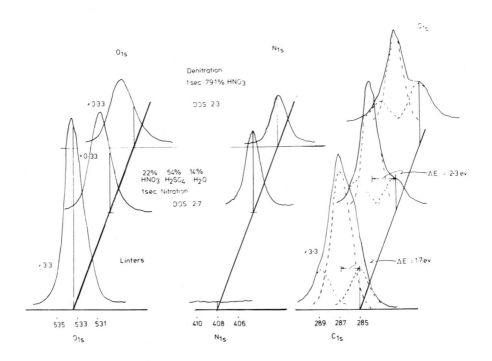

*Figure 40.   Comparison of surface nitration and denitration of cellulose*

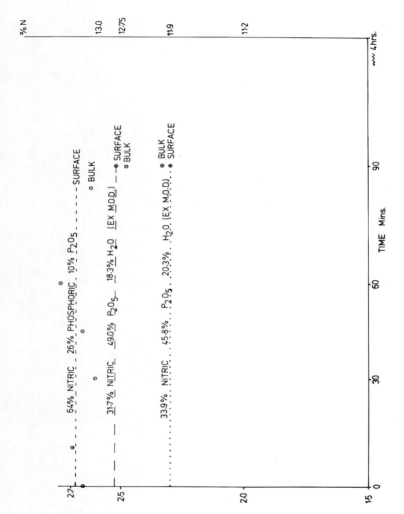

*Figure 41.  Comparison of surface and bulk DOS as a function of time for the nitration of cellulose in nitric–phosphoric acid mixes*

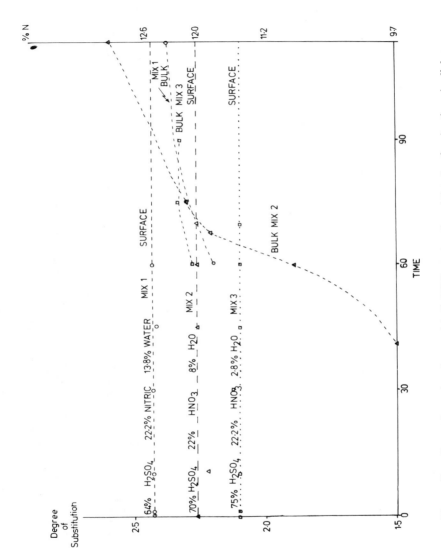

Figure 42.    Comparison of surface and bulk DOS as a function of time for the nitration of cellulose in nitric–sulfuric acid mixes

nitration. Since equilibration in the bulk requires diffusion of reagents then naturally establishment of this equilibrium occurs on a longer time scale than for the surface. The important feature however is that the DOS obtained from bulk analyses is essentially the same as for the surface regions (Figure 41). The situation with regard to nitrations in mixed acid is considerably more complex and representative data for bulk and surface nitrations in mixed acid are given in Figure 42.

Considering firstly the data for the 64% sulphuric mix the surface DOS as assessed by ESCA utilizing at $MgK_\alpha$ source is essentially established after 1 sec. exposure and remains constant thereafter. The diffusion of nitrating mix into the bulk leads to a time dependent DOS which tends to approach that of the surface. As the sulphuric content is increased and water content decreased the surface DOS decreases. In terms of competitive equilibria we might anticipate that as the sulphuric content increases the formation of sulphate esters becomes relatively more favored. The suggestion has been made that in mixed acid nitrations the sulphuric acid does not penetrate the bulk (14) and on this basis we might anticipate significant differences in surface and bulk chemistries irrespective of time scale. ESCA has demonstrated this in detail for the first time. Thus for both the 70% mix and the 75% sulphuric mixes the DOS of the surface is actually lower than for the bulk and work recently completed has shown that with a $TiK_\alpha$ X-ray source the DOS is also close to that for the bulk.

It is clear therefore that ESCA provides a new dimension to this complex problem of the nitration-denitration of cellulose materials and the work described here provides a strong basis for the study of the even more complex systems represented by double and triple based propellant formulations.

## Acknowledgements

Thanks are due to SRC for provision of equipment. The work described in this article is largely based on the research programmes in Durham of Alan Harrison, Zaki Abrahman and Peter Stephenson.

## Literature Cited

1.    Clark, D.T., "ESCA Applied To Polymers" in Advances in Polymer Science, 1978, 24, 126, Springer Verlag, Heidelberg.

2.    Cf. Clark, D.T., "Chemical Aspects Of ESCA" in Electron Emission Spectroscopy, Ed. W. Dekeyser, D. Reidel, Dordrecht, Holland 1973.

3.    Clark, D.T., "ESCA Applied To Organic and Polymeric

Systems" in Handbook of X-ray and Ultra-Violet Photoelectron Spectroscopy, Ed. D. Briggs, Heydon & Son Ltd., London
1977.

4.   Clark, D.T., Cromarty, B.J. and Dilks, A., J. Polym. Sci.,
     Polym. Chem. Ed., 1978, 16, 3173.

5.   Clark, D.T. and Harrison, A., J. Polym. Sci., to be submit-
     ted.

6.   Clark, D.T. and Thomas, H.R., J. Polym. Sci., Poly. Chem.
     Ed., 1978, 16, 791.

7.   (a) Clark, D.T., Dilks, A. and Shuttleworth, D., J. Elec.
         Spec., 1978, 14, 247.

     (b) Clark, D.T. and Shuttleworth, D., J. Elec. Spec.,
         1979, 17, 15.

8.   (a) Clark, D.T., Dilks, A. and Thomas, H.R., J. Polym.
         Sci., Polym. Chem. Ed., 1978, 16, 1461.

     (b) Clark, D.T., Dilks, A., Thomas, H.R. and Shuttleworth,
         D., J. Polym. Sci., Polym. Chem. Ed., 1979, 17, 627.

9.   Clark, D.T., Dilks, A. and Harrison, A., J. Polym. Sci.,
     Polym. Chem. Ed., to be submitted.

10.  Cf. Yasuda, H. and Hirotsu, T., J. Polym. Sci., Polym.
     Chem. Ed., 1978, 16, 743.

11.  Clark, D.T., Dilks, A. and Shuttleworth, D., "The
     Application of Plasmas to the Synthesis and Surface Modifi-
     cation of Polymers" in Polymer Surfaces, Ed. D.T. Clark and
     W.J. Feast, J. Wiley, London 1978.

12.  (a) Clark, D.T. and Shuttleworth, D., J. Polym. Sci.,
         Polym. Chem. Ed., 1978, 16, 1093.

     (b) Clark, D.T. and Shuttleworth, D., J. Polym. Sci.,
         Polym. Chem. Ed., 1979, 17, 1317.

     (c) Clark, D.T. and Shuttleworth, D., European Polym. J.,
         1979, 15, 265.

     (d) Clark, D.T. and Shuttleworth, D., J. Polym. Sci.,
         Polym. Chem. Ed., 1980, 18, 27.

     (e) Clark, D.T. and Shuttleworth, D., J. Polym. Sci.,
         Polym. Chem. Ed., 1980, 18, 407.

13. Clark, D.T. and Abrahman, Z., J. Polym. Sci., Polym. Chem. Ed., to be submitted.

14. Miles, F.D., "Cellulose Nitrate", Oliver and Boyd, London 1955.

15. Braconnot, H., Ann. Chim., 1819, 12, 172 and subsequent papers.

16. Cf. MacDonald, A., "Historical Papers on Modern Explosives", Oliver and Boyd, London 1912.

17. Clark, D.T. and Stephenson, P.J., in "Nitrocellulose Characterization and Double-Base Propellant Structure", Ed. T.J. Lewis, to be published.

18. Lewis, T.J., MOD Report PERME, TR109, 1979.

19. Wu, T.K., Macromolecules, 1980, 13, 74.

20. Murray, G.E. and Purves, C.B., J.A.C.S., 1940, 62, 3197.

21. Svetlow, B.S., Kinet Katal., 1972, 13, 792.

22. Mitchell, R.L., Ann. Chem., 1949, 21, 1496.

RECEIVED January 28, 1981.

# X-Ray Photoelectron Spectroscopy for the Investigation of Polymer Surfaces

A. DILKS[1]

Department of Chemistry, University of Durham, South Road,
Durham, DH1 3LE, England

Since the binding energies of electrons in core levels are characteristic of a particular element, X-ray photoelectron spectroscopy (XPS or ESCA) in its simplest form can be employed to produce an elemental analysis of a polymer surface. However a more detailed consideration of the information levels available from XPS has led to more comprehensive structural determinations for polymers, in terms of both the structural features present at the surface, and how the outermost few layers of the material differ from the subsurface and bulk sample. In this article a description of the application of XPS to the investigation of polymer surfaces is presented with the aid of some current examples. An emphasis is placed on the measurement of absolute and relative binding energies and, absolute and relative signal intensities, which is often sufficient to characterize a given polymer surface. For certain exceptions however it is also necessary to consider the extra information which may be derived from the observation of shake-up and sample charging phenomena. It is beyond the scope of this article to discuss these latter phenomena but detailed descriptions can be found elsewhere ($\underline{1}$, $\underline{2}$).

## Absolute and Relative Signal Intensities

Analytical Depth Profiling. Figure 1 shows data for four photon sources available for use in XPS, along with their photon energies and natural linewidths. The two most commonly used are $MgK\alpha_{1,2}$ and $AlK\alpha_{1,2}$ with photon energies of $\sim 1254$ and $\sim 1487 eV$ respectively. Although these X-rays may penetrate many thousands of Ångstroms into the sample, the mean free paths of the

[1] Current address: Xerox Corporation W-114, 800 Phillips Road, Webster, New York 14580 U.S.A.

0097-6156/81/0162-0293$06.25/0

| Photon Source | Energy (eV) | Linewidth (eV) | K.E.($C_{1s}$) (eV) | $\lambda(C_{1s})$ (Å) | Sampling Depth (Å) |
|---|---|---|---|---|---|
| $Mg_{K_\alpha}$ | 1254 | 0.7 | 969 | ~14 | ~40 |
| $Al_{K_\alpha}$ | 1487 | 0.85 | 1202 | ~23 | ~70 |
| $Ti_{K_\alpha}$ | 4510 | 2.0 | 4225 | ~100 | ~300 |
| $Cr_{K_\alpha}$ | 5417 | 2.1 | 5132 | ~200 | ~600 |

*Figure 1. Photon sources of use in XPS and typical electron mean-free paths*

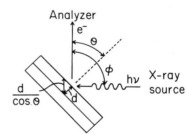

*Figure 2. Substrate/overlayer model and schematic of angles involved in XPS*

photoemitted electrons are rather short and it is therefore only
those from the outermost layers of the material which escape
without energy loss. Taking the C1s core level as an example for
polymers, the kinetic energy of the photoemitted electron is
shown in Figure 1 for each photon source. For $MgK\alpha_{1,2}$ and $AlK\alpha_{1,2}$
radiation, the mean free paths of these emitted electrons,
through thin films of polyparaxylyene, have been measured as $\sim 14\text{Å}$
and $\sim 23\text{Å}$ respectively ($\underline{3}$). These correspond to sampling depths
of $\sim 40\text{Å}$ and $\sim 70\text{Å}$ respectively, where the sampling depth is
defined as the depth from which 95% of the signal intensity
derives. The mean free path of a photoemitted electron is a
strong function of its kinetic energy, and this depends on the
photon source employed, as well as the core level being studied.
Clearly the sampling depth may be increased, if desired, by
employing a more energetic photon source, although the larger
natural linewidths of these sources result in decreased resolution in the spectrum.

The signal intensity derived from a given core level is
therefore dependent upon the mean free path for the photoemitted
electron and for a thick homogeneous sample is given by equation
(1).

$$I = f(\theta)F\alpha K\lambda N = f(\theta)I^{o}N \tag{1}$$

F is the X-ray flux, $\alpha$ is the cross section for photoionization in
the direction of the analyzer, K is a spectrometer dependent
factor, $\lambda$ is the mean free path of the photoemitted electrons and
N is the number of atoms per unit volume on which the core level
resides. $I^{o}$ is defined as F K . $f(\theta)$ is included in equation (1)
to describe how the absolute signal intensity varies with $\theta$, the
electron take-off angle, which is defined as the angle between the
normal to the sample surface and a line joining the sample and the
analyzer entrance slit, (see Figure 2). $f(\theta)$ depends upon
geometric factors ($\underline{1}$) such as the spectrometer configuration and
sample size and shape, and is often termed the angular response
function of the configuration.

For an inhomogeneous sample the situation which is most
often encountered is that shown in Figure 2, in which a modified
surface layer of thickness d overlays an unmodified bulk. This is
usually referred to as the substrate/overlayer model. The signal
intensity for a core level in the surface layer is given by
equation (2).

$$I_{s} = f(\theta)I^{o}_{s}N_{s}(1 - e^{-d/\lambda\cos\theta}) \tag{2}$$

while that for the bulk material is given by equation (3).

$$I_{b} = f(\theta)I^{o}_{b}N_{b} . e^{-d/\lambda\cos\theta} \tag{3}$$

The factor of $1/\cos\theta$ in the exponents of equations (2) and (3) arises as a consequence of the fact that, as can be seen from Figure 2, the effective overlayer thickness, as seen by the analyzer, increases as $1/\cos\theta$ as $\theta$ is increased from $0^{\circ}$ to $90^{\circ}$. This has the effect of relatively enhancing surface features as $\theta$ approaches $90^{\circ}$ (i.e. at grazing electron take-off). In terms of relative signal intensities, ratioing the intensities at a given angle $\theta$ of either two signals, i and j, deriving from the surface or two signals, i and j, deriving from the bulk allows d to be determined if the stoichiometry $Ni/Nj$ is known. ($f(\theta)$ cancels and the total relative sensitivity of i and j, $I^{\circ}i/I^{\circ}j$ can be determined for the particular spectrometer being used from standard homogeneous samples containing i and j). This is particularly useful for surface modifications involving fluorine where i and j can be the F1s and F2s levels - the F2s levels being essentially corelike - therefore the "stoichiometry" is unity. For other systems however, (for example surface oxidation of polymer films), the stoichiometry in the surface layer is not straight-forwardly obtainable and measurement of relative intensities at more than one take-off angle is necessary to obtain d <u>and</u> the stoichiometry. Measurement of the absolute signal intensities as a function of electron take-off angle can also provide valuable information in this respect

Figure 3 shows the absolute C1s signal intensity derived from a homogeneous polymer film as a function of the electron take-off angle, $\theta$, in an AEI ES 200B X-ray photoelectron spectrometer (4). For this spectrometer the angle between the X-ray source and analyzer, $\phi$ is fixed at $90^{\circ}$. It is clear from Figure 2 that if the sample is turned to face away from either the X-ray source or analyzer all signal intensities will fall to zero. In Figure 3 therefore, at $\theta \sim 90^{\circ}$ the sample faces away from the analyzer and at $\theta \sim -25^{\circ}$ the sample has completely eclipsed the X-ray source. The maximum signal intensity for this particular configuration and a homogeneous polymer occurs at $\theta \sim 35^{\circ}$ (4). This value of $35^{\circ}$ is rather important and the overall shape of the curve in Figure 3 is the angular response function described by $f(\theta)$.

In contrast to a homogeneous sample, where the signal intensity arising from a given core level is described by equation (1), the signal intensities arising from core levels in either the surface or bulk of an inhomogeneous sample, given by equations (2) and (3) depend on a product of two terms involving $\theta(\underline{1})$. Figure 4 shows plots of these expressions for the surface (overlayer) and bulk (substrate) intensities respectively, using C1s levels studied with $MgK\alpha_{1,2}$ radiation as examples. In each case the curves are calculated for overlayer thicknesses of 1, 2 and 3 monolayers. Although the data in Figure 4 refers specifically to the spectrometer and sample configuration employed in this work, the major features are generally applicable to any similar arrangement.

Considering firstly the substrate or bulk levels, we have

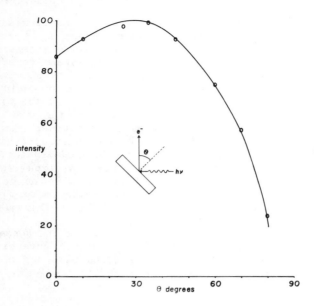

*Figure 3.   Variation of the absolute C-1s signal intensity of a homogeneous sample with θ*

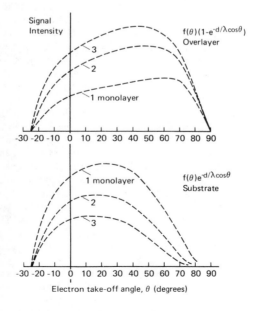

*Figure 4.   Variation of the absolute signal intensities of signals in the substrate and overlayer with θ*

seen from Figure 3 that with no overlayer the maximum intensity
occurs at $\theta \sim 35^{\circ}$. As the overlayer thickness is increased
however, in Figure 4, the angle giving a maximum intensity from
the substrate decreases such that for 3 monolayers coverage the
signal maximizes at $\theta \sim 10^{\circ}$. For the surface overlayer on the
other hand we know that an infinitely thick layer would give a
maximum intensity at $\theta \sim 35^{\circ}$. As the thickness of the overlayer
decreases however in Figure 4 the angle giving the maximum
intensity from the overlayer increases such that for 1 monolayer
the signal maximizes at $\theta \sim 65^{\circ}$. Figure 4 demonstrates therefore
how we can determine whether a core level is predominately in the
surface or the bulk by establishing whether its signal intensity
is a maximum at an electron take-off angle above or below the
value for a similar homogeneous sample. Furthermore, in principle it is possible to determine the overlayer thickness from
this angle alone (5).

It is convenient at this stage to illustrate the potential of
the measurement of absolute and relative signal intensities by
reference to Figure 5. The following discussion has been kept
deliberately qualitative, although accurate analysis of the
signal intensities given do indeed provide information relating
to the depth of reaction and stoichiometry of the modified surface
layer. Figure 5 compares signal intensity data for three polymer
films oxidized in different ways (6). The first is polyethylene
treated in a corona discharge in air. The second is hot-pressed
polyethylene and the third is polystyrene oxidized in a low-
power, inductively coupled radiofrequency plasma excited in a low
pressure of oxygen. For each is given the absolute C1s and O1s
intensities and the relative O1s/C1s intensity, at electron take-
off angles of $35^{\circ}$ and $70^{\circ}$, respectively. The latter set therefore
correspond to a near grazing electron take-off angle at which
surface features will be relatively enhanced. The final column of
numbers in Figure 5 compares the O1s/C1s intensity ratio at $\theta =
70^{\circ}$ and $\theta = 35^{\circ}$.

In each case the behavior of the absolute C1s intensity is
similar since the C1s signal derives from both surface and bulk,
and will therefore tend to follow $f(\theta)$. The behavior of the O1s
signal intensity however differs for the three samples. For the
corona treated sample the O1s/C1s intensity ratio is independent
of the take-off angle suggesting the degree of oxidation is
homogeneous within the XPS sampling depth ( $\sim 50\text{Å}$). This is
illustrated schematically as a plot of degree of oxidation versus
depth into the film on the far right of the figure. We can
therefore propose that oxidation in the corona proceeds via a
mechanism involving rather energetic species which may penetrate
at least $100\text{Å}$ into the polymer.

For the hot-pressed polyethylene the O1s signal intensity
remains approximately the same in going from $\theta = 35^{\circ}$ to $\theta = 70^{\circ}$
and the relative O1s/C1s intensity increases by $\sim 30\%$. It can be
shown therefore that oxidation for this sample is confined to the

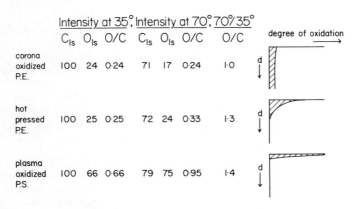

| | Intensity at 35°, | | | Intensity at 70°, | | | 70°/35° | degree of oxidation |
|---|---|---|---|---|---|---|---|---|
| | $C_{1s}$ | $O_{1s}$ | O/C | $C_{1s}$ | $O_{1s}$ | O/C | O/C | |
| corona oxidized P.E. | 100 | 24 | 0·24 | 71 | 17 | 0·24 | 1·0 | |
| hot pressed P.E. | 100 | 25 | 0·25 | 72 | 24 | 0·33 | 1·3 | |
| plasma oxidized P.S. | 100 | 66 | 0·66 | 79 | 75 | 0·95 | 1·4 | |

*Figure 5.   Signal intensity data comparing three polymer films oxidized in different ways*

top two or three monolayers of material, as illustrated by the schematic on the far right of the figure.

Finally, for the plasma oxidized polystyrene film, the absolute Ols signal intensity increases quite dramatically with electron take-off angle in going from $\theta = 35^{\circ}$ to $\theta = 70^{\circ}$. By comparison with Figure 4 this behavior indicates that the oxygen atoms are located in the outermost monolayer of the sample. This conclusion is reinforced by the $\sim$ 40% increase in the relative Ols/Cls intensity on going to the near grazing electron take-off angle. On the right of the figure is a plot of degree of oxidation versus depth into the film and this is consistent with a reaction mechanism involving oxygen atoms, which have an extremely short mean free path in hydrocarbon polymers.

Kinetic Studies. The possibility of employing XPS for depth profiling and for investigations of the kinetics of surface modifications of polymers has been demonstrated in a study of the interaction of polymers with inductively coupled radiofrequency plasmas excited in a variety of inert gases (4, 7). For example, Figure 6(a) shows the Fls and Cls spectra of a largely alternating ethylene-tetrafluoroethylene copolymer (52% TFE), while Figure 6(b) shows similar data for the sample after treatment in an argon plasma, (0.1 torr, 0.2 watts for 25 secs). The spectra are shown at electron take-off angles of $\theta = 18^{\circ}$ and $\theta = 80^{\circ}$ respectively. The changes in the spectra caused by plasma treatment reveal a decrease in the fluorine content of the surface region of the material, by the large relative decrease in intensity of the component arising from $\underline{CF_2}$ structural features and Fls levels, and a concomitant increase in the intermediate binding energy component associated with $\underline{CF}$ environments. The integrated intensities of the component signals for the treated sample as a percentage of the total Cls signal are $CF_2 : CF : \underline{C} = 32 : 8 : 60$ at $\theta = 18^{\circ}$ and 20 : 18 : 62 at $\theta = 80^{\circ}$. The difference strongly demonstrates the inhomogeneity of the outermost few monolayers of the sample, and can be understood in terms of the substrate/overlayer model. The $CF_2$ structural features are contained in the "substrate" (unreacted polymer) and the signal arising from these features relatively decreases in going from $\theta = 18^{\circ}$ to $\theta = 80^{\circ}$. The CF structural features are contained in the "overlayer" (modified polymer) and the signal intensity relatively increases on going to grazing electron take-off angle. The third component has contributions from both surface and bulk and is therefore much less dependent on $\theta$.

The kinetic model (4) developed for analysis of the XPS data is based on a system in which the modification of a surface layer of thickness d occurs via both direct and radiative energy transfer processes, while beneath this layer only radiative energy transfer processes are considered to be important. This assumption derives from the fact that the U.V. and vacuum U.V. radiation, emitted from the plasma, is expected to penetrate the

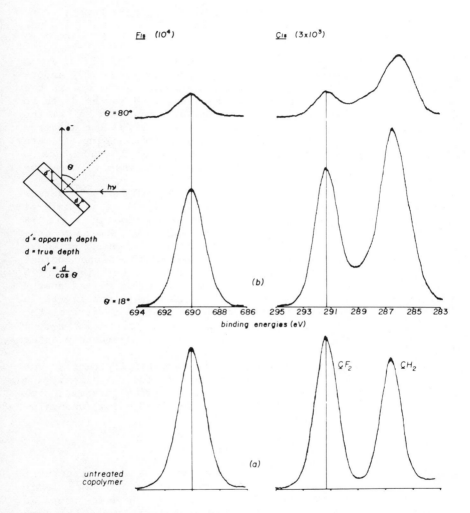

*Figure 6.   Core-level spectra of the ethylene–tetrafluoroethylene copolymer and of a sample treated in an argon plasma ($\theta = 18°$ and $80°$)*

sample further than active species such as argon ions and meta-
stables. The kinetic equation (4) can be summarized as equation
(4):

$$\frac{I^t}{I_o}\text{tot} = (1-e^{-d/\lambda\cos\theta}).e^{-K_s t} + e^{-d/\lambda\cos\theta}.e^{-K_b t} \qquad (4)$$

where $I_o^{tot}$ is the total intensity of the core level signal for a
given structural feature in the initial polymer which will undergo modification (i.e. $CF_2$ or Fls for the copolymer). $I^t$ is the
intensity after time t exposed to the plasma. d is the depth to
which direct energy transfer processes are important, $\lambda$ is the
mean free path of the photoemitted electrons and $\theta$ is the electron
take-off angle. $K_s$ is a composite rate constant encoding all rate
processes in the surface layer of thickness d and $K_b$ is a
composite rate constant encoding all rate processes in the sub-
surface and bulk of the polymer.

Figure 7 shows a semi-logarithmic plot of $I^t/I_o^{tot}$ versus
time of exposure to the plasma, for the Fls signal of a sample of
the copolymer treated in an argon plasma. The analysis is
straightforward, since as t becomes large the dominant contribution is from the term of small exponent (see equation (4)).
The plot therefore approaches linearity with the slope corresponding to $-K_b$. The intercept of this line with the axis at t = 0
corresponds to $-d/\lambda\cos\theta$ giving the depth, to which direct energy
transfer processes are important, of $\sim 12\overset{\circ}{A}$ or $\sim 2$ monolayers in
the argon plasma. Replotting the difference of this extrapolated
line from the experimental data, at low t, yields a second
straight line of slope $-K_s$. $K_s$ is an order of magnitude greater
than $K_b$.

This type of XPS analysis has been successfully carried out
over a wide range of operating parameters (4) for the plasma, and
a variety of inert gases (7), and clearly provides information on
the kinetics and depth of reaction at a level of detail unobtain-
able by any other technique.

## Absolute and Relative Binding Energies

In this section the identification of various structural
features from the measured binding energies of the core level
electrons is discussed. Examples have been chosen in the areas of
(a) plasma polymerization of fluorinated materials and (b) surface oxidation of polymers, to encompass both fluorine-containing
and oxygen-containing systems.

Plasma Polymerization of Fluorinated Materials. Figure 8
shows the Cls and Fls binding energies measured for the linear
fluoropolymers (8). While the shifts in binding energy of the Fls
levels are relatively small, the shift in the binding energy of
the Cls levels induced by fluorine as a substituent is relatively

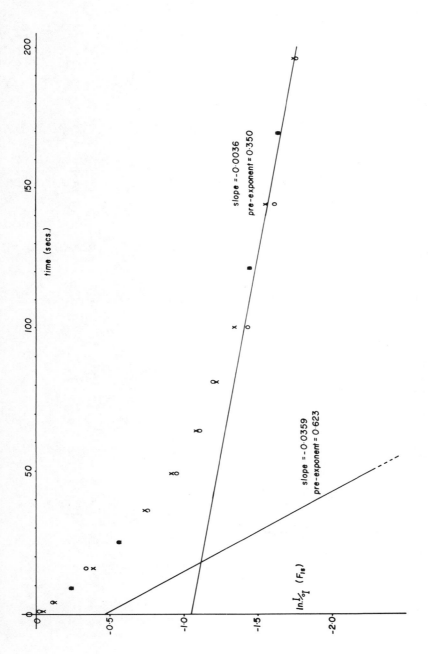

*Figure 7. Semi-logarithmic plot of $I^t/I_o^{tot}$ for the F-1s levels of the argon plasma treated copolymer vs. t (($\bigcirc$) experimental data; ($\times$) calculated points)*

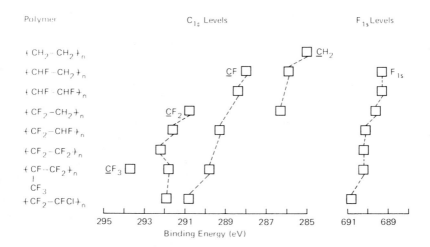

*Figure 8.    The C-1s and F-1s binding energies measured for the linear fluoropolymers*

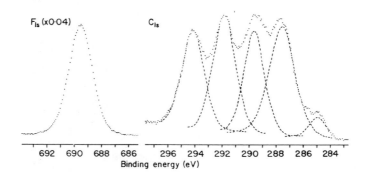

*Figure 9.    Core-level spectra of plasma polytetrafluoroethylene*

large.  A simple additive model may be constructed to describe the primary substituent effect of fluorine on carbon, where the shift in binding energy of a given C1s core level electron subsequent to replacing hydrogen or carbon by fluorine is $\sim$ 2.7eV per fluorine substituent.  It is also apparent from Figure 8 that substitution of fluorine on the next carbon atom along the chain also produces a significant, secondary shift, which has been shown to be $\sim$ 0.4eV per beta fluorine substituent (8).  While the C1s spectra of simple linear fluoropolymers are well resolved due to the large primary shifts, the corresponding spectra for more complicated fluoropolymer structures (e.g. as found in plasma polymerization) tend to be rather unresolved due to the relatively large range of secondary shift effects on a given structural feature which may occur in a variety of chemically different environments (9).  (For example, a $\rightarrow$CF structural feature bonded to three $\rangle CF_2$ groups will have a C1s binding energy $\sim$ 2eV higher than for a $\rangle CF$ structural feature having no beta fluorine substituents).

Figure 9 shows the F1s and C1s spectra of a plasma polymerized tetrafluoroethylene film produced in a radiofrequency capacitively coupled diode reactor configuration (9).  The C1s spectrum clearly exhibits several distinct types of carbon environment and comparison with previously well characterized systems allows the individual components to be assigned to particular structural features.  Thus the peak centered at $\sim$ 294.1eV on the binding energy scale is attributable to carbon attached to three fluorine atoms in $CF_3$ while those at $\sim$ 291.9eV and 289.7eV are due to $CF_2$ and CF structural features respectively.  The signal centered at $\sim$287.6eV can be assigned to carbon atoms not directly attached to fluorine but having fluorine substituents in a beta position, while the component at $\sim$ 285.0eV arises from either carbon atoms in a highly crosslinked environment, having no alpha or beta fluorine substituents, or from a small amount of hydrocarbon contamination.  The full width at half maximum (FWHM) of each component is relatively large, indicating a variety of environments for each structural type.

Similar data for plasma polytrifluoroethylene, plasma poly 1,1 difluoroethylene and plasma polyfluoroethylene are presented in Figure 10 (9).  The components of the C1s spectra of the series of fluorinated ethylenes are assigned as in Table 1, where subscripts a and b refer to the structural type being in a largely fluorinated or largely hydrocarbon environment respectively.  A full description of the lineshape analysis can be found elsewhere (1, 9).

The XPS data reveal that a considerable degree of rearrangement of the injected fluorocarbon is involved in the plasma polymerization process.  The relative quantities of $CF_3$, $CF_2$, CF and non-fluorine substituted features are readily monitored by XPS and show a strong dependence on the injected fluorocarbon.  As the fluorine content of the injected material decreases so does that of the polymer.  Table 2 lists the measured F/C stoichio-

Table I

Cls and Fls peak assignments for the plasma polymers derived from
ethylene and the series of fluorinated ethylenes.

| Injected | Binding Energies (eV) | | | | | | | |
|----------|------|------|------|------|------|------|------|------|
| Material | $CF_3$ | $CF_2$ | $CF_a$ | $CF_b$ | $C_a$ | $C_b$ | CH | F |
| $CH_2CH_2$ | | | | | | | 285.0 | |
| $CH_2CHF$ | | 290.0 | | 287.4 | | 285.9 | 285.0 | 687.3 |
| cis $CHFCHF$ | 292.9 | 290.0 | | 287.7 | 287.0 | 285.8 | 285.0 | 687.6 |
| trans $CHFCHF$ | 292.7 | 290.0 | | 287.8 | 286.8 | 285.8 | 285.0 | 687.5 |
| $CF_2CH_2$ | 293.4 | 290.6 | | 288.2 | 287.0 | 286.0 | 285.0 | 688.4 |
| $CF_2CHF$ | 294.0 | 291.6 | 289.9 | 289.0 | 287.2 | | 285.0 | 689.1 |
| $CF_2CF_2$ | 294.1 | 291.9 | 289.7 | | 287.6 | | 285.0 | 689.5 |

Table II

Fluorine/Carbon stoichiometries derived from the XPS data for the
series of plasma polymers studied.

| Injected | Stoichiometry (F/C) | |
|----------|--------------|----------------|
| Material | From Fls/Cls | From Cls comps. |
| $CH_2CH_2$ | 0.00 | 0.00 |
| $CH_2CHF$ | 0.14 | 0.26 |
| cis $CHFCHF$ | 0.40 | 0.52 |
| trans $CHFCHF$ | 0.40 | 0.51 |
| $CF_2CH_2$ | 0.51 | 0.61 |
| $CF_2CHF$ | 0.99 | 0.99 |
| $CF_2CF_2$ | 1.33 | 1.33 |

metries of the films derived from the series of ethylene and the
fluorinated ethylenes. The first column indicates the injected
material. The second is the F/C stoichiometry determined from the
Fls/Cls total signal intensity ratio corrected for the relative
sensitivity of these core levels, while the final column provides
the F/C stoichiometry determined from the relative intensities of
the Cls components. For plasma polytetrafluoroethylene and
plasma polytrifluoroethylene the two values agree precisely.
However, it is clear that this is not the case for the other
plasma fluoropolymers, for which the derivation from the Cls
levels always gives a higher figure. This can be attributed to
the presence of vinylic CH which is accompanied by a small shake-
up satellite positioned ∿7eV higher on the binding energy scale.
This adds a few percent to the measured intensities for the $\underline{CF_2}$
and $\underline{CF_3}$ components and a high F/C stoichiometry is calculated.
The greater the discrepancy between the two columns in Table 2,
the greater is the vinylic CH content. These XPS observations,
along with supporting data from the mass spectrometric analysis
of the low molecular weight, neutral species in the plasma
effluents, have led to the identification of the most likely
precursor to plasma polymerization in these systems ($\underline{9}$).

In an investigation of metal containing plasma polymers XPS
has been used to determine both the carbon-fluorine structure of
the polymer matrix and the quantity and form in which the metal is
present ($\underline{10}$, $\underline{11}$). Figure 11 shows the core level spectra of
polymers produced at the grounded electrode in a radiofrequency
capacitively coupled diode reactor system using perfluoropropane
as the injected gas ($\underline{10}$). The data pertain to three separate
experiments in which germanium, molybdenum and copper were used
as the excitation electrode material. The XPS data immediately
affirm that both molybdenum and copper are incorporated into the
polymer films whereas germanium is not under the conditions used.
Analysis of the relative signal intensities allow the metal
contents to be estimated as ∿20% and ∿14% by weight for Mo and Cu
respectively. In the Fls spectra, in addition to the intense peak
at ∿ 689.2eV associated with fluorine attached to carbon there is
a smaller component at lower binding energy ( ∿685.5eV), for the
$Mo-C_3F_8$ and $Cu-C_3F_8$ systems, due to metal fluoride. This,
however, is insufficient to account for the high oxidation states
of the metals observed in the $Mo_{3p}$ and $Cu_{2p}$ spectra. In fact, as
is apparent from the Ols regions, the metals are present predominately as oxides and hydroxides which are formed on air exposure
($\underline{10}$).

The similarity of the Cls spectra for the three films in
Figure 11 suggests that the polymer structure is the same irrespective of the excitation electrode material used. It has been
noted that the Cls spectra in Figure 11 are also similar to that
of plasma polytetrafluoroethylene formed under similar conditions
($\underline{9}$, $\underline{12}$). This observation, coupled with mass spectrometric
analysis of the low molecular weight neutral species in the plasma

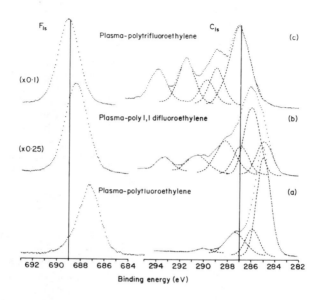

*Figure 10.   Core-level spectra of the plasma polymers derived from trifluoro-
ethylene,1,1-difluoroethylene, and fluoroethylene*

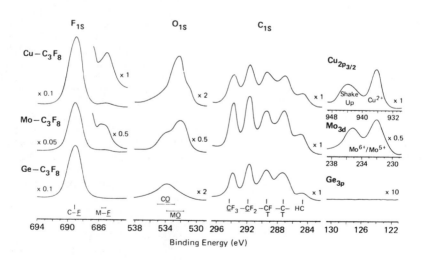

*Figure 11.   Core-level spectra of metal containing plasma fluoropolymers*

effuents suggested a likely polymerization mechanism for these systems involving difluorocarbene and tetrafluoroethylene as the primary precursors (10).

Surface Oxidation of Polymers. The investigation of oxygen-containing polymers and more particularly surface oxidized polymers, by XPS has been greatly aided by several recent publications providing the necessary experimental (1, 2, 6, 13) and theoretical (13, 14) reference data. This has been seen to be of great importance since, for the majority of cases, modification of a polymer involves the formation of oxygenated structural features at the surface. In contrast to the situation found for fluorine as a substituent, the secondary shift exerted by oxygen is small (< 0.2eV). This allows us to summarize shifts in Cls levels for oxygen-containing structural features, in terms of the number of carbon-oxygen bonds, as follows (6, 15):

$$O=\underline{C}\diagdown^{O}_{O} \qquad O=\underline{C}-O \quad > \quad \underline{C}=O \quad \backsim \quad O-\underline{C}-O \quad > \quad \underline{C}-O$$

$$\backsim290.6eV \qquad \backsim289.0eV \qquad \backsim287.9eV \qquad\qquad \backsim286.6eV$$

The reference binding energy for carbon not attached to oxygen (e.g. in polyethylene) is 285.0eV and, as found in the case of fluorine as a substituent, the primary substituent effect of oxygen can be described in terms of a simple additive model. The shift in binding energy, of a Cls core level electron, subsequent to replacing carbon or hydrogen by oxygen is $\backsim$ 1.5eV per carbon-oxygen bond. This is only half the shift induced by fluorine, and coupled with the typical full width at half maximum (FWHM) of a Cls level of $\backsim$1.4eV with the resolution of the spectrometer employed here, manifests itself in broad unresolved Cls profiles for oxidized polymers. Because the secondary shift of oxygen is small however, lineshape analysis of the Cls spectrum is possible, employing components of FWHM only slightly greater than 1.4eV and at binding energies appropriate to the structural types summarized above (6, 15).

The binding energies of the Ols core levels, although somewhat less predictable, can be approximately grouped as follows:

$$\underline{O}-C \quad > \quad \underline{O}-C \quad > \quad \underline{O}-C \quad \backsim \quad \underline{O}=C \quad > \quad \underline{O}=C$$

| $\underline{O}-C$ in carbonate | $\underline{O}-C$ in acid, ester | $\underline{O}-C$ in alcohol, ether | $\underline{O}=C$ in ketone | $\underline{O}=C$ in acid ester, carbonate |
|---|---|---|---|---|
| $\backsim$535.0eV | $\backsim$534.3eV | $\backsim$533.6eV | | $\backsim$532.8eV |

However, lineshape analysis of complex Ols band profiles can only usually be accomplished at a confirmatory rather than diagnostic level.

Figure 12 shows the Cls spectra for samples of polyphenylene

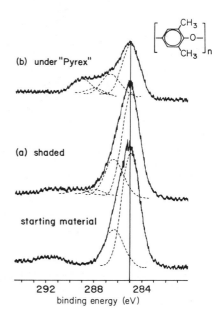

*Figure 12. The C-1s spectra of poly-phenylene oxide: (a) sample exposed in a shaded configuration and (b) sample exposed under a Pyrex glass slide (3 months)*

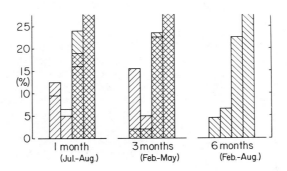

*Figure 13.    Summary of polyphenylene,   oxide weathering data in the form of a bar chart ((☐) full; (▨) Pyrex; (◩) shaded)*

oxide used in an investigation of the natural weathering of
polymer surfaces (16). The bottom spectrum is for the starting
material, (a) is for a sample exposed for 3 months shaded from the
sun and rain and (b) is for a sample exposed under a 1mm Pyrex
glass slide. The weathering site was in San Jose, California.

The C1s spectrum of the starting material in Figure 12 shows
two major components at binding energies of ∿285.0 and ∿286.6eV.
These can be assigned to carbon atoms not attached to oxygen and
carbon atoms in the ether linkage, respectively. A shake-up
satellite, shifted ∿7eV higher on the binding energy scale, is
also apparent due to π* ← π excitation accompanying photoionization in the aromatic structure. After 3 months exposure in the
shaded configuration the well between the direct photoionization
peaks and the shake-up satellite has filled with components due to
C=O and O-C=O type carbon which have been formed by a thermal
oxidation process. For the sample which has been exposed under a
Pyrex glass slide, the extra heat and light available has clearly
caused a much greater degree of oxidation in the sample surface.
From the relative intensities of the C1s components it can be seen
that ∿ 50% of the carbon atoms in the surface region are attached
to oxygen, and although a low concentration of carbonyl features
is present, the spectrum corresponds essentially to a 1:1:2 ratio
of O=C-O, C-O and CH features. It would, therefore, seem that
weathering in this configuration has resulted in almost 100%
conversion of the methyl groups at the surface to carboxylic
acids, with little other change.

A more convenient way to display data such as that shown in
Figure 12 is in the form of a bar chart, in which each component
in the spectrum is represented by a bar whose height corresponds
to the area under the curve. Figure 13 summarizes the data for
polyphenylene oxide, in the form of bar charts, for samples
exposed for 3 months and 6 months and for 1 month in the height of
the summer. For each chart the bars represent, from left to
right, carbon not attached to oxygen (off scale), C-O, C=O and
O=C-O. The conclusions drawn from Figure 12 are markedly emphasized by the center chart in Figure 13. It is interesting to
note, however, that if a fresh sample is exposed for only 1 month
mid-summer the C-O component decreases for the "Pyrex" and fully
exposed samples providing evidence that, under these conditions,
the ether linkage is cleaved.

XPS is able to answer the question of how the weathering at a
polymer surface differs from that in the bulk material and is able
to detect changes in the surface chemistry of weathered polymers
at a stage where the degree of bulk oxidation is small.

Modifications of polymer surfaces by exposure to electrical
plasmas and discharges have also been subjected to XPS examination in several recent articles (4, 6, 7). An example is the
plasma oxidation of polyethylene, polypropylene and polystyrene
in a radiofrequency inductively coupled system (6). Figure 14
shows the C1s and O1s spectra of a polyethylene film after

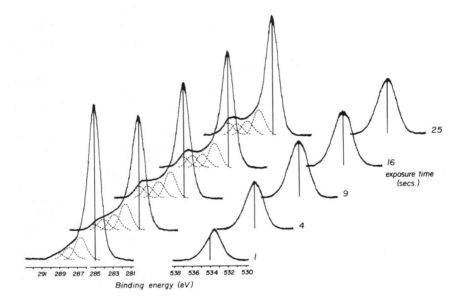

Figure 14.    Core-level spectra of a polyethylene film after various times of exposure
to an oxygen plasma

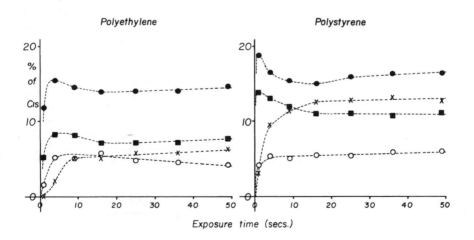

Figure 15.    Plots of the intensities of the C-1s components vs. t for polyethylene
and polystyrene ((● $\underline{C}$—O; (■) $\underline{C}$=O (O—$\underline{C}$—O); (○) $\underline{C}$=O; (×) $\underline{C}$OOO)

exposure to a pure oxygen plasma at reduced pressure, as a function of time of exposure.   Although untreated polyethylene shows only a single peak, in the Cls region at 285.0eV, the sample treated for just 1 sec. shows a well developed tail to higher binding energy and a broad envelope appears in the Ols region. The tail in the Cls spectrum is associated with the oxidation products and can be resolved into three components due to C-O, C=O + O-C-O and O=C-O, from low to high binding energy.   Longer exposures to the plasma increases the degree of oxidation and an additional component appears in the Cls spectrum due to the formation of carbonate structural features.   By following the intensities of the various components as a function of time of exposure to the plasma it is possible to obtain mechanistic and kinetic information on the processes involved.

The Ols spectra in Figure 14 are not so readily resolved but are entirely consistent, in terms of band profile and intensity, with the findings from the Cls spectra.   Thus as the sample is exposed to the plasma for longer periods the Ols intensity increases and the centroid of the Ols envelope shifts to higher binding energy.   This shift can be associated with the appearance of a component at $\sim$535.0eV due to the singly bonded oxygen in the carbonate functionality.

It is quite straightforward to monitor the various components of the Cls spectra versus time of exposure to the plasma (6). Figure 15 exhibits the relevant plots for polyethylene and polystyrene.   The oxidation for polystyrene is very much more rapid than for polyethylene but both tend to reach their final surface compositions after $\sim$16 secs in the plasma.   The polysytrene also exhibits a greater uptake of oxygen in the form of carbonate and, C=O and/or O-C-O functionalities, the amounts of C-O and O=C-O being approximately the same for the two samples after extended exposure times.   The general shapes of the curves in Figure 15, in particular the initial peak in the concentration of C-O and O-C-O has led, when compared ot other data, to the conclusion that oxidation in the oxygen plasma occurs via the initial crosslinking of the polymer chains by either linkages followed by a subsequent formation of more complex oxygen-containing features (6).

The analytical depth profiling for these systems (e.g. the polystyrene data is shown in Figure 5) revealed that the reaction is essentially confined to the topmost monolayer of material (6). This is entirely reasonable in terms of the plasma chemistry since the most prominent reactive species is atomic oxygen which is expected to have an extremely short mean free path in hydrocarbon polymers.   This serves as a very good example of the powerful nature of XPS when applied to the study of the surface modification of polymers.

In addition to the relatively simple oxygen-containing structural feature already discussed, to be able to obtain a complete picture of an oxidized polymer surface it is also

| Compound | Binding energies (eV) | | |
|---|---|---|---|
| | $\underline{C_{1s}}$ (CH) | $\underline{C_{1s}}$ (CO) | $\underline{O}_{1s}$. |
| Ph(CH$_3$)$_2$O-OH | 285.0 | 286.7 | 533.9 |
| (CH$_3$)$_3$CO-OH | 285.0 | 286.6 | 534.1 |
| ((CH$_3$)$_3$CO)$_2$ | 285.0 | 286.6 | 533.8 |
| (PhC$\begin{smallmatrix}O\\ \\O-\end{smallmatrix}$)$_2$ | 285.0 | 289.3 | 533.0<br>535.5 |

*Figure 16. The C-1s and O-1s binding energies observed in peroxy compounds*

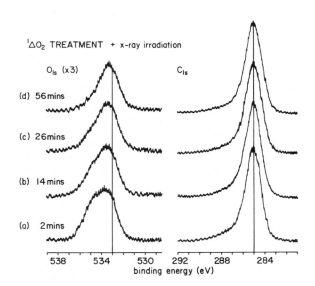

*Figure 17. Core-level spectra of a high-density polyethylene sample after expo-sure to a stream of oxygen rich in singlet molecular oxygen vs. time of x-ray irra-diation*

important to be able to recognize and identify peroxy features when they occur. As with the assignment of any feature in the XPS spectrum the most logical place to start is with detailed experimental and theoretical studies of model compounds. Figure 16 therefore shows the Cls and Ols binding energies observed for a series of peroxy compounds, studied as thin condensed films (17). The Cls binding energies fall into line with those previously discussed. If we consider the Ols binding energies an interesting situation emerges. The binding energies for singly bonded oxygen in a simple hydroperoxide or dialkyl peroxide is $\sim$ 533.9eV, only slightly higher than for the corresponding ether or alcohol. For benzoylperoxide on the other hand, while the doubly bonded oxygen is at a similar binding energy to that found for the corresponding oxygen in an acid or ester, the singly bonded oxygen has the highest Ols binding energy observed so far, at $\sim$ 535.5eV. This investigation has been accompanied by theoretical calculations on a much wider range of peroxy features and it is generally found that while simple peroxides exhibit an Ols binding energy similar to alcohols and ethers, peroxy features in highly oxygenated environments exhibit a high Ols binding energy. This latter category includes features such as oxetanes and dioxiranes as well as peracids and peresters.

In this context it is interesting to consider the data for a high density polyethylene sample whose surface has been exposed to the effluent of a microwave singlet molecular oxygen ($^1\Delta O_2$) generator (15, 17). The bottom spectra in Figure 17 pertain to a sample exposed for 1 hour in such a system. The Cls spectrum possesses a tail to high binding energy owing to the presence of predominately carbon singly bonded to oxygen, although components are also present due to $\underline{C}$=O + O-$\underline{C}$-O, and O=$\underline{C}$-O. The Ols band profile is at a relatively high binding energy and clearly contains a component at $\sim$ 535eV. The Ols and Cls spectra of the sample were repeated after various times of exposure to the X-ray radiation in the spectrometer and it is apparent from Figure 17 that X-ray decomposition of the sample is occuring, until after $\sim$ 1 hour the high binding energy Ols components have completely disappeared. X-ray decomposition of polymer samples is not generally observed for the low dose rates used in XPS. These data are therefore indicative of relatively labile structures, i.e. peroxy functionalities. Construction of a difference spectrum of the first and last Ols spectra reveals that the disappearing portion contains two components at $\sim$ 534.0eV and $\sim$ 535.2eV which can be assigned to simple peroxides and peroxy features in highly oxygenated environments, respectively.

## Conclusion

It should be clear, even from this short account, that although the analysis of XPS data pertaining to polymeric materials is relatively straightforward with the necessary background data, the information content per experiment is large. Thus, elemental analysis, structural elucidation and depth profiling can, in most cases, be accomplished to a high degree of detail.

## Acknowledgements

Thanks are due to the Science Research Council, U.K. for provision of financial support and instrumentation, without which this work would not have been possible.

## Literature Cited

1.  Dilks, A., "Electron Spectroscopy," Ed. C. R. Brundle and A. D. Baker, Academic Press, London, Vol. 4, in press.

2.  Clark, D.T., "Handbook of X-ray and Ultraviolet Photoelectron Spectroscopy," Ed. D. Briggs, Heyden, London, 1977.

3.  Clark, D.T. and Thomas, H.R., J. Polym. Sci., Polym. Chem. Ed., 1977, 15, 2843.

4.  Clark, D.T. and Dilks, A., J. Polym. Sci., Polym. Chem. Ed., 1977, 15, 2321.

5.  Clark, D.T., Dilks, A., Shuttleworth, D. and Thomas, H. R., J. Elect. Spectr., 1978, 14, 247.

6.  Clark, D.T. and Dilks, A., J. Polym. Sci., Polym. Chem. Ed., 1979, 17, 957.

7.  Clark, D.T. and Dilks, A., J. Polym. Sci., Polym. Chem. Ed., 1978, 16, 911.

8.  Clark, D.T., "Advances in Polymer Friction and Wear," Ed. L.M. Lee, Plenum Press, New York, Vol. 5A, 1974.

9.  Dilks, A. and Kay, E., submitted 1980.

10. Dilks, A. and Kay, E., "Plasma Polymerization," Ed. M. Shen and A.T. Bell, American Chemical Society Symposium Series, Washington, D.C., 1979, 108.

11. Dilks, A., Kay, E. and Seybold, J. Appl. Phys., in press.

12. O'Kane, D.F. and Rice, D.W., J. Macromol. Sci., Chem., 1976, A10, 567.

13. Clark, D.T. and Thomas, H.R., J. Polym. Sci., Polym. Chem. Ed., 1976, 14, 1671.

14.  Clark, D.T., Cromarty, B.J. and Dilks, A., J. Polym. Sci., Polym. Chem. Ed., 1978, 16, 3173.

15.  Dilks, A., "Developments in Polymer Characterization," Ed. J.V. Dawkins, Applied Science, London, 1980.

16.  Clark, D.T. and Dilks, A., in preparation.

17.  Dilks, A., submitted for publication.

RECEIVED January 15, 1981.

# Surface Studies on Multicomponent Polymer Systems by X-Ray Photoelectron Spectroscopy

H. R. THOMAS and J. J. O'MALLEY[1]

Xerox Webster Research Center, Xerox Square W-114, Rochester NY 14644

Multicomponent polymers systems such as polyblends, and block copolymers often exhibit phase separation in the solid state which results in one polymer component dispersed in a continuous phase of a second component. The morphological properties of these systems depend upon a number of factors such as the molar ratios of the components, the molecular weights, the thermal history of the system and, for solvent cast films, the solvent and drying conditions.

Although a number of techniques have been devised to investigate the bulk domain structure of multicomponent polymer systems the detailed structure of the surface, i.e., the outermost few tens of angstroms, has been studied in much less detail. Since many of the important properties of a polymeric solid are dependent upon the surface structure and since the surface can differ considerably from the bulk a technique which can differentiate the surface from bulk properties is likely to be of considerable importance.

In this paper we describe the application of X-ray photoelectron spectroscopy (XPS) (2) to the quantitative and qualitative investigation of the surface properties of several multicomponent polymer systems in the solid state. We have divided the polymer systems into three categories, 1) regular block copolymers, 2) random block copolymers, and 3) physical blends. In each category representative examples are presented of polymeric systems studied by XPS to delineate the surface topography, morphology, structure and bonding. In all these studies we have used angular dependent XPS XPS($\theta$) (3) to depth profile the compositional variation within the outermost few tens of angstroms of the polymer-air-interface.

## EXPERIMENTAL

In Table 1 are shown the systems reported in this paper. The synthesis and characterization of the polymer samples are

reported elsewhere and the references for each system are found on
the right hand side of Table I.  The samples were all studied as
thin ( ⩰ 1-10 μ ) films cast from the appropriate solvents for
each system, also found in Table I.  (In this paper we describe
only systems cast from solvents where both polymer components
were soluble, more detailed studies on selective solvents can be
found in the papers referenced).  All solvents were spectroscopic
grade and the samples were studied as films cast onto flat
aluminum substrates by dip coating.  All films were dried in an
argon atmosphere to reduce surface contamination and were sufficiently thick ( > 100 Å) to mask the Al $_{2p}$ core level signal
from the substrate.

Table I - Bulk Compositions, Solvents and Synthesis
References for Polymer Systems Studied

| System | %PS | | % PDMS | | Solvent | Synthesis Reference |
|---|---|---|---|---|---|---|
| | wt. | mole | wt. | mole | | |
| a) PS-PEO diblock | 19.6 | 9.6 | | | CHCl$_3$ | 19 |
| | 39.3 | 21.4 | | | | |
| | 70.0 | 49.5 | | | | |
| b) PEO-PS-PEO triblock | 23.5 | 11.4 | | | CHCl$_3$ | 19 |
| | 38.5 | 21.0 | | | | |
| | 70.3 | 49.8 | | | | |
| c) HMS-DMS Random block | | | 27.0 | 58.0 | CHCl$_3$ | 19 |
| | | | 57.2 | 84.0 | | |
| | | | 72.5 | 91.0 | | |
| d) PS-PEO physical blend | 20.0 | 9.8 | | | CHCl$_3$ | * |
| | 40.0 | 22.0 | | | | |
| | 70.0 | 49.5 | | | | |

*Commercial homopolymers Scientific Polymer Products

Spectra were recorded with an AEI ES 200 B spectrometer by
using Mg$_{K\alpha \, 1,2}$ exciting radiation.  Typical operating conditions
were X-ray gun, 12 Kv, 15mA; pressure in the sample chamber ca.
$10^{-8}$ torr.  Under the experimental conditions employed, the gold
4f$_{7/2}$ level at 84eV used for calibration had a full width at half
maximum (FWHM) of 1.2 eV.  No evidence was obtained for radiation
damage to the samples from long-term exposure to the X-ray beam.
Overlapping peaks were resolved into their individual
components by use of a DuPont 310 curve resolver (an analog
computer).  The detailed deconvolutions were based on a knowledge
of line widths determined from the homopolymers and model compounds studied previously.(4)  Studies have shown that for indi-

vidual components of the core-level spectra for the $C_{1s}$ and $O_{1s}$ levels, the line shapes approximate fiarly closely to Gaussian.

## Results and Discussion

A)  Regular Block Copolymers.
1)  Poly(ethylene oxide)/Poly(styrene) Diblock Copolymer. Previous studies (5,6) on bulk morphology of the PEO/PS copolymers indicated that the individual components are incompatible and that they undergo microphase separation and domain formation. We can anticipate based upon the differences in solid state surface tension between the two polymers (7) that the surface of the solid copolymer may well vary from the bulk.  In order to establish a firm basis for the interpretation of the diblock copolymer data, it is necessary to study the component homopolymers, polystyrene (PS) and poly(ethylene oxide) (PEO), and determine their absolute and relative binding energies and relative peak intensities.  The XPS core level spectra for PS and PEO are shown in Figure 1 and the experimental binding energies and peak intensity ratios are tabulated in Table II.

The PS spectrum has a strong peak centered at 285 eV associated with the direct photoionization of the $C_{1s}$ core levels and a low-energy satellite peak at 291.6 eV arising from shake-up transitions ( $\pi^* \leftarrow \pi$ ) accompanying core ionization.  These low-energy shake-up transitions are understood both theoretically and experimentally and the spectrum observed in Figure 1 is entirely consistent with previous work.(8-13)  The spectra for PEO show a single peak for the $C_{1s}$ levels at 286.5 eV, referenced to hydrocarbon at 285 eV, and a single peak for the $O_{1s}$ core levels at 533.3 eV.  The peak area ratio of the $C_{1s}$ to $O_{1s}$ core levels is 0.73; however, when these peak areas are corrected for the different theoretical photoionization cross sections (14), electron mean free paths (15) relative to the $C_{1s}$ and $O_{1s}$ core level electrons, and instrumental factors the correct stoichiometry for PEO is obtained.

The chemical shift of ∿ 1.5 eV for the $C_{1s}$ core level peak in PEO, relative to PS, can be attributed to each carbon being attached to an oxygen atom and is consistent with theoretical predictions and experimental results on related low-molecular-weight model compounds. (4, 16) As expected, there is no shake-up peak in The $C_{1s}$ core level spectrum for PEO since the polymer is fully saturated. (17)  These significant differences in XPS spectra of PS and PEO, i.e., the 1.5 eV chemical shift in the $C_{1s}$ core levels, the uniqueness of the $\pi^* \leftarrow \pi$ shake-up peak associated with the PS component, coupled with the peak intensity ratios from Table II, enable a unique analysis of the surface composition of the PS/PEO system.

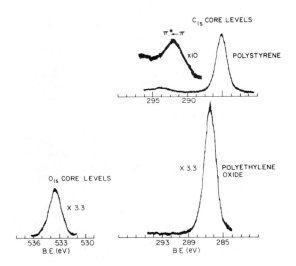

*Figure 1.    The C-1s and O-1s core-level spectra for polystyrene and poly(ethylene oxide) homopolymers*

Table II
Experimental Binding Energies and Peak Area Ratios for the Reference Homopolymers, Polystyrene
and Poly(ethylene oxide)

| | binding energy,[a] | | | peak area ratios |
|---|---|---|---|---|
| | $C_{1s}$ | $C_{1s}$ ($\pi * + \pi$) | $O_{1s}$ | |
| polystyrene | 285.0 | 291.6 | | $C_{1s}(PS/C_{1s}(PEO) = 1.60 \pm 0.1$ |
| poly(ethylene oxide) | 286.5 | | 533.3 | $C_{1s}(PEO)/O_{1s}(PEO) = 0.73 \pm 0.05$ |
| | | | | $C_{1s}(\pi * + \pi)(PS)/C_{1s}(PS)=0.081 \pm 0.005$ |

[a]Referenced to hydrocarbon at 285.0 eV.

   Films of the three block copolymers were cast from chloroform, a mutual solvent for PS and PEO,(6) and the measured $C_{1s}$ and
$O_{1s}$ core level spectra are shown in Figure 2. The spectra show the
characteristic $O_{1s}$ peak of PEO, the shake-up satellite of PS, and
an easily deconvoluted doublet for the $C_{1s}$ core levels in PS and
PEO. It is apparent from the spectra that the PS concentration at
the copolymer surface increases as the PS in the copolymer increases. More importantly, however, an analysis of the spectral data
clearly shows that the surface compositions are significantly
richer in PS than would be predicted based on a knowledge of the
bulk compositions of the block copolymers. In Figure 3 is shown a
plot of the surface-vs-bulk compositions for the diblock copolymers.
   Up to this point all the XPS measurements have been made by
analyzing the photoemitted electrons normal to the surface of the
sample under investigation. This experimental arrangement is the
most commonly used in XPS studies, and with it the effective
sampling depth is maximized. For systems such as PS/PEO, the
effective sampling depth is about 50 Å; i.e., about 95% of the
signal comes from the outermost ∿ 50Å, based upon a previous
knowledge of the electron mean free paths appropriate to photoemitted electrons from the $C_{1s}$ ( ∿ 960 eV) and $O_{1s}$ ( ∿ 720 eV) core
levels. (5)  Therefore, the surface composition data shown in
Figure 3 are the average composition on this effective sampling
depth of ∿ 50 Å.
   These measurements can be further refined by controllably
decreasing the effective sampling depth. This will, in effect,
allow for depth profiling the composition of the upper 50 Å of the
surface and provide us with a means to explore the molecular
organization of the surface macromolecules. The method for varying the effective sampling depth is shown schematically in Figure
4. The sample is rotated relative to the fixed position energy
analyzer by angle θ , designated as the angle between the normal
to the sample and the slits in the analyzer. It is readily seen
that electrons collected at grazing exit angles relative to the
surface ( θ → 90° ) will enhance surface features relative to
electrons collected normal to the surface. (15,17,18)

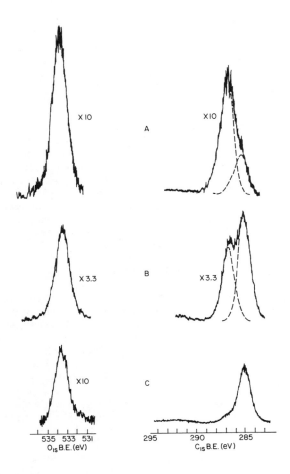

*Figure 2.    The C-1s and O-1s core-level spectra for the three PS/PEO diblock co-polymers cast from chloroform*

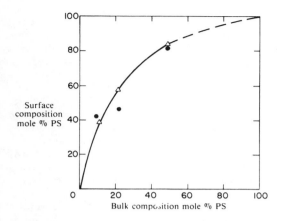

*Figure 3.    Surface vs. bulk compositions for PS/PEO diblock (●) and triblock (△) copolymers (XPS (θ) = 0°)*

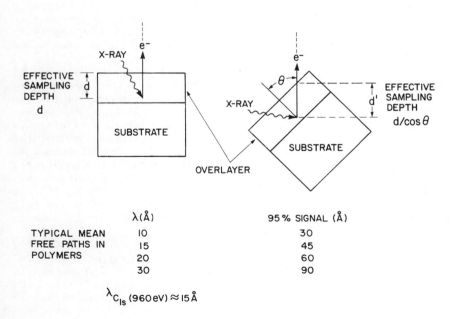

*Figure 4.    Angularly dependent studies for polymer samples in which spectra are studied as a function of electron take-off angle θ with respect to the sample surface*

In Figure 5 is shown the $C_{1s}$ and $O_{1s}$ core level spectra from the angular dependent XPS studies, XPS( $\theta$ ) , on copolymer B cast from chloroform. Spectra were recorded at three different angles, $\theta$ = 0, 45, and 80°, in order to achieve effective sampling depths of $\sim 50$, $\sim 25$, and $\sim 10$ A, respectively. Both the $C_{1s}$ and $O_{1s}$ core level spectra clearly show that each component of the copolymer is present at all sampling depths. Furthermore, from the relative intensity ratios of the deconvoluted $C_{1s}$ peaks associated with the PS and PEO components, we find that in particular cases, there is a concentration gradient in the top 50 A of the surface and that the relative concentration of PS and PEO increases as the air-solid polymer interface is approached.

Although the finding on the compositional variation between the surface and the bulk is interesting the XPS( $\theta$ ) studies reveal a much more interesting phenomena in the topography. Through the use of differences in the electron mean free paths(15) for photo-emitted electrons from $C_{1s}$ and $O_{1s}$ core levels and the relative core level intensity ratios for $C_{1s}/O_{1s}$ associated with the PEO component a unique surface structure emerges for the topography, and the models are illustrated in Figure 6. The models a-c for the surface topography can be eliminated from a careful examination of the experimental results. However the experimental data supports model d and for these diblock copolymers cast from chloroform, a mutual solvent for PS and PEO, there appears to be a tendency to form vertical isolated domain structures at all the copolymer compositions studied, with a non-planar surface of PS protruding above the PEO domain.

2) Poly(ethylene oxide)/Poly(styrene) Triblock (PEO/PS/PEO) Copolymers. Applying the XPS ( $\theta$ ) technique to these samples cast from chloroform results in the surface-vs-bulk composition for the triblock copolymers illustrated in Figure 3, along with the di-block copolymers discussed previously. The striking similarity in the data for the diblock and triblock copolymers indicate that the surface topography is similar in both cases. The experimental evidence indicates that the PS and PEO components in the copolymers are both exposed at the surface and that they are organized into domains which are thick compared to the XPS sampling depth. (3,15) A slight angular dependence is observed as $\theta$ is varied and this again points to a non-planar surface topography in which PS domains are elevated slightly above the PEO domains.

There have been numerous studies employing calorimetric(19), dynamic mechanical,(20) dielectric,(21) and morphological(23,24) techniques to elucidate the solid-state behavior of styrene-ethyl-ene oxide block copolymers. These measurements have focused on transition-temperature phenomena, and they have provided reference data on the bulk properties of the copolymers. The evidence accumulated to date indicates that PS and PEO are incompatible in the bulk. While this appears true, in general, one cannot rule out the possibility that PS and PEO have some limited degree of miscibility in the copolymers. It is also unknown, at this time, what influence an interface (e.g., the air-polymer interface) has

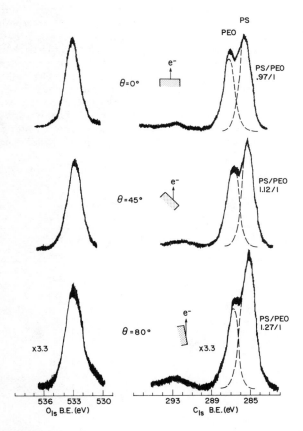

*Figure 5.    The C-1s and O-1s core-level spectra for XPS(θ) studies on PS/PEO diblock copolymer (B) cast from chloroform.   The experimental intensity ratios are corrected for absolute signal from Table II to obtain the molar ratios.*

on polymer–polymer compatibility and whether the degree of phase separation is the same or different in the bulk and in the surface. The present study, in part, addresses these issues by uniquely focusing on the surface properties rather than on the bulk properties of the copolymers. Our data suggest that, indeed, there is some compatibilization of PS and PEO.

In this section, we will discuss the interesting results on the $C_{1s}/O_{1s}$ intensity ratios for the PEO component and the $C_{1s}( \pi * \leftarrow \pi )/C_{1s}$ intensity ratios for the PS component in the triblock copolymers. Table III contains the measured intensity ratios for the two homopolymers and the three triblock copolymers. The triblocks deviate from the homopolymers in two ways: First, they have unusually high $C_{1s}/O_{1s}$ ratios for the PEO component compared to the PEO homopolymer, and second, they have unusually low $C_{1s}( \pi * \leftarrow \pi )/C_{1s}$ intensity ratios compared to the PS homopolymer. We will consider both of these significant deviations, in turn, and we will propose that these data are evidence of phase mixing in the solvent-cast PEO/PS/PEO triblock copolymer films.

Table III
XPS( $\theta$ ) Data for PS, PEO, and the
PEO/PS/PEO Triblock Copolymers

| sample | angle( $\theta$ ), deg | $C_{1s}/O_{1s}$ (PEO) | $\pi * \leftarrow \pi$ / $C_{1s}$ (PS) |
|---|---|---|---|
| PEO | 0, 45, 80 | 0.73 | |
| PS | 0, 45, 80 | | 0.080 |
| PEO–PS–PEO | 0 | 1.22 | 0.016 |
| (11.4 mol % PS) | 45 | 1.23 | 0.023 |
| | 80 | 1.22 | 0.032 |
| PEO–PS–PEO | 0 | 1.20 | 0.043 |
| (21.0 mol % PS) | 45 | 1.30 | 0.043 |
| | 80 | 1.40 | 0.049 |
| PEO–PS–PEO | 0 | 1.23 | 0.065 |
| (49.8 mol % PS) | 45 | 1.23 | 0.072 |
| | 80 | 1.23 | 0.075 |

Let us consider now the suggestion that the high $C_{1s}/O_{1s}$ intensity ratio for the PEO component is due to partial phase mixing of PS and PEO. We can ask the question: "How much hydrocarbon would have to be intimately mixed with the PEO component to attenuate the oxygen signal of the PEO component and thereby effectively increase the $C_{1s}/O_{1s}$ intensity ratio from 0.73 (PEO homopolymer) to the 1.2–1.4 range of values shown in Table IV for the copolymers?" In other words, what carbon to oxygen stoichiometries in a mixed phase of amorphous PEO and amorphous PS will result in measured $C_{1s}/O_{1s}$ intensity ratios of 1.2–1.4?

Using $C_{1s}/O_{1s}$(PEO in copolym) $= x_c\ C_{1s}/O_{1s}$ (unmixed PEO) $+$

$$(1 - x_c)\ C_{1s}/O_{1s} \text{ (mixed PS-PEO)} \qquad (1)$$

where the term on the left-hand side of eq. 1 is the measured intensity ratio for the copolymer (values ranging from 1.2 to 1.4), $x_c$ is the degree of crystallinity of the PEO component measured by calorimetry(19) and assumed to be the same for the surface region as for the bulk, and the term in the first bracket is the measured value of 0.73 for PEO homopolymer, we can calculate the term $C_{1s}/O_{1s}$-(mixed PS-PEO) , which is the intensity ratio for an intimate mixture of amorphous PEO and PS. To convert the calculated intensity ratios of the mixed phases in the copolymers into chemical compositions of the mixed phases, we need an experimental calibration curve relating these two quantities. This calibration curve is shown in Figure 7, in which we report $C_{1s}/O_{1s}$ intensity ratios for a series of related homopolymers with known carbon to oxygen stoichiometries ranging from 1:1 in the case of poly(oxymethylene) to 2.5:1 in the case of poly(methylmethacrylate). In using these polymers as model systems, we make the reasonable assumptions that the electron mean free paths differ insignificantly for these polymers at the same kinetic energy for photoemitted electrons and that the carbon and oxygen species are randomly arranged in the surface regions of these polymers.

The above procedure was used to calculate the composition of the mixed phase in each of the triblock copolymers. The results of the calculations are tabulated in Table IV, and they show that the molar ratio of ethylene oxide to styrene in the mixed phase varies from 0.6:1 to 1.5:1 in going from sample A to C. Mixing of PS with amorphous PEO is most prevalent in sample A, which has the smallest concentration of PS and the shortest PS block length ($M_n$ = 5.1K). Almost 50% of the PS in the surface region of sample A films is mixed with PEO, whereas only 6% of the surface PS is mixed in sample C. It seems clear that PS and PEO are partially miscible in the surface regions of these triblock copolymers. Our finding that PS

Table IV

Phase Mixing in PEO-PS-PEO Triblock Copolymers from Analysis of $C_{1s}/O_{1s}$(PEO) Intensity Ratios

| surface comp (mol %) | | PEO/PS comp (mixed phase) | mol % PS (mixed) | mol % PS (unmixed) |
|---|---|---|---|---|
| PS | PEO | | | |
| 39.0 | 61.0 | 0.6/1 | 18.3 | 20.7 |
| 58.5 | 41.5 | 1.0/1 | 12.5 | 46.0 |
| 83.8 | 16.2 | 1.5/1 | 5.0 | 78.8 |

is most miscible with PEO when the molecular weight and concentration of PS are low is qualitatively consistent with current concepts of phase-separation behavior in block copolymers(24-26) and polyblends(27).

Let us now turn our attention to other evidence bearing on

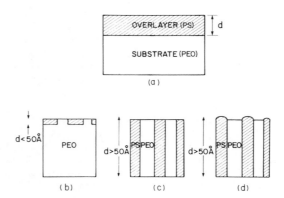

*Figure 6.   Models for the surface topography of the PS/PEO diblock copolymers*

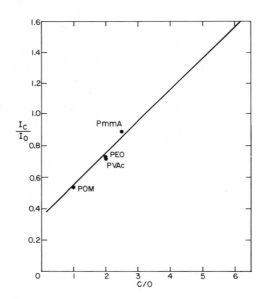

*Figure 7.   Calibration curve relating experimentally measured C-1s/O-1s intensity ratios to the known carbon-to-oxygen stoichiometries of a series of oxygen-containing homopolymers: POM, PVAc, PEO, and PMMA.*

the issue of phase mixing in PEO/PS/PEO copolymers. Inspection of the $C_{1s}$( $\pi * \leftarrow \pi$ )/$C_{1s}$ intensity ratios in Table III reveals that these ratios decrease with decreasing PS concentration in the surface regions of the copolymers and that, for samples A and B, the ratios are significantly lower than the value of 0.075 for PS homopolymer. It is proposed that the low $C_{1s}$( $\pi * \leftarrow \pi$ )/$C_{1s}$ intensity ratios are a result of PS mixing with amorphous PEO in the surface and that the mixed PS and PEO components are electronically interacting. The effect of this interaction is to decrease the probability, relative to PS homopolymer, of a $\pi * \leftarrow \pi$ transition occurring simultaneously with photoejection of a $C_{1s}$ core level electron from a given aromatic ring in the copolymer.

Although the precise nature of the electronic interaction between PS and PEO is unknown at this time, we speculate that it involves the interaction of the electron-rich oxygen atoms in the PEO block with the aromatic rings in the PS portion of the copolymers. Based on the work of Dilks,(28,29,30) we can anticipate that an electron donor/acceptor interaction between the ether oxygen in PEO and the phenyl rings in PS would result in (1) a shift in the PEO $C_{1s}$ peak to lower binding energy because the influence of ether oxygen electrons on the adjacent carbon atoms in PEO is effectively being modulated by the interaction of these electrons with the phenyl ring in PS and (2) a change in the line shape for the $\pi * \leftarrow \pi$ shake-up satellite peak. Both of these anticipated spectral changes are observed, as discussed below.

Figure 8 shows the $C_{1s}$ core level spectra for a triblock copolymer and a diblock copolymer, both of which contain 21 mol % PS. Deconvolution and line-shape analysis of the main $C_{1s}$ envelope for each copolymer indicate that the full width at half-maximum of the individual component peaks does not vary with copolymer structure. However, there is an obvious "filling in" of the region between the two $C_{1s}$ component peaks in the triblock copolymer spectrum. We attribute this to an additional PEO $C_{1s}$ peak, shifted by about 0.3 eV to lower binding energy, that arises from an electronic interaction between PEO and PS.

B)  Random Block Copolymers
1)    Poly(hexamethylene sebacate)/Poly(dimethyl siloxane) Block Copolymer.

An example of a random multiblock copolymer is found in the study of HMS/DMS copolymers where the DMS bulk content was varied from 27%, to 57.2% to 72.5% by weight. XPS ( $\theta$ ) studies established that trends in the surface composition and morphology were similar in nature to both the PS/PEO diblock and triblock systems. In other words, the polymer components are segregated into isolated domains and the lower surface energy component dominates the immediate surface. By plotting the results from the XPS ( $\theta$ ) studies on compositional variations with depth and extrapolating to an angle of $\theta = 90°$ , or just the immediate outer surface layer, reveals the outermost surface composition. When plotted against the composition determined by contact angle experiments, as shown in Figure 9, the striking agreement between the two

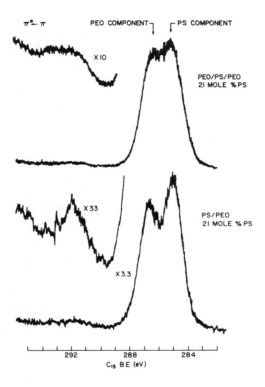

*Figure 8.    Comparison of the C-1s core-level spectra and shake-up satellites found in PEO/PS/PEO triblock and PS/PEO diblock copolymers containing about 21 mol % PS*

techniques is apparent. The XPS ( $\theta$ ) studies however not only reveal the immediate surface composition but indicate a heterogeneous distribution of components within the XPS sampling depth of $\sim$ 50 A which varies with the bulk composition. As the DMS component increases in the copolymer the thickness of the DMS segment at the surface increases, ranging from $\sim$ 7A , $\sim$ 10 A to $\sim$ 15 A in going from 27%, 57.2% to 72.5% DMS respectively. However it is important to note that even at 72.5% DMS the DMS component never totally dominates the surface layer and the segmented domain structure at the surface is still intact to some degree.

C)   Physical Blends
1)   Poly(ethylene oxide)/Poly(styrene)
An interesting comparison can be made between the results on the PEO/PS blends and the PEO/PS diblock and triblock copolymers. In comparing the three blends prepared in molar ratios equivalent to the diblock and triblock copolymers the results for the blends are surprisingly similar to the copolymers. As the PS increases in the bulk the PS concentration at the surface increases, however, as in the diblock and triblock copolymer systems, the PS ratio at the surface always exceeds that in the bulk. A comparison of the PS surface content in the diblock copolymers and the blends at equivalent bulk concentrations is shown in Figure 10. The data indicate that the blends have lower concentrations of PS at the surface than do the diblock copolymers. This is a surprising result because the thermodynamic driving force is for the PS, the lower surface energy component, to reside at the surface. At this time we have not satisfactorily explained this experimental observation and work in this area is continuing.

The $C_{1s}/O_{1s}$ intensity ratios for the blends, are similar to those for the triblocks, in that they are somewhat higher than the value for the homopolymer. Considering that the higher $C_{1s}O_{1s}$ intensity ratio results from phase mixing of the PEO and PS homopolymers, the small increase from the value for the homopolymer of $\sim$ 0.73 to $\sim$ 0.85 for the blends indicates substantially less mixing than in the triblock system when the value increased for $\sim$ 0.73 to $\sim$ 1.3. If we assume the PS only mixes with the amorphous phase of the PEO component and that it is distributed homogeneously throughout we calculate that one mole of PS mixes with 16 moles of PEO repeat units. From calorimetry measurements on bulk samples, we know the degree of crystallinity of PEO is $\sim$ 70% for all the polyblends.(9) The remaining 30% is amorphous and available for mixing with the PS component and the calculated ratios for the three polyblends for the mixed phases would contain amounts ranging for 2-0.5% PS. Calculations at all angles result in similar PS mixing as a function of depth and therefore within the XPS sampling depths the samples are homogeneously mixed.

PEO/PS Layers

The surprising results on the surface composition of the

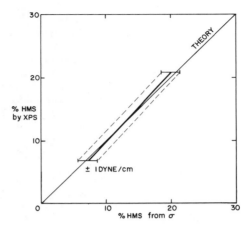

*Figure 9.    Plot of the % HMS at the surface determined by XPS vs. the % HMS
determined by contact-angle measurements*

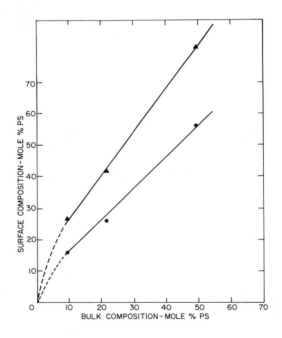

*Figure 10.    Comparison of the surface-vs.-bulk composition for the blends (●)
and the diblock copolymers (▲) (from CHCl₃)*

blends of PS and PEO, where the PS component did not dominate the surface and the domains did not appear to be discrete, prompted us to investigate polymer "diffusion" for the blends.  A sample was prepared where a PS film was overcoated with a layer of PEO and the sample heated to 130° C, above the $T_g$ of both homopolymers.  The $C_{1s}$ core levels were monitored, Figure 11, as a function of temperature at constant angle $\theta$ =25° (where the $C_{1s}$ core level signal maximized).  It can be seen at ambient temperature the PEO component dominated the surface, whereas and at 130° C the mole % PS relative to PEO had increased to $\sim$ 18%.

Upon reaching 130° C, which is designated as t=0 in Figure 12, a time study was carried out for a total of 18 hours to establish the surface homogeniety of the layed sample.

The XPS ( $\theta$ ) studies indicate significant angular dependence on the $C_{1s}$ core level spectra, where the PS component increases in intensity relative to the PEO as grazing angles are approached.  This effect persists throughout the time involved in the study and is attributed to a similar topography as assigned to the diblock copolymer and blend surfaces.

## Conclusions

The experimental results indicate that these blends are similar in surface topography to PS/PEO diblock copolymers cast from chloroform.  However, for identical bulk compositions, the concentration of PS on the surfaces of the blends is significantly less than on the surfaces of the copolymers.  In general, we can conclude that:

1) In the blends, the concentration of PS at the polymer-air interface is higher than the bulk concentration of PS.  However, the surface excess of PS in the polyblends is less than that found in the diblock copolymers of comparable bulk composition.

2) The surfaces of the blends are laterally inhomogeneous in PS and PEO.  In a manner similar to the diblock copolymers, we find the PS component raised somewhat above the PEO component.  However, our results suggest there is some phase mixing of the PS and PEO components in the blends.  In view of the propensity of PEO towards crystallization and the well known incompatibility of PS and PEO in the bulk, we suspect the degree of mixing to be quite small.  Nevertheless, it is surprising, in view of current theories(26) on phase mixing in multicomponent systems, that evidence for phase mixing would be found in the polyblend system but not in the diblock copolymers.  More work is obviously needed in this area for a better understanding of these mixing phenomena.

3) As in the diblock system, the molar composition at the homopolymer blend surface determined by XPS corresponds to the surface area occupied by each copolymer component, with the slight deviation accounted for by the mixing.

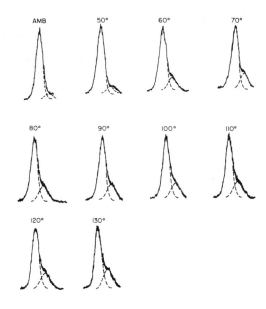

*Figure 11.    The C-1s core-level spectra for the PEO/PS layered structure as a function of temperature*

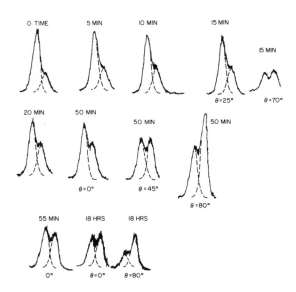

*Figure 12.    The C-1s core-level spectra for a layered PEO/PS structure as a function of time at 130°C (time zero taken when sample reached 130°C, from Figure 11)*

## Acknowledgements

We wish to extend our thanks to Drs. W. Salaneck, T. Davidson, C. Beatty, D. Shuttleworth and A. Dilks for helpful discussions.

## Literature Cited

1. Present address: Exxon Chemical Company, P.O. Box 4255, Baytown, Texas 77520.

2. Siegbahn, K., Nordling, C., Fahlman, A., Nordberg, R., Hamrin, K., Hedman, J., Johansson, G., Berkmark, T., Karlsson, S.E., Lindgren, I., and Lindberg, B., "ESCA, Atomic, Molecular and Solid State Structure Studied by Means of Electron Spectroscopy", Almquist and Wiksells, Uppsala, 1967.

3. Thomas, H.R., and O'Malley, J.J., Macromolecules, 12 323 (1979).

4. Clark, D.T. and Thomas, H.R., J. Polym. Sci., Polym. Chem. Ed. 14, 1671 (1976).

5. Crystal, R.G., Erhardt, P.F. and O'Malley, J.J., in "Block Copolymers:" S.L. Aggarwal, Ed., Plenum Press, N.Y., (1970) pp. 179-193 and references therein.

6. Short, J.M., and Crystal, R.G., Appl. Polym. Symp., 16, 137 (1971).

7. Lee, L.H, "Adv. Chem. Ser. 87", p. 106 (1968).

8. Clark, D.T., in "Progress in Theoretical Organic Chemistry", Vol. 2, I.G. Csizmadia, Ed., Elsevier, Amsterdam, 1976.

9. Clark, D.T., Adams, D.B., Scanlan, I.W., and Woolsey, I.S., Chem. Phys. Lett., 25, 263 (1974).

10. Clark, D.T. and Adams, D.B., J. Electron Spectrosc. Relat. Phenom., 7, 401 (1975).

11. Clark, D.T., Dilks, A., Peeling, J. and Thomas, H.R., Faraday Discuss. Chem. Soc., 60, 183 (1975).

12. Clark, D.T., Adams, D.B., Dilks, A., Peeling, J., and Thomas, H.R., J. Electron Spectrosc. Relat. Phenom., 8, 51 (1976).

13. Manne, R., and Åberg, T., Chem. Phys. Lett., 7, 282 (1970).

14. Scofield, J.H., J. Electron Spectrosc. Relat. Phenom., 8, 129 (1976).

15. Clark, D.T. and Thomas, H.R., J. Polym. Sci., Polym. Chem. Ed., 15, 2843 (1977).

16. Clark, D.T., and Thomas, H.R., J. Polym. Sci., Polym. Chem. Ed., 16, 791 (1978).

17. Clark, D.T., "Advances in Polymer Sciences", Vol. 24, H.-J. Cantow et al., Eds. Springer-Verlag, Berlin, 1977, p. 126.

18. Fadley, C.S., Baird, R.J., Siekhaus, W., Novakov, T., and Biegstrom, S.A.A., J. Electron Spectrosc. Relat. Phenom., 4, 93 (1974).

19. O'Malley, J.J., Crystal, R.G., and Erhardt, P.F., in "Block Polymers", S.L. Aggarwal, Ed., Plenum Press, New York, 1970, pp. 163-178.

20. Erhardt, P.F., O'Malley, J.J., and Crystal, R.G., ref 19, pp. 195-211.

21. Pochan, J.M. and Crystal, R.G., in "Dielectric Properties of Polymers", F.E. Karasz, Ed., Plenum Press, New York, 1972, pp. 313-327.

22. Crystal, R.G., Erhardt, P.F., and O'Malley, J.J., ref. 8, pp. 179-193.

23. Crystal, R.G., in "The Colloidal and Morphological Properties of Block and Graft Copolymers", G. Molav, Ed., Plenum Press, New York, 1971, pp. 279-293.

24. Meier, D.J., J. Polym. Sci., Part C, 26, 81 (1969).

25. Meier, D.J., Polym. Prepr., Am. Chem. Soc., Div. Polym. Chem., 11, 400 (1970).

26. Krause, S., Macromolecules, 3, 84 (1970).

27. Sanchez, I., in "Polymer Blends", Vol. 1, D. Paul, Ed., Academic Press, New York, 1978, pp. 115-139.

28. Dilks, A., Ph.D. Thesis, University of Durham, U.K., 1977.

29. Clark, D.T., and Dilks, A., J. Polym. Sci., Polym. Chem. Ed., 14, 533 (1976).

30. Clark, D.T. and Dilks, A., J. Polym. Sci., Polym. Chem. Ed., 15, 15 (1977).

RECEIVED March 19, 1981.

# Metal–Polymer Interfaces:
# Studies with X-Ray Photoemission

J. M. BURKSTRAND

Physics Department, General Motors Research Laboratories, Warren, MI 48090

Polymer substrates are often coated with metals for a wide variety of reasons. In many cases, the chemical condition of the surface has been found (1) to alter the adhesion of the metal film to the substrate. In particular, oxygen plasma treatment of polymer surfaces before metal deposition has been found (2, 3, 4) to increase the adhesion of the metal. We have been able to identify (5) hydroxyl, carbonyl and ester groups which were created on the polymer surfaces during oxygen plasma treatment. We have also identified (6) the formation of metal–oxygen–polymer complexes at an oxygen treated polymer surface and correlated their presence with an increase in the adhesion of the metal film.

In complimentary experiments, a number of other workers (7–14) have studied with photoemission the electronic structure of thin metal overlayers on inorganic substrates. In particular, Tibbetts and Egelhoff found for small metal clusters on a clean, amorphous carbon substrate that all the metal electronic binding energies increased by about 0.6eV with respect to the bulk values. They attributed this to either decreased extra–atomic relaxation energies or to an atomic renormalization (expansion) of the valance orbitals. We have attributed (15) analogous changes in the core binding energies of copper atoms on polystyrene to changes in both extra–atomic and intra–atomic relaxation energies.

We briefly describe here the results obtained from studies of copper, nickel and chromium overlayers deposited on polystyrene, polyvinyl alcohol, polyvinyl methyl ether, polyethylene oxide, polyvinyl acetate and polymethyl methacrylate. Using X-ray photoemission spectroscopy we measured significant variations in the core binding energies and lineshapes as we varied both the metal and the substrate atoms. These changes can be related to both differences between the intrinsic properties of the metal atoms as well as to differences in the interactions with the substrates. In the following sections we describe the details of

0097-6156/81/0162-0339$05.00/0

the experimental preparation, the data acquisition, the experimental results, and the significance of the measurements.

## Experimental

The interactions taking place on the surface were monitored with X-ray photoemission spectroscopy (XPS) using the same system as described previously (5, 6). This consisted of a commercial (16) double pass cylindrical mirror analyzer and a Mg $K_\alpha$ X-ray source in an ultra-high vacuum system whose operating pressure was about $2.6 \times 10^{-8}$ Pa ($2 \times 10^{-10}$ Torr). The cylindrical mirror analyzer was used with a constant pass energy of 50 eV or .08 eV energy resolution.

The polymer films were solvent cast on stainless steel substrates and air dried at 22C; their final thickness was about 0.001 mm. After insertion into the ultra-high vacuum chamber through a load-lock chamber, the polymers were warmed to temperatures above their respective glass transition temperatures for the time needed to remove the remaining solvent from the bulk of the film.

Copper and nickel were deposited from metal foil wrapped around a hot tungsten filament. Chromium was evaporated from a chrome plated tungsten wire. XPS measurements were made to determine the metal coverage as well as the electronic structure at the interface. The metal coverage was determined by substituting the experimentally measured areas under the XPS curves, core hole cross sections (17), and electron mean free path in both the metal (18) and the polymer (19) and an instrument response function into the equations for emitted electron intensity (20). At coverages near one monolayer, these values were checked with results from a quartz crystal thin film monitor.

During photoemission, a small buildup of positive charge occurred on the surface, resulting in a shifting of the energy scale by 1.0 - 2.5eV. This charging was accounted for by setting the binding energy relative to the fermi level of the C-1s electrons to 285.0eV for carbon atoms involved in $CH_2$ bonds and referencing the other levels to that value, a method which has been successfully used in the examination of other polymer surfaces. (5, 6, 12, 21).

## Results

Copper, nickel, or chromium was deposited on the clean polymers at coverages from 0.002 to 10.0 monolayers. At each coverage, the metal $2p^{3/2}$ core electron binding energy was measured with XPS. The peak positions of $2p^{3/2}$ lineshapes for a number of coverages of copper, nickel and chromium on polystyrene and on polyvinyl alcohol are plotted in Figs. 1-3 respectively. The data for the other metal/oxygen containing polymer systems are similar to that shown for polyvinyl alcohol (if all the metal binding energies are plotted as a difference from the bulk value).

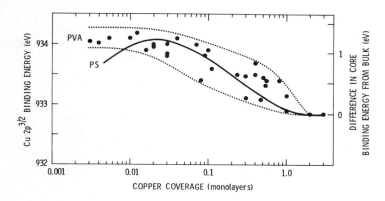

*Figure 1.    Variation in the Cu-2p$^{3/2}$ electron core binding energy as a function of coverage for vapor-deposited Cu on PS and PVA.  The individual data points for PS are not shown.*

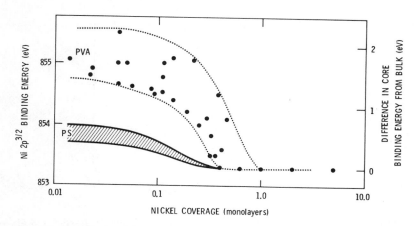

*Figure 2.    Variation in the Ni-2p$^{3/2}$ electron core binding energy as a function of coverage for vapor-deposited Ni on PS and PVA*

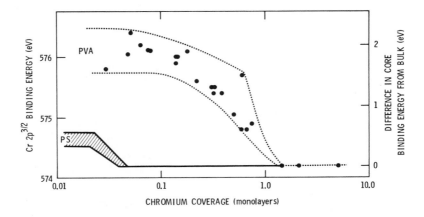

*Figure 3.    Variation in the Cr-2p$^{3/2}$ electron core binding energy as a function of coverage for vapor-deposited Cr on PS and PVA*

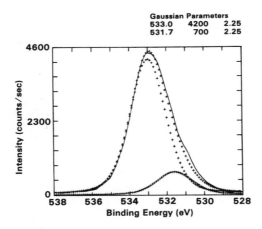

*Figure 4.    The O-1s XPS spectrum (———) from PVA following deposition of 0.3 monolayer of Ni. The inner curves (+), having the parameters as shown (peak center, height, FWHM), were fit to the data as described in the text. The outer curve (+) is the sum of the inner curves and shows the accuracy of the fit to the experimental data.*

In comparing the different sets of data, a number of overall differences are striking. First, the data for Cu, Ni and Cr on polystyrene were all quite reproducable, within 0.2eV. On the other hand, when these same metals were deposited on the oxygen containing polymers, the scatter was often as large as 1.0eV, with the data forming a broad band. Second, when nickel and chromium were deposited on polystyrene, the bulk value of the $2p^{3/2}$ core energy was reached when the metal coverage was well less than one monolayer. However, when these same metals were deposited on polyvinyl alcohol, the metal coverage needed to be about one monolayer before the bulk value was obtained. Third, the largest difference in the $2p^{3/2}$ core binding energy from the bulk value for Cr and Ni on PS is about 0.5eV while for these metals on polyvinyl alcohol it is as large as 2.2eV.

In addition to the information obtained from the core lineshapes of the overlayer atoms, there is much information to be derived from the lineshapes of substrate polymer atoms. For Cu, Ni and Cr deposited on polystyrene, there were no measurable changes in the C-1s core level except for a decrease in intensity caused by overlayer attenuation. However, the addition of these metal atoms to the surface of the oxygen containing polymers modified the lineshapes of the core levels of the surface carbon and oxygen atoms.

On the clean polyvinyl alcohol surface, the C-1s core lineshape consists of a single broad peak centered at 285.8eV and the O-1s lineshape of a single narrow peak at 533.0eV binding energy. The carbon line can be decomposed into two components (22), one with a peak at 285.0eV associated with a $CH_2$ group and one with a peak at 286.6eV associated with a HCOH group. After Cu, Ni, or Cr atoms were deposited, the overall intensity decreased and the lineshape altered. This alteration can best be described as a decrease in the 286.6eV component relative to that at 285.0eV. The overall intensity decrease of both the carbon and oxygen lines can be accounted for by simple overlayer attenuation. The relative decrease of the carbon 286.6 eV component implies a modification of the original carbon–oxygen bond.

Following deposition of Cu, Ni or Cr on polyvinyl alcohol, the oxygen 1s core level spectrum was broadened on the low binding energy side by the addition of a new peak. Figure 4 shows an example of this structure for coverage of about 0.3 monolayer of Ni. In all cases, the curve decomposition was made by a nonlinear least squares fitting of the data by lineshapes each composed of a Gaussian and a Lorentian, the relative weighting being 0.7 and 0.3 respectively. The instrumental function was not deconvolved out, and the new additional lineshape was assumed to have the same shape as the original oxygen 1s spectrum. With these assumptions, the mathematical uncertainty of the various component energy positions is about $\pm$ 0.1 eV. If the restrictions on lineshapes are reduced, this uncertainty becomes somewhat larger.

The data curve shown in Fig. 4 is comprised of a large peak at 533.0eV and one at 531.7eV. The higher binding energy component is that from the oxygen atoms in polyvinyl alcohol which have not been affected by the presence of Ni atoms. The new oxygen signature is shifted 1.3eV to lower binding energy. For this case, and in general, the size of this shift in binding energy was independent of metal coverage. The intensity of the new peaks scale approximately with coverage for metal coverages less than one monolayer. These same results are true for polyvinyl methyl ether and polyethylene oxide -- both polymers with single bonded oxygen.

The analysis of the data from polyvinyl acetate and polymethyl methacrylate is not as straightforward. Figure 5 shows the oxygen 1s spectrum from clean polymethyl methacrylate which is easily decomposed into two components: a peak at higher binding energy from the single bonded oxygen and a peak at lower binding energy from the double bonded oxygen. When 0.6 monolayer of Ni is deposited on this surface, the oxygen 1s spectrum changes to that shown by the solid curve in Fig. 6. This can be deconvolved in two ways. First, we can assume that the original components are fixed in energy and that only a third peak need be added, as is illustrated in Fig. 6. While this is a straightforward decomposition, we believe that both the single and double bonded oxygen atoms should interact with the adsorbed Ni atoms. Thus there should be another peak "hidden" beneath that at 532.2eV. This fourth peak would be the XPS signature of the single bonded oxygen atoms which interacted with the Ni. If we assume that this peak is shifted in binding energy the same as those single bonded oxygen atoms in polyvinyl alcohol and polyvinyl methyl ether (-1.3eV), then the resulting curve decomposition is shown in Fig. 7. Similar procedures were carried out on the other metal/polymer systems.

The changes in the oxygen 1s core binding energies are summarized below. For polyvinyl acetate and polymethyl methacrylate, the first value is for the single bonded oxygen, the second for the double bonded oxygen.

|                            | Cu          | Ni          | Cr          |
| -------------------------- | ----------- | ----------- | ----------- |
| Polyvinyl alcohol          | -2.7        | -1.3        | -1.7        |
| Polyvinyl methyl ether     | -1.7        | -1.3        | -1.3        |
| Polyethylene oxide         |             | -1.2        | -1.3        |
| Polyvinyl acetate          | -1.3, -1.4  | -1.3, -0.9  | -1.3, -1.5  |
| Polymethyl methacrylate    |             | -1.3, -1.0  | -1.3, -1.6  |

Discussion

After a careful analysis of the data, it is clear that the

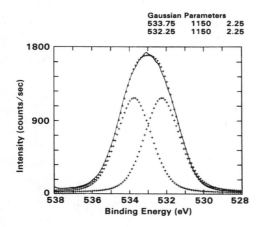

Figure 5.   *The O-1s XPS spectrum (──) from clean polymethyl methacrylate. The curves comprised of (+) are as described in Figure 4.*

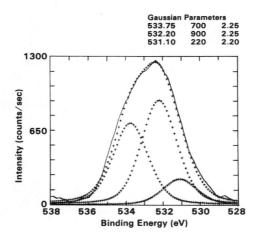

Figure 6.   *The O-1s XPS spectrum (──) from polymethyl methacrylate following deposition of 0.6 monolayer of Ni. The curves comprised of (+) are as described in Figure 4.*

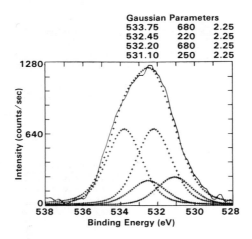

Figure 7.    The O-1s XPS spectrum (———) from polymethyl methacrylate following deposition of 0.6 monolayer of Ni. The curves comprised of (+) are as described in Figure 4.

substrate plays a significant role in determining the properties of the metal overlayer. This is not entirely surprising in light of previous work using metal and semiconductor substrates. When a metal atom first bonds to an oxygen-containing polymer surface, it is able to bond rather strongly with an oxygen atom already present, forming a metal-oxygen-polymer complex (6). This changes both the initial electronic state of the metal atom as well as the atomic and extra-atomic screening contributions to the measured photoemission energy.

It has previously been pointed out (8, 9, 11, 12, 15, 23) that changes in screening can account for the general changes in binding energy as a function of coverage such as are observed in Figures 1-3.

As the size of the metal cluster increases, the amount of extra-atomic screening of the core hole left after photoemission increases until the screening equals that of the bulk. The differences between Cu, Ni and Cr on polystyrene can be attributed (23) to differences in the mobilities of the metal atoms. Both the Ni and Cr $2p^{3/2}$ core levels reach the bulk value before one monolayer, indicating that they form larger clusters with bulk screening values more rapidly than does Cu.

But when these metals are deposited on polyvinyl alcohol, the shapes of the curves change dramatically. In addition, the FWHM for Ni and Cr also change. This suggests a rather different state for the metal atoms on the surface in which the initial state and/or atomic and extra-atomic screening is different.

We can obtain another clue to the makeup of this new state by examining the changes in the substrate core lineshapes, Figure 4. The changes in the carbon (not shown) and the oxygen lineshapes indicate that the metal atoms have perturbed the original C-O bonds and formed (partial) bonds with the oxygen atoms. The relative intensity of the change is a measure of the strength of the new bonds. The copper, not shown, produces the smallest change in both the carbon and oxygen spectra -- the Cu coverage is almost twice as large as that for Ni and Cr. This is not unexpected, as copper does not react as strongly with oxygen as does nickel and chromium (24, 25, 26, 27). The initial sticking coefficient of oxygen on Cu is lower than it is on Cr or Ni. In addition, the dissociation energy of copper oxide is much lower than that of chromium or nickel oxide. For all three metals, the additional structure in the O-1s spectra lies between the binding energy of the oxygen in the clean polymer and the appropriate bulk metal-oxide binding energy (24, 25) or the binding energy for chemisorbed oxygen (25, 26, 27). This implies that the oxygen is still partially bonded to the polymer, and indeed a metal-polymer complex has been formed at the interface.

The existence of these metal-oxygen-carbon complexes can account for many of the observed changes already discussed. We expect the metal to be more tightly bonded to the polymer (polyvinyl alcohol) through such a complex than if one was not present, as on clean polystyrene. Indeed, quantitative adhesion results confirm this -- copper, nickel and chromium adhere better to polyvinyl alcohol than to clean polystyrene and better to oxygen treated polystyrene than to clean polystyrene (6, 28).

A stronger bond between the metal and the polymer would reduce the migration of metal atoms on the surface. This means smaller clusters would be formed at any given coverage, implying smaller relaxation energies and larger chemical shifts with a resulting larger shift in core binding energy from the bulk value, as evidenced in Figs. 2-3. This would also account for the larger scatter in the data on polyvinyl alcohol, for the metal atoms would tend to stay near where they first contacted the surface. Thus a series of different depositions on polyvinyl alcohol could easily give a series of different results. However, on polystyrene, where the binding of the metal atoms is relatively weak, the memory of the arrival site tends to be erased.

Unfortunately, the exact configuration of this complex cannot be deduced from the energy values of the XPS structure. There is simply not enough information. However, it is possible that a vibrational spectroscopy such as infrared, Raman or electron energy loss would yield the molecular configuration of such a polymer-metal-complex.

On the other hand, the XPS data we do have gives us a clue to the atomic makeup of these complexes. At the lowest coverages, the $2p^{3/2}$ core binding energies of Ni and Cr on polyvinyl alcohol are about the same as those for bulk NiO and $Cr_2O_3$, 854.6eV (25) and 576.6eV (26) respectively. This suggests that these polymer metal complexes contain one or two metal atoms per oxygen atom. In addition, the value of the oxygen 1s core binding energy is also approaching that for these metal oxides. The oxygen 1s core shift obtained after depositing Cu is, however, larger than one first expects. The lower electronegativity of Cu implies that a Cu atom would transfer a smaller electronic charge than Ni or Cr, and this would produce a smaller core shift in the oxygen. This implies that more Cu atoms are involved in the complexes than Ni or Cr atoms. This is confirmed by the relative coverage data -- it takes about twice as much Cu than Ni or Cr to produce an equal number of new metal-oxygen complexes. If the Ni and Cr complexes contain one metal atom, then this suggests that the Cu complexes on polyvinyl alcohol contain two Cu atoms.

The data for the other oxygen-containing polymers is a little less clear. The sizes of the oxygen 1s core shifts following deposition of Ni suggests that the Ni atoms interact with the single bonded oxygens to form complexes which are similar to one another and approximately independent of the kind of polymer. This is probably the case with Cr also. However, the oxygen 1s core shift following Cr deposition is smaller for polyvinyl methyl ether and polyethylene oxide than it is for the other polymers. This is not totally unreasonable, as the oxygen in these two polymers is bonded between two carbon atoms, and one might expect that the Cr could not interact as strongly with it.

In general then, it appears as if these metal-polymer complexes have a similar nature. The data indicates that the metal
atoms bond through the oxygen atoms to the polymer chains. The
ratio of Ni or Cr atoms to interacted oxygen atoms (using peak
areas) is about 1:2. We suggest, then, that the simplest model
for these complexes which is consistent with the data and with
bulk metal-polymer chemistry is one in which a one or two atom
cluster of metal atoms forms a chelate (29) structure with the
polymer. In polyvinyl alcohol this would have the form of a metal
atom or cluster "crosslinking" adjacent polymer chains through
four oxygen atoms. On the other polymers, the arrangement of the
atoms in the complexes would be similar, but not as simple;
especially in the cases where both single and double bonded oxygen
atoms are present in the pendant groups.

## Abstract

The interfaces formed by evaporating copper, nickel and
chromium layers on polystyrene, polyvinyl alcohol, polyethylene
oxide, polyvinyl methyl ether, polyvinyl acetate and polymethyl
methacrylate have been studied with X-ray photoemission spectroscopy (XPS). At submonolayer coverages of the metals, the peak
positions and widths of the metallic electron core levels vary
significantly from one polymer substrate to another. Most of
these variations can be accounted for in terms of changes in the
atomic and extra-atomic relaxation energies during the photoemission process. Much of this change is brought about when the metal
atom deposited on an oxygen containing polymer interacts with the
substrate oxygen and forms a metal-oxygen-polymer complex. The
presence of this complex is verified by changes in the photoemission lineshapes of the substrate carbon and oxygen atoms. The XPS
signature of these various complexes are quite similar and suggest that they are chelate-like complexes.

## Literature Cited

1.  Mittal, K.L., *J. Vac. Sci. Technol.*, 1976, <u>13</u>, 19 and
    references therein.

2.  Thayer, D.W. and Wedel, R., (unpublished).

3.  Hall, J.R., Westerduhl, C.A., Devine, A.T. and Bodnar, M.J.,
    *J. Appl. Polym. Sci.*, 1969, <u>13</u>, 2085.

4.  Hall, J.R., Westerduhl, C.A., Devine, A.T. and Bodnar, M.J.,
    *J. Appl. Polym. Sci.*, 1972, <u>16</u>, 1465.

5.  Burkstrand, J.M., *J. Vac. Sci. Technol.*, 1978, <u>15</u> 223.

6.   Burkstrand, J.M., Appl. Phys. Lett., 1978, 33, 387.

7.   Kiang, K.S., Salaneck, W.R. and Aksay, I.A., Solid State
     Commun., 1976, 19, 329.

8.   Ascarelli, P., Cini, M., Missoni, G. and Nistico, N., J.
     Phys., (Paris) Colluque, 1977, 38, 125.

9.   Takasu, Y., Unwin, R., Tesche, B., Bradshaw, A. and Grunze,
     M., Surface Science, 1978, 77, 219.

10.  Tibbetts, G.G. and Egelhoff, W.F., Jr., Phys. Rev. Lett.,
     1978, 41, 188.

11.  Tibbetts, G.G. and Egelhoff, W.F., Jr., (a) J. Vac. Sci.
     Technol., (in press); (b) Solid State Commun., (in press);
     (c) Phys. Rev., (to be published).

12.  Mason, M.G., Gerenser, L.G. and Lee, S.T., Phys. Rev. Lett.,
     1977, 39, 288.

13.  Kim, K.S. and Winograd, N., Chem. Phys. Lett., 1975, 30, 91.

14.  Mason, M.G., Baetzold, R.C., J. Chem. Phys., 1976, 64, 271.

15.  Burkstrand, J.M., J. Vac. Sci. Technol., 1978, 15, 658;
     Surface Science, 1978, 78 513.

16.  Physical Electronics Industries, Eden Prairie, Minn.

17.  Scofield, J.H., Lawrence Livermore Laboratory Report No.
     UCRL 51326.

18.  Tracy, J.C. and Burkstrand, J.M., CRC Crit. Review, Solid
     State Sci., 1974, 4, 381.

19.  Evans, S., J. Phys. C., Solid State Phys., 1977, 10 2483;
     Clark, D.T., Thomas, H.R. and Shuttleworth, D., J. Polym.
     Sci. Polym. Letters Ed., 1978, 16, 465.

20.  See for example: Joshi, A., Davis, L.E. and Palmberg, P.W.,
     Czanderna, A.W. Ed. "Methods of Surface Analysis," Elsevier
     Scientific Publishing Company, Amsterdam, 1975, p. 159.

21.  Clark, D.T. and Thomas, H.R. J. Polym. Sci. Chem. Ed., 1976,
     14, 1671.

22.  Clark, D.T. and Thomas, H.R., J. Polym. Sci. Chem. Ed., 1978,
     16, 791.

23. Burkstrand, J.M., J. Appl. Phys., 1979, 50, 1152.

24. Haber, J., Machej, T., Ungier, L. and Ziolkowski, J., Journal of Solid State Chemistry, 1978, 25, 207.

25. Fleisch, T., Winograd, N. and Delgass, W., Surf. Sci., 1978, 78, 141.

26. Allen, G.C., Tucker, P.M. and Wild, R.K., Faraday Trans. II., 1978, 74, 1126.

27. Tibbetts, G.G., Burkstrand, J.M. and Tracy, J.C., Phys. Rev. B., 1977, 15, 3652.

28. Burkstrand, J.M., to be published.

29. For the bulk analog, see Tsuchida, E. and Nishide, H., "Advances in Polymer Sciences," Vol. 24, Springer-Verlag, New York, 1977, p. 1.

RECEIVED January 21, 1981.

# Surface Characterization of Plasma-Fluorinated Polymers

M. ANAND, R. E. COHEN, and R. F. BADDOUR

Department of Chemical Engineering, Massachusetts Institute of Technology, Cambridge, MA 02139

Previous Work. In previous publications (1, 2, 3) we have reported on the surface fluorination of low density polyethylene using a low pressure atmosphere of a dilute mixture of fluorine in helium and using a cold plasma generated from this same gas mixture. It was found that in the glow discharge, perfluorination of the surface was accomplished readily, whereas with elemental fluorine treatment, the reaction was slow and did not lead to complete surface fluorination under the conditions employed (1). In dilute elemental fluorine reactions, several partially fluorinated species, such as $CH_2$-CHF, CHF-CHF, were formed on the surface as determined by X-ray photoelectron spectroscopy measurements (XPS), while in the case of plasma treatment, largely-$CF_2$ type groups were generated. Presence of $CF_2$ groups on the surface was also shown by Fourier Transform Multiple Internal Reflectance (FTMIR) infrared studies and by advancing contact angle measurements. Previous experiments also showed that, for reaction conditions under which complete surface fluorination was not obtained, a post-reaction with oxygen and oxygen-containing species (largely moisture from the air) occurred, resulting in large oxygen signals in XPS measurements and in sharp decreases in contact angles (1, 3). However, for samples which were fully fluorinated at the surface, very small oxygen levels were found, and this favorable condition persisted indefinitely. No reaction or post-reaction with oxygen or water is observed for reactions that go to completion. In previous fluorination studies in which elemental fluorine was employed, several oxygen-containing species including acylfluoride groups, carboxyl and hydroxyl groups, and hydrogen-bonded water have been found in the surface and sub-surface layers (4, 5, 6, 7).

Reaction conditions which influence the fluorination of LDPE in the fluorine plasma have also been discussed previously (2). High pressures and flow rates favored the fluorination reaction kinetics. There was a competition between ion-assisted etching

0097-6156/81/0162-0353$05.00/0

and chemical reaction resulting in small (~40A) depths of fluorination.  However, the depth of fluorination was increased in
reactions carried out under ion-depleted conditions.  This was
achieved at the expense of the reaction rate which was reduced by
a factor of about two (3).  Even under these reaction conditions,
an upper limit on the depth of fluorination (~60A) was observed
possibly due to a diffusion-controlled reaction.

    Present Work.  In this paper we report the results of
reactions carried out under conditions of higher fluorine concentrations with the aim of increasing the depth of fluorination.
Results for reactions with several other polymers are also reported.  Further, a discussion of the fluorination of LDPE powders
in a fluidized bed reactor and the molding properties of these
powders is also included.
    The apparatus being used to treat polymer films is shown in
Fig. 1(a) and for treatment of powders in Fig. 1(b).  The surface
analytical tools employed for characterizaton are X-ray photoelectron spectroscopy (~50A), FTMIR ( 10000A) and contact angle
measurements (first molecular layer), the details of which have
been previously described (3).  In the results being presented
here, 5% and 15% premixtures of fluorine with helium are used in
the fluorine glow discharge.
    The gases, obtained from Matheson Gas,were passed through a
sodium fluoride column to remove trace amounts of hydrogen fluoride.  No effort was made to remove small amounts of water and
oxygen that may have been present in the incoming gases.  The
polymer films were treated at the plasma bulk gas temperature
which was close to ambient for all reactions.  Visual observation
of the films did not show any melting or deformation and hence it
was assumed that if the surface temperatures did vary, the
fluctuations were small.

## Results and Discussion

    Reactions Under Ion-Shielding Conditions.  Fluorine-contain-
ing plasmas are known to cause etching in both organic and
inorganic systems (8, 9, 10, 11).  In the case of hydrocarbon
polymers, ion assisted etching is apparently caused by fluorine
ions such as $F_2^+$, $F_2^-$, $F^-$ in the plasma.  Our experiments in the
glow region of the plasma have clearly shown etching taking place
as evidenced by the apparent upper limit on depth of fluorination
(3).  The amount of etching could be reduced by elimination of
ions from the reaction zone.  This can be achieved either by
isolating the samples from the plasma by enclosing them in a
Faraday cage (an electrically grounded metal screen) inside the
glow region or by carrying out the reactions in the dark region
downstream of the plasma.  In our study we have used a modified
Faraday cage in which the metal screen was not grounded.  The
metal acts as a sink for the ions and permits only free radicals

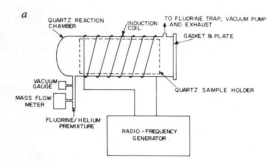

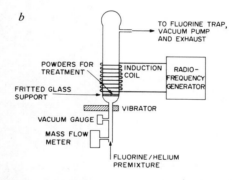

*Figure 1.   Schematic of experimental setup: (a) reactor for film treatment; (b) fluidized bed for treatment of powders*

with mean free paths smaller than the screen opening into the reaction zone. This was clearly observed by a glow outside the screen but a dark region inside it.

In the results published earlier (3), we showed that with $F_2$/He mixtures containing 5% fluorine, the depth of fluorination was increased from 40A in the case of reaction in the glow to about 60A for reactions in the dark region. However, we could not achieve treatments to depths much beyond 60A possibly due to inadequate concentration of fluorine species at the solid substrate or due to extremely slow propagation reaction into the bulk. To test this hypothesis, experiments were carried out with a $F_2$/He mixture containing 15% $F_2$ in the dark region inside the metal cage. The results of these experiments are shown in Figures 2, 3, 4, 5 and Tables 1 and 2.

TABLE I

DEPTH OF FLUORINATION FOR VARIOUS PLASMA TREATMENTS

| Conditions | Time Of Treatment, min | Depth, A |
|---|---|---|
| 3.0 mm, 40 cc/min, 50W | 2 | 30 |
| Aluminum Cage | 5 | 38 |
| (5% $F_2$) | 15 | $>$60 |
|  | 60 | $\sim$60 |
| 3.0 mm, 40 cc/min, 50W | 1 | 22 |
| No Cage | 2 | 30 |
| (5% $F_2$) | 60 | 35 |
| 3.0 mm, 40 cc/min, 50W | 2 | $>$60 |
| Aluminum Cage | 5 | $>$60 |
| (15% $F_2$) | 15 | $>$60 |

Table 1 compares depths of fluorination calculated (4) from XPS measurements for reactions in the glow and in the dark region with both 5% and 15% $F_2$ premixtures. In the case of 15% fluorine, the depth of fluorination is beyond that which can be measured by XPS even after only 2 minutes of treatment. An estimate of the depth fluorination was obtained by assuming that the increase in weight of the treated polymer was due to substitution of hydrogens in the polymer by fluorine. The results of these calculations are shown in Table 2. The exact values of the depths of fluorination must be viewed cautiously because of complicating factors such as adsorption and chemisorption (especially for polymers not per-fluorinated) of water on the surface in the time between taking the sample out of the reactor and weighing it. However, trends are clear especially for the case of 15% $F_2$ treatment where large increases in weight are observed. These results suggest much greater depths of fluorination with increased treatment time, supporting the hypothesis that observed limitations in the depth of fluorination arise from a diffusion controlled reaction. Additional evidence in support of this increase in fluorination depth with treatment time is provided by the observed increase with time in the intensity of the infrared band corresponding to $-CF_2$ species, as shown in Fig. (2).

TABLE II

DEPTH OF FLUORINATION ESTIMATED FROM WEIGHT INCREASES

Reaction conditions: 3.0 mm, 40 cc/min, 50W, Aluminum Cage

| Time of Treatment | % of Increase in Weight | | Depth of Fluorination, A | |
|---|---|---|---|---|
| | $5\%F_2$, 95%He | $15\%F_2$, 85%He | $5\%F_2$, 95%He | $15\%F_2$, 85%He |
| 2 min | 0.136 | 0.113 | 235 | 195 |
| 5 min | 0.096 | 0.170 | 160 | 290 |
| 15 min | 0.155 | 0.276 | 260 | 475 |

The increase in concentration of fluorine from a 5% to a 15% mixture should accelerate the reaction because of increased reactive species available for reaction. This is clearly seen in Fig. (3) in the plot of contact angle against time of treatment. Curve (b) for treatment with 15% $F_2$ suggests that the surface layer is perfluorinated (formamide has an advancing contact angle of 92° with PTFE) within 5 minutes of treatment, which is not the case for treatment with the 5% fluorine plasma treatment. Higher oxygen content is also seen for the 5% $F_2$ plasma at 5 minute treatment time, suggesting a case in which the surface is not fully fluorinated and is susceptible to chemisorption of water. However, Fig. 4, a plot of atomic percent fluorine vs. time of treatment does not suggest a faster reaction with the 15% $F_2$ mixture. This result may be explained if the active fluorine first attacks the amorphous zones of the polyethylene and the reaction propagates beyond the first 60A before the crystalline regions are fluorinated. In such a case, the XPS signals from $F_{1s}$ and $F_{2s}$ would suggest a depth of fluorination of greater than 60Å because the calculation assumes a homogeneous material. Such a situation possibly exists in our case because the $C_{1s}$ XPS spectra of Fig. 5 clearly show a residual $CH_2$ signal for the case of the 2 minute treatment. The reaction in the glow region of the plasma is clearly faster than in the dark region of the plasma as seen

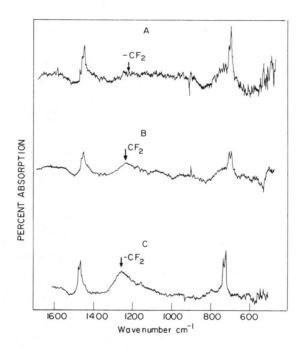

*Figure 2. FTMIR spectra of LDPE treated with 15% fluorine mixture with helium under ion-depleted conditions: (A) 2-min, (B) 5-min, and (C) 15-min treatments. Reaction conditions: 40 cc/min, 3.0 mm, 50 W; spectra at 45° incidence.*

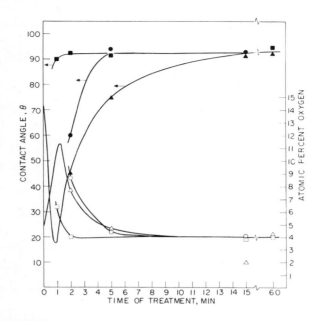

*Figure 3. Variation of contact angle and atomic percent oxygen as a function of time of treatment ((■, □) 5% F₂, no cage; (▲, △) 5% F₂, aluminum cage; (●, ○) 15% F, aluminum cage; reaction conditions: 3.0 mm, 40 cc/min, 50 W).*

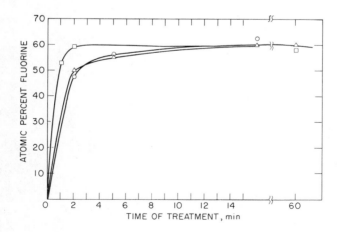

*Figure 4. Build-up of fluorine on the surface of LDPE with treatment time ((□) 5% F₂, no cage; (△) 5% F₂, aluminum cage; (○) 15% F₂, aluminum cage; reaction conditions: 3.0 mm, 40 cc/min, 50 W).*

PLASMA TREATMENT UNDER ION-DEPLETED CONDITIONS

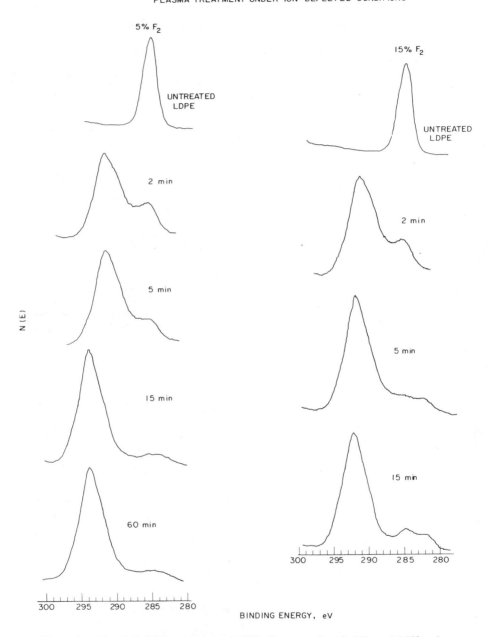

*Figure 5.   The C-1s XPS spectra for LDPE film treated with 5% and 15% mix-tures of fluorine in helium.  Reactions carried out in the dark region of the plasma.*

from both Figures (3) and (4) and from the $C_{1s}$ XPS spectra reported earlier ($\underline{1}$, $\underline{3}$).

Another important feature of the reactions in the ion-depleted zone is that $-CF_2$ type groups are formed predominantly as in the case of reactions in the glow region of the plasma. This is seen by the chemical shift or about 6eV from the original $-CH_2CH_2-$ peak in the $C_{1s}$ XPS spectra (Fig. 5). The details of such assignments have been discussed earlier ($\underline{3}$). This is very different from reaction with elemental fluorine ($\underline{3}$, $\underline{4}$) where several partially fluorinated species are formed before genera-tion of $-CF_2$ groups. Based on XPS data, it appears that the $-CF_2$ groups formed in the case of treatment with and without the metal screen are identical. However, FTMIR spectra reveal that there are differences in the type of $-CF_2$ groups as shown in Figs. (2) and (6). The predominant $-CF_2$ species formed from treatment in the aluminum cage appear at 1250 cm$^{-1}$ as opposed to buildup of $-CF_2$ species at 1090 cm$^{-1}$ for treatment in the glow region. The bands associated with fluorine-containing species appear in our spectra in the range of 1000 cm$^{-1}$ to 1300 cm$^{-1}$. The bands at 1090 cm$^{-1}$ and 1250 cm$^{-1}$ are possibly associated with $-CF_2$ symmetrical $(E_1)$ and $-CF_2$ assymetrical $(E_1)$ stretches respectively ($\underline{12}$, $\underline{13}$, $\underline{14}$). (The band due to $-CF_2$ symmetrical stretch is actually re-ported to occur at 1141 cm$^{-1}$ ($\underline{12}$, $\underline{13}$, $\underline{14}$) but it seems to mix with FCF bend, CC stretch and CCC bend as a consequence of which the position might be shifted to a lower value. Fluorine contain-ing CF unsaturates and CF saturates are also known to occur in the range of 1000 cm$^{-1}$ - 1200 cm$^{-1}$. It is possible that in the depth being sampled, some of these groups maybe present especially if there is a gradient in the polymer composition from a completely $-CH_2 - CH_2-$state in the bulk to a completely $-CF_2 - CF_2-$state on the surface). This suggests subtle differences in the two fluorinated polymers.

Further, infrared spectra reveal that in the case of plasma treatment in the glow, there is some carbon-carbon unsaturation present on the surface evidenced by band at 890 cm$^{-1}$ (Fig. 6). These are not present in the case of treatment in the dark region of the plasma. Also, the amorphous nature of the fluorinated polymer layer is exhibited by the small band around 740 cm$^{-1}$ ($\underline{12}$).

## Fluorination of Other Polymers

Surface fluorination of several polymers other than poly-ethylene has been attempted to show the effectiveness of this technique in generating perfluorinated surfaces. Results of formamide contact angle experiments are summarized in Table (3). It appears from these results that under the conditions studied, polypropylene and nylon 6,6 can be perfluorinated but polymethyl-methacrylate does not perfluorinate in the time scale shown. The $C_{1s}$ XPS spectra and the respective chemical compositions are shown in Figure (7) and Table (3). Even though PMMA shows high

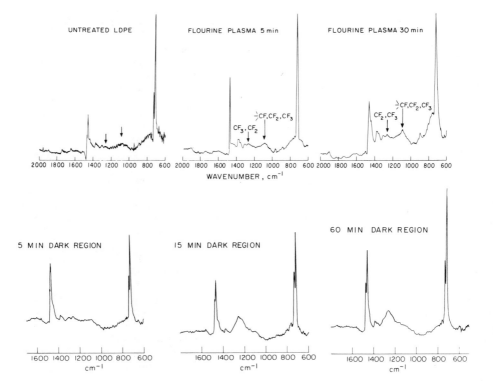

*Figure 6.    FTMIR spectra comparing treatments in the glow and the dark region of the plasma (spectra at 30° incidence)*

fluorine contents and reasonable depths of fluorination, the
contact angle with formamide is low, suggesting that the oxygen
must reside in the first molecular layer. It is possible that
oxygen in the polymethylmethacrylate molecule is disassociated
from the molecule chain and post-reacts with the surface after the
plasma is turned off. Results of contact angle measurements from
nylon 6,6 and polypropylene are surprising because the $C_{1s}$ XPS
spectra suggests partially fluorinated surfaces. However, the
first molecular layers must be perfluorinated as suggested by
contact angle measurements and calculations of depth of fluorination.

TABLE III

COMPOSITION OF FLUORINATED FILMS OF VARIOUS POLYMERS

| Polymer | Treatment* Time | F | Atomic % O | C | N | $d_F$, $\overset{o}{A}$ | Contact Angle, ** |
|---|---|---|---|---|---|---|---|
| Polymethylmethacrylate | 0 | – | 25.2 | 74.8 | – | – | 56$^o$ |
| Polymethylmethacrylate | 2 | 44.6 | 13.2 | 42.2 | – | 20 | 15$^o$ |
| Polymethylmethacrylate | 5 | 52.7 | 7.8 | 39.5 | – | 29 | 40$^o$ |
| Nylon 6,6 | 0 | – | 15.4 | 81.6 | 3.0 | – | 55 |
| Nylon 6,6 | 3 | 49.6 | 9.3 | 39.4 | 1.7 | 5 | 92–93$^o$ |
| Polypropylene | 0 | – | 7.0 | 93.0 | – | – | 78$^o$ |
| Polypropylene | 2 | 49.3 | 7.2 | 43.5 | – | 26 | 90 |
| LDPE | 0 | – | 5.0 | 95.0 | – | – | 71 |
| LDPE | 2 | 56.5 | 5.2 | 38.3 | – | 37 | 92–93$^o$ |

*  All with 5% $F_2$ mixture at 2.0 mm, 40 cc/min., 50W power

** with Formamide

## Treatment Of Polymer Powders

     Low density polyethylene was cryogenically ground to yield a
powder with a specific surface area of about 0.15 m$^2$/g as determined by BET measurements. We found that for powders which were
not fully fluorinated (short reaction times) significant amounts
of water were adsorbed on the surface. As a consequence of this,
the powders seemed similar in appearance to moist salt. Further,
a decrease in pH was observed when these powders were suspended in
distilled water, suggesting the presence of acidic groups. Also,
when these samples were mounted for XPS experiments, the water
apparently did not completely desorb; when the X-ray source was

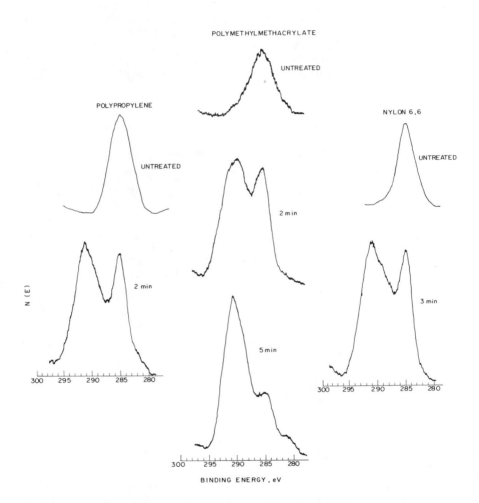

*Figure 7.  The C-1s XPS spectra for polypropylene, polymethylmethacrylate, and Nylon 6,6 treated in the plasma*

moved to such a sample, outgassing was observed as indicated by an increase in pressure in the XPS chamber. Thus we concluded that water was chemisorbed to the acidic sites on the polymer surface. Most significantly, for these specimens, very large amounts of oxygen were found on the surface with very small carbon signals. Furthermore, the sample charging properties were altered so dramatically that the peaks were shifted to lower binding energies by as much as 20 - 30eV. The signal strengths and widths varied during the experiments which made it impossible to take any meaningful spectra. It is possible that if 2-3 layers of water are chemisorbed, little or no carbon signal will be seen and when the X-ray source is directed on the sample, the energy may be high enough to debond some water resulting in a continuous change of the surface. This explanation is consistent with our observations. However, for powders which had undergone significant extents of fluorination, XPS spectra could be obtained easily. The results presented below are for such cases.

As shown in Fig. (8), fluorination reactions of polyethylene powders in the fluidized bed were slower compared to those of films in the plasma. This may be related to the greater surface area involved in the treatment of powders ($\sim 0.15$ $m^2$/g for powder compared to $0.03$ $m^2$/g for film). Further, for treatment of powders, the overall extent of fluorination is lower compared to that of films - 52% F as opposed to 60% for films (Fig. 8). This is also supported by the $C_{1s}$ spectra (Fig. 9) in which residual -$CH_2$ signals are present at 285.0eV. This raises questions about uniformity of the treatment and the depth to which fluorination has progressed. On the basis of calculations similar to those used for films, we found that the depth of fluorination (in the cases where atomic percent fluorine was of the order of 52%) was about 40-50A which suggested that if the treatment were uniform, the residual -$CH_2$ signal should have been very small. It is possible that the treatment is not uniform, which could be related to poor mixing due to inadequate fluidization or due to zones which might not be accessible for fluorination because of formation of stagnant layers, especially of HF and $H_2$, which prevent further reaction. Figure (9) is also useful in identifying chemical species that are formed on the polymer powder. On the basis of chemical shifts as in the case of polymer films, the species generated are largely -$CH_2$.

Differential scanning calorimetry ($20^{\circ}$K/min over the range $350^{\circ}$K to $650^{\circ}$K) of the treated powders showed no change in the melting point of polyethylene, and did not reveal any significant features attributable to the fluorinated polymer. The treated powders could not be molded under conditions which were suitable for the untreated polyethylene ($150^{\circ}$, 30 MPa). However, the treated powders could be preformed at $95^{\circ}$C and 60 MPa to allow for cold compacting in an isostatic press. The cold compacting improved the mechanical strength of the molded polymers but not to an extent where they could be practically used. The conditions

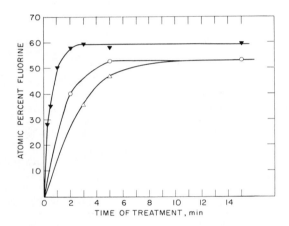

*Figure 8.    Comparison of fluorination of LDPE films and powders ((▼) 40 cc/ min, 2.0 mm, 50 W, film; (○) 40 cc/min, 9.0 mm, 25 W, powder; (△) 30 cc/min, 9.0 mm, 25 W, powder)*

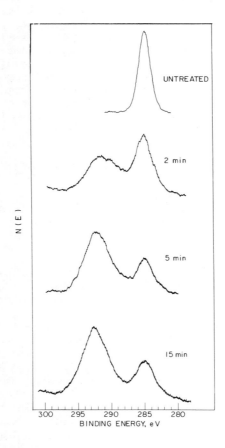

*Figure 9. The C-1s XPS spectra for fluorinated powders (reaction conditions: 9.0 mm, 50 W, 40 cc/min)*

for sintering of these powders is being investigated with the aim to improve the mechanical properties without damaging the fluorinated layer.

## Summary

Reactions in the plasma can be carried out in the glow region in the presence of ions or in the dark region in the absence of ions. In the fluorine plasma treatments we found that for experiments carried out in the glow, there was a competition between chemical reaction and ablation of material from the surface of the polymer thus limiting the depth of fluorination. An increase in the depth of fluorination was achieved by shielding the specimens from ions. This resulted in a drop in the rate of chemical reaction on the surface. In both these cases $-CF_2$ groups were rapidly formed on the surface as determined by XPS measurements. FTMIR experiments suggest that some differences do exist in the organization of the $-CF_2$ groups in the fluorinated polymer surfaces obtained from these reactions.

An apparent upper limit on the depth of fluorination was observed even for reactions under ion-depleted conditions. This was attributed to an insufficient fluorine gradient from the surface of the polymer into the bulk. An increase in this gradient obtained by an increase of fluorine concentration in the plasma resulted in greater depths of fluorination suggesting a mechanism controlled by diffusion of fluorine species into the polymer.

Perfluorination on the surfaces of polypropylene and nylon 6,6 could be achieved but not with polymethylmethacrylate under the reaction conditions employed. It is possible that the oxygen from the PMMA chain is dislodged from the molecule and post-reacts with the surface when the plasma is turned off.

LDPE powders treated with the fluorine plasma in a fluidized bed did not show perfluorination on the surface even though the chemical groups generated were largely $-CF_2$. This could be associated with non-uniform treatment caused by inadequate mixing in the fluidized bed or due to formation of stagnant HF and $H_2$ layers in areas of the bed. The pre-forming conditions for molding of these powders were identified; sintering conditions without disrupting the fluorinated layer are being investigated.

## Acknowledgement

This research is supported by National Science Foundation, Division of Engineering.

## Abstract

A study of fluorination of low density polyethylene in the glow and dark regions of a low temperature plasma is reported.

The gas feed to the plasma consisted of dilute mixtures of fluorine in helium. Reactions which were carried out in the glow region of the plasma and which employed a fluorine/helium ratio of 0.05 revealed that the depth of fluorination was limited to about 40A owing to ion assisted etching. This depth was increased to about 60A by carrying out the reactions under ion-depleted conditions but the small fluorine concentration retarded the advance of the fluorinated layer into the bulk. This process was accelerated by increasing the fluorine/helium ratio to 0.15. The kinetics of these fluorination processes are discussed qualitatively. Results of plasma fluorinations of polymers other than low density polyethylene are also reported. Fluorine plasma treatment of low density polyethylene powders in a fluidized bed reactor is also described along with the molding properties of these powders.

## Literature Cited

1.  Anand, M., Cohen, R.E. and Baddour, R.F., ACS Polymer Preprints, 1979, 20(2), 507.

2.  Anand,M., Cohen, R.E. and Baddour, R.F., ACS Polymer Preprints, 1980, 21(1), 139.

3.  Anand, M., Cohen, R.E. and Baddour, R.F., Polymer, in press.

4.  Clark, D.T., Feast, W.J., Musgrave, W.K.R. and Ritchie, I., J. Polym. Sci., Polym. Chem. Ed., 1975, 13, 857.

5.  Florin, R.E. and Wall, L.A., J. Chem. Phys., 1972, 57(4), 1791.

6.  Shinohara, H., Iwasaki, M., Tsujimura, T., Watanabe, K., and Okazaki, S., J. Polymer Sci., 1972, A-1, 10, 2129.

7.  Shimada, J. and Hoshino, M., J. Appl. Polymer Sci., 1975, 19, 1439.

8.  Masuoka, T., Yasuda, H. and Morosoff, N., ACS Polym. Preprints, 1978, 91(2), 498.

9.  Yasuda, H., ACS Polymer Preprints, 1978, 19(2), 491.

10. Hatsuo, Seitaro and Takehara, Yumido, Japan J. Appl. Phys., 1977, 16, 175.

11. Coburn, J.W., Winters, H.F. and Chuang, T.J., J. Appl. Phys., 1977, 48, 3532.

12. Giegengack, H. and Hinze, D., Phys. State Sol. (A), 1971, 8, 513.

13.  Hannon, M.J., Boerio, F.J. and Koenig, J.L., _J. Chem._ _Phys._,
     1969, 50(7), 2829.

14.  Liang, C.Y. and Krimm, S., _J. Chem. Phys._, 1956, 25(3), 503.

RECEIVED January 27, 1981.

# ESCA and SEM Studies on Polyurethanes for Biomedical Applications

B. D. RATNER

Department of Chemical Engineering and Center for Bioengineering, BF-10,
University of Washington, Seattle, WA 98195

Polyetherurethanes (PEU's) have long been considered for use in biomedical applications because of their excellent mechanical properties, their resistance to hydrolysis and degradation, and in some instances, their good biocompatibility.

Assuming the absence of toxic, leachable, low molecular weight components, the biocompatibility (and, in particular, the blood compatibility) of PEU's will be strongly influenced by the surface properties of the polymers. Techniques which have been used to study polyurethane surfaces include contact angle measurements (2), attenuated total reflectance IR (ATR-IR) (1, 2, 3), Auger spectroscopy (4), and electron spectroscopy for chemical analysis (ESCA) (5-9). One of the most important observations from previous characterization studies is that for solvent-cast PEU films, the air and the casting surface sides differ significantly in average chemical structure (3-7).

The significance of differences in surface structure of polyurethanes on biological reactivity has received relatively little study. Many vena cava ring test studies were performed on PEU's (10), but no clear-cut structure-blood compatibility relationships were arrived at. Lyman, et al. (7), noted differences in the ratio of adsorbed albumin to other plasma proteins on two PEU's differing in polyether chain length. They also observed that platelet adhesion to these materials decreased with increasing albumin fractions at the surface. Stupp, et al. (2), found differences in the amount of adsorbed fibrinogen and possibly in the conformation of the adsorbed fibrinogen on PEU's cast against glass and poly(ethylene terephthalate). They related this to differences in the aromatic and polyether contents of the surface as observed by ATR-IR.

This study represents a preliminary investigation on the chemistry and morphology of polyurethane surfaces with long term goals directed toward relating these factors both to the bulk structure of the polyurethanes and to their blood and tissue

compatibility. The primary analytical technique used was ESCA
because of its surface sensitivity (10-100 Å) and because of the
high information content from high resolution Cls spectra.

## Methods

The ESCA analysis of the polymers was performed on a Hewlett
Packard Model 5950B ESCA system. A 0.8 kwatt monochromatized X-
ray beam from an aluminum anode was used for all spectra. An
emission from an electron flood gun was used to neutralize charge
build-up. Cls peaks associated with hydrocarbon-like environments were assigned a binding energy of 285.0 eV to correct for
the energy shift resulting from the electron flood gun. Areas
under the various overlapping peaks in the Cls spectra were
determined using a Dupont 310 curve resolver.

All polyurethane materials used in this study were either in
the form of extruded tubes (Tygothane, Superthane, Pellethane) or
as films cast on clean glass from reagent grade dimethylacetamide
(DMAC). Only the luminal surfaces of the tubes and the glass-
facing sides of the cast films were observed by ESCA.

## Results and Discussion

The approach in this study to the analysis of polyurethane
surfaces was based upon the idea that the surface properties
represent a summation of the effects of all chemical groups at the
surface. The various structural units which might be expected in
a polyurethane were represented by model compounds. ESCA spectra
of these model compounds revealed the chemical shifts for each
type of group and the peak widths to be expected (Figure 1).
Unambiguous curve resolution can be performed using these two
pieces of information (11).

Initial ESCA analysis of a series of commercially available
polyurethanes revealed a wide range of surface structures with
large variations in the proportions of the various important
structural groups at the surface of the materials (Figure 2).
Since the possibility of surface contaminant films obscuring the
true polyurethane surface was considered, and also since leachable components from polymeric materials frequently induce un-
desirable blood and tissue responses, various extraction procedures were investigated. Figure 3 shows Cls ESCA spectra of the
polyesterurethane Tygothane (Norton Plastics) after a series of
washes and extractions. Figures 4 and 5 show similar results with
the polyetherurethanes Superthane (Newage Industries, Inc.) and
Pellethane 2363-80A (Upjohn, Inc). Data on peak areas and shifts
are tabulated in Table I for Tygothane.

The Ivory soap solution wash (with sonication) probably
begins to remove only a surface contaminant film. Evidence for
this is suggested by the peak shift to higher binding energies of
the highest binding energy component of the Cls spectra for the

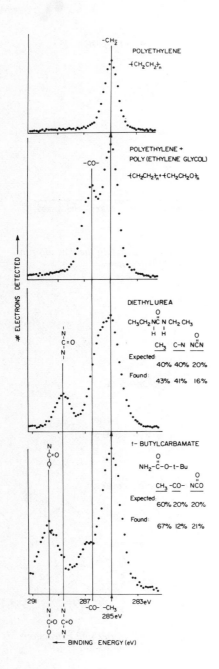

*Figure 1. The C-1s ESCA spectra of model compounds used to obtain peak widths and peak shifts for functional groups expected in polyurethanes*

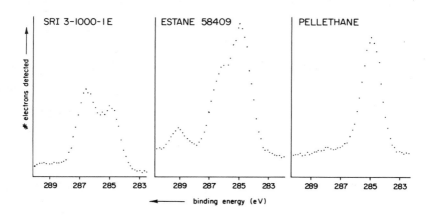

*Figure 2.   The C-1s ESCA spectra of three polyurethanes showing the variation in peak shapes that can be expected for polyurethanes*

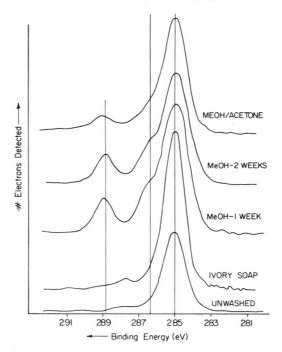

*Figure 3.   The C-1s spectra for Tygothane after a series of washes and extractions*

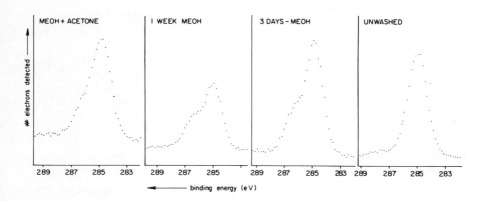

*Figure 4.   The C-1s spectra for Superthane after a series of extraction procedures*

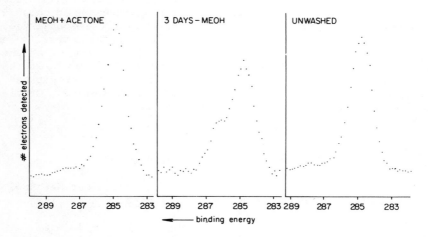

*Figure 5.   The C-1s spectra for Pellethane after a series of extraction procedures*

Table I

Relative Concentrations of Carbon Species at the Surface of Tygothane Tubing After Various Cleaning Procedures

Percent of Cls Spectrum

| | | $-NH$<br>\|<br>$-O-C=O$ | $O$<br>\|\|<br>$-C-N-$ | $-C-O$ | $-C-H$ |
|---|---|---|---|---|---|
| Tygothane: | Unwashed | -- | 5 | 26 | 69 |
| Tygothane: | Ivory | -- | 5 | 11 | 84 |
| Tygothane: | 1 Week − MeOH | 16 | − | 25 | 59 |
| Tygothane: | 2 Week − MeOH | 16 | − | 25 | 59 |
| Tygothane: | 2 Week − MeOH<br>+ 3 Days − Acetone | 9 | − | 20 | 71 |

unwashed and Ivory soap washed Tygothanes (Table I) after solvent treatment. The peak position prior to solvent treatment is indicative of an amide group (288 eV) rather than the expected carbamate linkage (289.2 eV). Thus, the surface observed may be that of a relatively low molecular weight surfactant/amide of the type frequently used as an extrusion lubricant. After the methanol extraction (in a soxhlet extractor) increases in components associated with higher binding energy peaks are observed. Acetone extraction of the methanol-extracted Tygothane results in a partial loss of these higher binding energy components at the surface. Upon cooling, the methanol and acetone used in the extraction were found to contain white, floculant precipitates. The amount of precipiate was always greater with acetone. Gravimetric extraction data is listed in Table II.

Table II

Extraction Data For Polyurethanes

| | % Wt. Loss After<br>Methanol Extraction | % Wt. Loss After Methanol<br>+ Acetone Extraction |
|---|---|---|
| Tygothane | 1.31 + 0.01 | 2.54 + 0.08 |
| Pellethane | 1.24 + 0.02 | 2.49 + 0.12 |
| Superthane | 1.56 + 0.03 | 5.25 + 0.14 |

For Superthane, the extract precipitate from methanol had a C/O ratio (as determined by ESCA) of 3.71. This value should be compared to C/O values for the luminal surface of the tubing after methanol extraction (3.39-3.83) and after acetone extraction (4.47). The analysis of the Cls spectrum of the Superthane extract (Figure 6) indicates the possibility of a high proportion

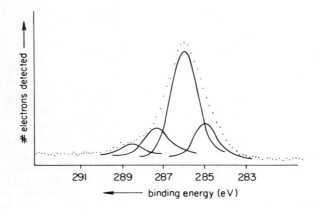

*Figure 6.   The C-1s spectrum of the precipitate from the methanol used to extract Superthane.  The component peaks were resolved using a peak width at half height of 1.2 eV.*

of C-O type bonds. The extract has been found by GPC to be of substantially lower molecular weight than the original polymer (Table III).

Table III

Gel Permeation Chromatographic Data for Polyurethanes and Extracts

| | RETENTION TIME (Minutes)* | |
| --- | --- | --- |
| | with LiBr | without LiBr |
| Untreated Pellethane | 32.4 | 31.7 |
| Extracted Pellethane | 31.3 | 32.3 |
| Methanol Extract of Pellethane | 44.1 | 44.2 |
| Acetone Extract of Pellethane | 35.0 | 36.0 |
| Polystyrene (MW = 3,000) | -- | 54.7 |
| Polystyrene (MW = 37,000) | -- | 43.5 |
| Polystyrene (MW = 80,000) | -- | 40.5 |
| Polystyrene (MW = 233,000) | -- | 34.2 |

* DMF, Flow rate = 0.75 ml/min, $10^3$ Å, $10^4$ Å, and $10^5$ Å Styragel columns.

A hypothesis concerning the surface structure of these materials has been constructed based upon this data. It would appear that the polyurethanes observed contain polyether or polyester enriched oligomers. These have relatively low solubility in warm methanol and migrate to the surface of the polymer tube where they slowly leach into solution. In acetone, these low molecular weight components are rapidly removed. Thus, the surface of the tube observed after (incomplete) methanol extraction is rich in the oligomer phase (i.e., the C/O ratio of the extract is approximately equal to that of the tubing surface), while after acetone extraction, the C/O ratio is higher indicating loss of material high in oxygen. The disappearance of C-O type bonds in the C1s spectra after acetone extraction also supports this hypothesis. Experiments with precipitation-purified polyetherurethanes of known structure prepared by casting on glass further support the hypothesis. These polymers, prepared by SRI International and containing poly(oxypropylene glycol) blocks of various molecular weights always showed, by ESCA, a surface enrichment of the polyether component compared to what would be expected based upon the stoichometry. Other investigators have also recently reported polyether surface enrichment or depletion for PEU's ([12], [13]). Detailed results with these polymers will be reported elsewhere.

Distinct morphological changes in the surface structure at each state of the washing/extraction process are observed by scanning electron microscopy. Figure 7 shows results for Tygo-

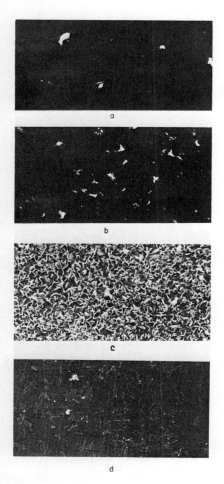

*Figure 7.    Scanning electron micrographs of Tygothane surfaces: (a) unwashed, 1500×; (b) Ivory soap washed with sonication, 750×; (c) extracted one week in methanol, 750×; (d) after methanol extraction, 3 days acetone extraction, 750×.*

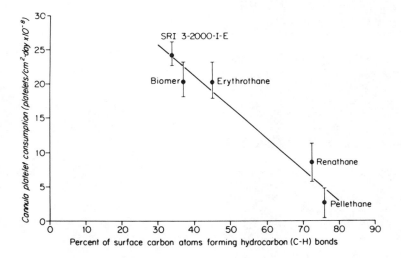

*Figure 8.   Platelet consumption of polyurethanes as a function of percent hydro-
carbon component in the C-1s spectra (from Ref. 9)*

thane. Similar observations were made for Superthane and Pellethane. In the earliest wash stages, inclusions, which appear a
light color, seem to be imbedded in a darker matrix. Extraction
with methanol may remove this "matrix" revealing the nature of the
substructure. One can easily speculate on the nature of these
morphological changes based upon the chemical evidence presented
previously. However, further experiments must be performed before the precise nature of these morphological changes can be
clarified.

## Conclusions

Using model compounds, the surface composition of polyurethane materials of unknown composition can be identified. Large
differences in surface structure are observed for polyurethanes.
Also, the surface structure is sensitive to extraction and cleaning procedures. Some of these changes may be related to processing additives in the commercial grade polyurethanes used or to low
molecular weight polyurethanes. Future experiments will look at
carefully synthesized polyetherurethanes of known composition to
relate bulk and surface structure. Also, extracts will be further
analyzed by ESCA, GPC and IR to determine their structure.

The biological significance of this work is revealed in
recent experiments in which the surface properties of various
polyurethanes were divided into non-dispersive and dispersive
force components by considering, as a simple first-order approximation, that the hydrocarbon-type C1s peak represents contributions to only the dispersive force component, and all other C1s
peaks indicate contributions to non-dispersive-type surface
forces. When platelet consumption as measured in a baboon A-V
shunt model is plotted against the fraction of the C1s ESCA
spectra representative of hydrocarbon-type groups (dispersive
force component), a linear relationship is obtained (Figure 8,
reference 9). Recently, ESCA and platelet consumption data from
extracted and unextracted Tygothane have been found to fit this
trend. This is particularly interesting since it indicates that
both polyesterurethanes and polyetherurethanes behave in a mecha-
nistically similar fashion with respect to their reaction with
blood. Also, further analysis of the ESCA data has indicated that
platelet consumption correlates strongly with the concentration
of C-O type linkages at the surface and not with the concentration
of carbamate-type groups. These conclusions do not agree with the
conclusions arrived at by another research group implicating the
hard segment (i.e. carbamate-type functionalities) as the initia-
tor of blood reaction (14). A detailed study on the platelet
consumption of polyurethanes will be published shortly.

## Acknowledgement

This work was supported by NHLBI Program Project Grant HL
22163.

Literature Cited

1.   Boretos, J.W., Pierce, W.S., Baier, R.E., Leroy, A.F.,
     Donachy, H.J., J. Biomed. Mater. Res., 1975, 9, 327.

2.   Stupp, S.I., Kauffman, J.W., Carr, S.H., J. Biomed. Mater.
     Res., 1977, 11, 237.

3.   Paik Sung, C.S., Hu, C.B., Merrill, E.W., ACS Polym. Pre-
     prints., 1978, 19(1), 20.

4.   Paik Sung, C.S., Hu, C.B., J. Biomed. Mater. Res., 1979, 13,
     45.

5.   Paik Sung, C.S., Hu, C.B., J. Biomed. Mater. Res., 1979, 13,
     161.

6.   Andrade, J.D., Iwamoto, G.K., McNeill, B., Shibatani, K.,
     ACS Organ. Coatings Plastics Chem. Preprints, 1976, 36(1),
     161.

7.   Lyman, D.J., Knutson, K., McNeill, B., Shibatani, K., Trans.
     Am. Soc. Int. Org., 1975, 21, 49.

8.   Lyman, D.J., Albro, D., Jr., Jackson, R., Knutson, K.,
     Trans. Am. Soc. Artif. Int. Org., 1977, 23, 253.

9.   Hanson, S.R., Harker, L.A., Ratner, B.D., Hoffman, A.S., J.
     Lab. Clin. Med., 1980, 95, 289.

10.  Gott, S.R., Baier, R.E., "Evaluation of Materials by Vena
     Cava Rings in Dogs," NHLI Report PH 43-68-84-3-1, National
     Institutes of Health, Bethesda, MD, 1972.

11.  Clark, D.T., Thomas, H.R., J. Polym. Sci., Polym. Chem. Ed.,
     1976, 14, 1671.

12.  Knutson, K., and Lyman, D.J., ACS Organic Coatings and
     Plastics Chemistry Preprints, 1980, 42, 621.

13.  Hu, C.B., Paik Sung, C.S., ACS Polymer Preprints, 1980,
     21(1), 156.

14.  DaCosta, V.S., Brier-Russell, D., Trudel, G., Waugh, D.F.,
     Salzman, E.W., and Merrill, E.W., J. Coll. Interf. Sci.,
     1980, 76, 594.

RECEIVED March 10, 1981.

# Surface Analysis of Silicon:
# Alloyed and Unalloyed LTI Pyrolytic Carbon

R. N. KING and J. D. ANDRADE

Department of Materials Science and Engineering, Department of Bioengineering and Surface Analysis Laboratory, University of Utah, Salt Lake City, UT 84112

A. D. HAUBOLD and H. S. SHIM

Carbo-Medics, Inc., 11388 Surrento Valley Road, San Diego, CA 92121

LTI pyrolytic carbon is one of the very few synthetic materials generally accepted as suitable for long-term blood contact applications (1). Although a number of hypotheses have been formulated with respect to the blood tolerability of materials, a general theory or mechanism is not yet available. Nyilas, et al., (2) have shown that in certain situations the local hemodynamics can play a predominant role, while in most cases the solid-blood interfacial properties have been shown to be equally important (2, 3). It is assumed that understanding the plasma protein adsorption processes on solids used for blood-contact applications will lead to a better understanding of solid-blood interactions (1, 2, 3).

In terms of LTI carbon surfaces, a number of preliminary studies of plasma protein adsorption are available (4, 5, 6, 7). Kim, et al., (4) have utilized radioiodinated ($I^{125}$) proteins to measure adsorption of individual proteins and protein mixtures on LTI carbon surfaces. Their results indicate a very rapid adsorption of albumin onto the LTI carbon surface, consistent with Kim's model of blood interactions via a platelet-adhesion mechanism (8). Microcalorimetric and electrophoretic mobility studies of protein adsorption on LTI carbon surfaces have been done by Chiu, et al., (5). The extension of the adsorbed protein layers have been directly measured by Fenstermaker, et al., (6) and Stromberg et al., (7) at NBS using ellipsometric methods.

Most of these studies have been performed on relatively uncharacterized LTI carbon surfaces. Since we assume that a large part of blood compatibility depends on the nature of the solid-plasma interface, particularly with respect to protein adsorption, we have elected to characterize some of the surface properties of LTI carbon in hopes of further understanding the solid-blood interaction mechanisms.

In this paper we concentrate on the nature of the LTI carbon surface as determined by X-ray photoelectron spectroscopy (XPS), and energy-dispersive X-ray analysis (EDAX). These data are discussed with reference to additional literature data on LTI carbon surfaces obtained by infrared spectroscopy, chemical reaction analysis, and electrochemical methods.

## Background

The lattice structure of most pyrolytic carbon crystallites is characterized by carbon atoms arranged in planar hexagonal arrays, with varying degrees of lattice perfection. X-ray diffraction results suggest that the planar arrays are either slightly "wrinkled," or contain single or multiple lattice vacancies (9), and that the layer spacings are found to be somewhat greater than those found in graphite. This slightly distorted lattice, in which the layers are arranged roughly parallel and equidistant but not otherwise mutually oriented, has been termed "turbostratic" by Biscoe and Warren (10). Hosemann (1), in more recent X-ray studies on the crystal lattices of polymers, has introduced the term "paracrystalline" in describing similar lattice distortions. The former terminology, however, has been generally adopted in the carbon literature.

On the microstructural level, several types of pyrolytic carbons may be deposited each with one of four distinctly different structures, ranging from layered, highly anisotropic forms to structures with very small, randomly-oriented crystallites with no preferred orientation. All of these structural variations are a result of modifications in processing conditions. In this particular study, only the isotropic forms of both pure LTI carbon and co-deposited LTI carbon-silicon alloyed carbon (Pyrolite[R] – registered trademark of Carbo-Medics, Inc., San Diego, California) were investigated.

In addition, it is important to note that the isotropic forms of pyrolytic carbon produce at least a two-phase microstructure during formation, consisting of the previously-described turbostratic microcrystalline phase along with an amorphous carbon phase presumably interspersed between the crystalline regions (12).

In activated carbons a high internal porosity is formed by removal of much of the amorphous carbon by elevated temperature treatment, leading to a carbon material with a very high surface area (13).

In the case of the alloyed carbon, microcrystalline silicon carbide particles are randomly interspersed in the carbon matrix. The presence of this SiC phase greatly contributes to the hardness and wear resistance of these alloys compared to pure pyrolytic carbon.

The surface chemistry of carbon has been extensively studied and reviewed (14-18). It is generally believed that a variety of carbon-oxygen functional groups are present on carbon surfaces. Their nature and concentration are dependent on the sample history and depend, for example, on processing variables.

Carbon-oxygen combinations appear to be crystallographically specific.  Mattson and Mark (15) report that molecular oxygen preferentially attacks graphite crystals at the edges of the layer planes at a rate nearly 20 times that of the atoms within the basal planes.  From these data it is reasonable to assume that the surface oxide concentrations on carbons which have been mechanically polished would contain more oxygen than samples which have not been surface finished.  It will subsequently be shown that the surface oxidation of LTI pyrolytic carbon is significantly increased by surface finishing.

While a large number of oxygen-containing functional groups have been reported on carbon surfaces, the most generally accepted are carboxyls, hydroxyls, and quinone-like carbonyls.  These are illustrated in Figure 1 as they might appear on the plane edges of a turbostratic crystallite.

A variety of other carbon-oxygen groups have been suggested, including lactones, anhydrides, peroxides, ethers, and esters (14-18).  These surfaces oxides have been studied by functional group reactions (18), titration, and infrared spectroscopy (15, 17).

The quinone-like carbonyl groups have received considerable attention in several electrochemical studies (15, 19, 21).  Such studies clearly show anodic and cathodic peaks which can be interpreted as arising from the formation and reduction of quinone-like groups on the surface.  Our laboratory has also verified these results using cyclic sweep voltammetry.  Similar quinones or quinone-like groups have been identified by infrared studies (17) and by specific wet chemical reactions (20).

The most characteristic feature of quinones is their ease of reduction and reoxidation; they play a part in the redox processes of many living systems (22).  Furthermore, they can interact with amino and sulfhydryl groups.  One might therefore expect quinone-like groups to participate in protein adsorption and perhaps cell adhesion although no direct evidence is available to support this hypothesis.  Studies of quinone-like groups on activated carbon surfaces (23) indicate such groups can participate as electron donors in donor-acceptor complexes with adsorbed molecules (17, 24).

Much of the above data on activated carbon surfaces can be correlated with the results on isotropic pyrolytic carbon surfaces due to the similarities in microstructure.  The advantage of the pyrolytic carbon is in the reproducibility of the carbon surface (15) utilizing carefully controlled processing conditions.

## Materials and Methods

Unalloyed and silicon alloyed LTI carbon samples were prepared at Carbo-Medics using the "steady-state" fluidized bed process developed by Akins and Bokros (25).

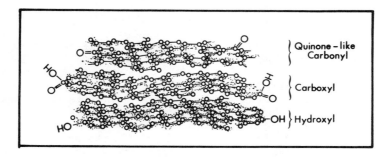

*Figure 1.   Oxygen-containing groups on turbostratic carbon plane edges (modified from Ref. 9))*

The unalloyed LTI pyrolytic carbon was prepared by introducing small graphite plates into a fluidized bed consisting of spherical $ZrO_2$ particles. Carbon deposition on the plates was accomplished by pyrolyzing a gas mixture of about 25 vol. % propane – 75 vol. % He (diluent gas) in a temperature range of $1300°$ to $1400°C$. The resulting plate coatings consisted of completely isotropic, turbostratic crystallites (approximately 24 Å to 40 Å crystals) (12). Densities range from $\sim 1.5$ to $1.8$ $gm/cm^3$, with an anisotropy ratio of less than 1.1 (9).

Pyrolytic carbon for some biomedical applications often requires a harder, more wear-resistant surface than is attainable with pure carbon. By alloying the pyrolytic carbon with a silicon carbide phase, the microhardness of the carbon surface can be roughly doubled (25). Preparation of the alloyed pyrolytic carbon is generally equivalent to that of the unalloyed carbon. The main processing difference is the addition of methyltrichlorosilane to the propane-He gas mixture. This pyrolyzing mixture produces an additional SiC phase in the final pyrolytic carbon. The alloyed microstructure is generally similar to pure LTI carbon in that a completely isotropic turbostratic structure results, with the added benefit of a silicon carbide phase for improved wear resistance. Densities are slightly higher ( $\sim 2.0$ to $2.2$ $gm/cm^3$) due to the additional SiC phase, with an anisotropy ratio again less than 1.1. Coating thickness for both types of carbon were on the order of 0.3 to 0.5 mm. For our studies, three lots of both the unalloyed and silicon-alloyed LTI pyrolytic carbons, all from the same runs, were prepared. Lots for the unalloyed LTI pyrolytic carbon (designated as LTI-A, LTI-B, and LTI-C) and silicon-alloyed LTI pyrolytic carbon (LTI/SI-A, LTI/Si-B, and LTI/SI-C) differed only by finishing operations. Cleaning and polishing procedures for both the unalloyed and silicon-alloyed lots were identical.

Lots LTI-A and LTI/SI-A were placed in clean glass vials immediately after coating and examined without any subsequent handling or treatment. Lots LTI-B and LTI/SI-B were ultrasonically cleaned in isopropyl alcohol, air dried, and stored in glass vials prior to study. Lots LTI-C and LTI/SI-C were ground and polished in several steps, and contacted various metal oxides, silicon carbide, diamond, water, detergent, isopropyl alcohol, and ethyl alcohol. Final cleaning was done in ethyl alcohol, air dried, and stored in clean glass vials. Several additional samples of LTI-C were packaged in a soft blue foam, representative of the surface which is normally delivered to medical device manufacturers for further processing and application in medical devices.

X-ray photoelectron spectra were obtained with a Hewlett-Packard 5950 B instrument utilizing monochromatic Al $K_{\alpha 1,2}$ radiation at 1487 eV. The samples were mounted in air, inserted into the spectrometer, and analyzed at ambient temperatures in a $10^{-9}$

torr vacuum. Power at the X-ray source was 800 watts. Instrument resolution in our spectrometer during this analysis series was measured as 0.76 eV for the full width at half maximum of the C-1s peak from spectroscopic grade graphite. An electron flood gun operating at 0.3 mA and 5.0 eV supplied a flux of low energy electrons to the carbon surface to minimize heterogeneous charging artifacts in the resulting spectra.

Wide scans (0 to 600 eV) were performed for surface elemental analyses as well as detailed 20 eV scans of the C-1s (275 to 295 eV) region. Several standards were also analyzed under the same scan conditions in order to obtain accurate chemical shift data for various carbon-oxygen functional groups. These included poly(ethylene terephthalate), poly(ethylene oxide), and anthraquinone. The latter was run at $-50^\circ$C in order to minimize volatility under our high vacuum conditions. Additional spectra were obtained on spectroscopic grade graphite for comparison purposes. All spectra were charge referenced to a C-1s line for an alkyl-like carbon at 284.0 eV.

Scanning electron micrographs were produced on a Cambridge Mark II Stereoscan SEM equipped with an EDAX 707B/NOVA accessory for elemental microanalysis of bulk regions. An edge of each sample was coated with silver paint in order to assure conductive contact between the samples and specimen holders. The analyzing vacuum was about $10^{-4}$ torr.

## Results and Discussion

SEM/EDAX. Figures 2a, b and 3a, b are representative of the surface morphologies and bulk elemental analyses of the "as formed" lots of LTI-A, B and LTI/SI-A, B pyrolytic carbons. As shown in Figures 2a and 3a, the surface of both the unalloyed and silicon-alloyed carbons is composed of disorganized particulates resulting in a rough and highly porous topography. The ultrasonic isopropyl alcohol cleaning process (LTI-B and LTI/SI-B) seems to remove many of the more loosely adherent carbon granules resulting in much finer surface debris for those materials compared to lots LTI-A and LTI/SI-A.

EDAX analysis of these materials, as illustrated in Figures 2b and 3b, show little difference between the samples with the exception of the silicon peak found in the carbon-silicon alloy. It should be noted that EDAX is inherently insensitive to the lower atomic number elements due to the low fluorescent yields of the lighter elements, internal absorption, and low transmission factors for these elements through the beryllium detector window of the instrument. Thus, carbon and oxygen are notably absent from the conventional EDAX spectra.

Figures 4a, b and 5a, b show the results of the surface finishing process on lots LTI-C and LTI/SI-C. The surface roughness and porosity are shown to be significantly decreased, with a corresponding disappearance of the granular structures

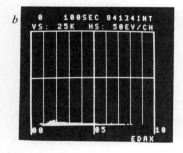

*Figure 2.    (a) Surface topography of "as formed" unalloyed pyrolytic carbon (LTI-A), 2600×. (b) EDAX analysis of "as formed" unalloyed carbon (LTI-A). Note absence of identifiable spectral lines. Vertical scale = 25,000 counts, data accumulation time = 100 s.*

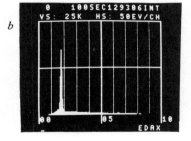

Figure 3. (a) Surface topography of "as formed" silicon-alloyed pyrolytic carbon (LTI/SI-A), 2600×. (b) EDAX analysis of "as formed" silicon-alloyed pyrolytic carbon (LTI/SI-A). Major spectral line is silicon. Vertical scale = 25,000 counts, data accumulation time = 100 s.

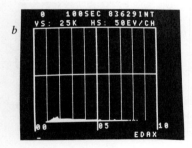

*Figure 4. (a) Surface topography of polished unalloyed pyrolytic carbon (LTI-C), 2600×. (b) EDAX analysis of particulate-free region of above polished unalloyed pyrolytic carbon. Identifiable spectral lines are absent. Vertical scale = 25,000 counts, data accumulation time = 100 s.*

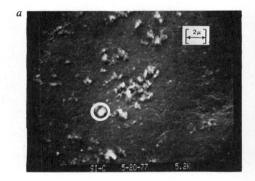

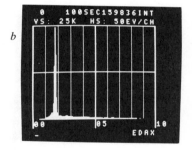

*Figure 5.   (a) Surface topography of polished alloyed pyrolytic carbon (LTI/SI-C), 2600×.  (b) EDAX analysis of particulate-free region of above polished silicon-alloyed pyrolytic carbon.  Major spectral line is silicon.  Vertical scale = 25,000 counts, data accumulation time = 100 s.*

noted in lots A and B. Particulate matter is still evident on both surface, however. These surfaces were examined "as received," and may represent either insufficient cleaning of the polished surface, or the presence of embedded particles from the grinding process. Evidence to support the latter conclusion is shown in Figure 6, which is the EDAX analysis of the particulate area circled in Figure 5a. The additional peaks in the spectrum are aluminum (indicating the particle may be $Al_2O_3$ which may be embedded during the finishing operation), and iron (which is the result of back-scattered electrons exciting Fe X-rays from the final lens assembly of the Stereoscan).

ESCA. The surface elements determined from wide-scan ESCA analysis for the various lots of unalloyed pyrolytic carbon (LTI-A, B, C) are given in Table I. We have adopted Scofield's theoretical cross-sections (26) for semi-quantitatively normalizing our ESCA spectra with respect to carbon.

Table I

ESCA Elemental Concentration for Unalloyed LTI Carbon Samples. Atomic Ratios Normalized to 100 Carbon Atoms

| LTI Lot Code | C | O | Si | Trace ($< 1.0$) |
|---|---|---|---|---|
| A | 100 | 1.8 | -- | -- |
| A | 100 | 1.2 | -- | -- |
| B | 100 | 1.4 | -- | -- |
| C-"as Received" in glass | 100 | 10.8 | -- | -- |
| C-packaged in Form | 100 | 8.5 | 1.9 | S, P, Cl, Al |
| C-packaged in Foam and Methanol Cleaned | 100 | 8.2 | -- | Cl |
| Foam Packing | 100 | 39.9 | 12.3 | N($\sim$ 2.8) |

As noted in the table, Lots A and B of the unalloyed LTI carbon were identical, with a carbon/oxygen ratio slightly less than 50:1. No other elements were detected on the surface. Detailed analysis of the C-1s region of these samples shows a small chemically-shifted peak in the major C-1s region indicating what appears to be an ether- or hydroxyl-like carbon-oxygen bond.

The "as received" LTI-C sample stored in glass shows a substantial increase in the oxygen concentration of the surface region. In contrast, the LTI-C sample packaged in blue foam again

*Figure 6. EDAX analysis of particulate area circled in Figure 5a. Note additional aluminum and iron lines in spectrum, assumed to result from the polishing process. Vertical scale = 25,000 counts, data accumulation time = 100 s.*

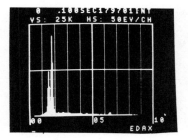

shows a much higher oxygen concentration and also a substantial quantity of silicon.   Other elements present on this surface include traces of sulphur, phosphorous, chlorine, and aluminum.

The foam packed lot LTI-C was then ultrasonically cleaned in absolute methanol for five minutes and rescanned under identical conditions.   As shown in Table I no silicon was evident in this spectrum.   The fact that the silicon spectral lines were readily removed by this washing process suggested a surface contaminant as the source of silicon, possibly a silicone release agent.   The only likely source of such a contaminant is the blue foam in which these specific samples were packaged.   ESCA examination of this foam packing material revealed relatively high concentrations of silicon on the surface as well as nitrogen and chlorine.   This suggests a polyurethane foam with a silicone-type release agent, and traces of possible NaCl from handling.   Studies of the methanol-cleaned LTI-C material (silicon-free) contacted with the foam confirmed that the silicon (silicone) on the foam can readily be transferred to the carbon surface.

The carbon/oxygen ratio on the surface of both the "as received" in glass and methanol-cleaned LTI-C is about 10:1, suggesting that the finishing operation on unalloyed pyrolytic carbon increases the oxygen concentration on the surface by a factor of 5 compared to the "as formed" Lots LTI-A and B.   It is assumed that the trace elements detected on the surface are also a result of the polishing operation.

Table II summarizes the ESCA elemental analyses for the three lots of silicon-alloyed pyrolytic carbon (LTI/SI-A, B, C). Again lots A and B are roughly similar, with a carbon/oxygen ratio of about 40:1, or a slightly higher oxygen content than that found for the unalloyed material.   The C-1s spectrum likewise shows an ether-like carbon-oxygen bond, similar to that observed in the unalloyed material.

Table II

Elemental Concentrations for Silicon-Alloyed LTI/SI Carbon Samples.   Atomic Ratios Normalized to 100 Carbon Atoms

| LTI/SI Lot Code | C | O | Si | Al | Cl |
|---|---|---|---|---|---|
| A | 100 | 3.4 | 3.0 | -- | -- |
| A | 100 | 2.9 | 1.6 | -- | -- |
| B | 100 | 2.7 | 1.2 | -- | -- |
| B | 100 | 2.2 | 1.2 | -- | -- |
| C | 100 | 9.6 | 0.9 | 1.2 | 1.0 |
| C-Methanol Cleaned | 100 | 12.2 | 0.7 | 1.6 | 0.7 |

The LTI/SI-C carbon results again illustrate that the car-
bon/oxygen ratio is significantly increased by the polishing
process. While the relative increase compared to the "as formed"
LTI/SI lots is slightly less than that found for the unalloyed
material (a factor of 3 to 4 compared to the five-fold increase
for the unalloyed), the final carbon/oxygen ratio is approximately 10:1 for the polished silicon-alloyed carbon.

A careful examination of the C-1s and O-1s regions of the
oxygen-containing compounds listed in Table III allow us to
deduce the nature of the carbon-oxygen functional groups on the
surface of the polished lots LTI-C and LTI/SI-C.

The    C-1s    spectra    of    poly(ethylene    terephthalate),
poly(ethylene oxide), and anthraquinone are shown in Figure 7.
All spectra were internally charge-referenced to an alkyl-like C-
1s line at 284.0 eV. As shown in the figure, the oxygen-
containing functional groups in these model compounds result in
pronounced chemical shifts in the C-1s spectra.

Note also that the area ratios of the various carbon peaks
shown in Figure 7 can be used to predict the stoichiometry of
these compounds. For example, from Table III the mer structure of
poly(ethylene terephthalate) shows two ether-like groups, two
ester-like groups, and six alkyl- or aromatic-like groups for a
ratio of 1:1:3, respectively. In Figure 7, the C-1s spectrum of
this polymer shows the predicted 1:1:3 ratio, the O-1s region (not
shown) gives a 1:1 ratio of the singly and doubly bonded oxygen.

Table III thus summarizes the chemical shift data of Figure
7, which was subsequently used to estimate the nature of the
oxygen-containing functional groups found on the LTI-C and
LTI/SI-C carbon surfaces.

The 1:1 silicon/aluminum ratio in Lot C persists even after
methanol cleaning. This would support the earlier observation
that a portion of the grinding media, in particular $Al_2O_3$, may be
firmly embedded in the surface during polishing.

The chlorine present on the surface may be a result of
surface contamination or possibly a chemically bound form in the
surface structure. Note that the SiC phase in these materials is
a pyrolysis product of methyltrichlorosilane, which conceivably
could introduce some chlorine in the final structure. Our
previous experience with a variety of surfaces has shown that
ultrasonically cleaning the surface in absolute methanol usually
eliminates most surface contaminants caused by handling. Unfortunately, no means were available for depth-profiling the alloyed
carbons to determine the presence of chlorine below the surface.

Figure 8 shows the C-1s spectra for "as received" LTI-C
unalloyed carbon stored in glass, LTI/SI-C silicon-alloyed carbon
and a spectroscopic grade graphite. The scale factors for the
spectra have been increased by a factor for five compared to those
in Figure 7 in order to enlarge the carbon-oxygen functional group
regions. While distinct peak separations cannot be observed in
the spectra due to the high intensity of the major C-1s line, the

Table III

Organic Models for Determination of Carbon-Oxygen Functional Groups by XPS; Oxygen-Containing Functional Group Data from the C-1s Line of Poly(ethylene) terephthalate), Poly(ethylene oxide), and Anthraquinone

| Compound and Mer Structure | Observed Binding Energy (eV)* | $\Delta$BE** | Fraction of Total Oxygen | | Functional Group |
|---|---|---|---|---|---|
| Poly(ethylene terephthalate) | 284.0 | 0.0 | -- | -C-C-or (benzene ring) | alkyl and/or aromatic |
| (structure) | 285.6 | +1.6 | .5 | -R-O-R- | ester-like |
| | 288.0 | +4.0 | .5 | R-C(=O)-O-R- | ester-like |
| Poly(ethylene oxide) | 285.8 | +1.8 | 1.0 | -R-O-R- | ether-like |
| $(CH_2CH_2O)_n$ | 284.0 | 0.0 | -- | -C-C-or | alkyl or aromatic |
| Anthraquinone | 286.5 | +2.5 | 1.0 | (=O quinone structure) | quinone-like |

* Charge-referenced to C-1s at 284.0 eV.

** Binding Energy -284.0 eV = $\Delta$BE

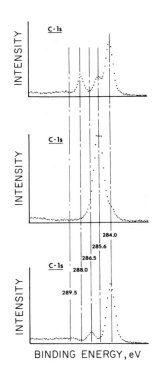

*Figure 7.  The C-1s ESCA spectra of poly(ethylene terephthalate) (top), poly-(ethylene oxide) (center), and antraqui-none (bottom).  All spectra charge-refer-enced to alkyl-like C-1s at 284.0 eV.*

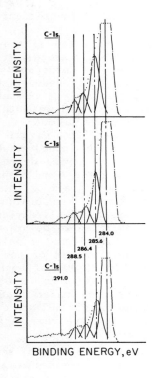

*Figure 8. The C-1s ESCA spectra of polished unalloyed (top) and alloyed (center) LTI carbon, compared to graphite (bottom). All spectra charge-referenced to alkyl-like C-1s at 284.0 eV.*

asymmetry evident on the high binding energy side of the peak is
indicative to several additional carbon species, presumably chemically bound to oxygen. Using the chemical shift data of Figure 7
and Table III, the carbon and oxygen regions were roughly resolved
(by maintaining the same full-width at half-maximum peak intensity) to form the peaks superimposed on the spectra in Figure 8.

These peaks are shown to reasonably account for the
asymmetry in the C-1s spectra of the carbon surfaces. These data
lead us to believe that the assigned peaks are reasonable to a
first approximation, and correspond to the carbon-oxygen functional groups indicated in Table III. Figure 8 also illustrates
that the graphite spectrum shows similar carbon-oxygen functionality.

In addition, a higher energy feature is noted in the C-1s
spectra which comprises less than 3% of the total carbon. If this
is assigned to a carbon-oxygen group, it roughly corresponds to a
carbonate-like carbon, as reported by Clark (29). However, Clark
(30) has also demonstrated that low kinetic energy $\pi \rightarrow \pi^*$ shake-up
transitions for aromatic polymers invariably lead to satellite
peaks located from 6 to 7 eV above the alkyl-like C-1s. One
would, therefore, expect to see such satellites in the C-1s
spectra of poly(ethylene terephthalate), anthraquinone, graphite,
and the pyrolytic carbons, since all exhibit varying degrees of
aromaticity. Such peaks are suggested in Figures 7 and 8 and are
shown to fall roughly within the same region for all of the C-1s
spectra. Of these two possible explanations, in our opinion, the
higher binding energy peak seen in these spectra is likely due to
the $\pi \rightarrow \pi^*$ shake up satellite, which is expected to be present.

Table IV summarized the ESCA data obtained on the polished
carbon surfaces along with estimates of percentages of the carbon-oxygen groups on the surface. Roughly 85-90% of both the
unalloyed and Si-alloyed carbon surfaces are shown to be oxygen-
free, with the remainder consisting of three major types of
oxygen-containing functional groups; approximately 60% are of the
ether-like variety, 25% are quinone-like, and the remaining 15%
are ester-like.

These data partially correlate with the XPS data of several
investigators. Marsh, et al., (27), using a non-monochromatized
X-ray source, have reported a 50% maximum oxygen coverage on the
surface of a variety of pyrolytic carbons reacted with oxygen. It
is important to note, however, that the carbons used in the
present study were prepared quite differently, and not specifically reacted with oxygen. They further reported deconvoluting
the O-1s spectra into five distinct peaks. They were unable,
however, to assign them to particular oxygen-containing functional groups. We cannot justify this large number, even though our
spectra appear to be better resolved. Recently Evans and Thomas
(28) studied single crystals of graphite and diamond surfaces and
suggested that >C-O-C< or >C-OH surface groups were probably
present in roughly equal concentrations on oxygenated carbon
surfaces.

Table IV

Estimations of Oxygen-Containing Functional Groups on Polished LTI Pyrolytic Carbon and Graphite Surfaces

| Material | Estimated Binding Energy (eV)* | BE** | Approximate % of Higher Energy Forms of Carbon | Functional Group |
|---|---|---|---|---|
| LTI-C unalloyed carbon | 284.0 | 0.0 | -- | alkyl or aromatic |
| | 285.76 | +1.76 | 63 | ether-like |
| | 286.34 | +2.34 | 23 | quinone-like |
| | 288.48 | +4.48 | 9 | ester-like |
| | 291.00 | +7.00 | 6 | $\pi \to \pi*$ satellite or carbonate-like |
| LTI/SI-C silicon-alloyed carbon | 284.0 | 0.0 | -- | alkyl or aromatic |
| | 285.60 | +1.60 | 65 | ether-like |
| | 286.40 | +2.40 | 23 | quinone-like |
| | 288.54 | +4.64 | 8 | ester-like |
| | 291.00 | +7.00 | 4 | $\pi \to \pi*$ satellite or carbonate-like |
| Graphite | 284.0 | 0.0 | -- | alkyl or aromatic |
| | 285.60 | +1.60 | 67 | ether-like |
| | 286.56 | +2.56 | 21 | quinone-like |
| | 288.16 | +4.16 | 7 | ester-like |
| | 291.00 | +7.00 | 5 | $\pi \to \pi*$ satellite or Carbonate-like |

The data also correlate with the electrochemical investigations of Epstein, et al., (21), i.e., that quinone-hydroquinone
groups are most likely present on LTI carbon surfaces.   They
further concluded that such groups can be readily converted
electrochemically from one to the other. As previously noted, our
data indicates that approximately 25% of the surface oxygen is
quinone-like in nature.

## Conclusions

At least three major types of carbon-oxygen functional
groups are present on polished LTI and LTI/SI carbon surfaces.
The origin of these oxygen-containing species is predominately
the polishing process, during which the oxygen content increases
nearly five-fold compared to the unpolished materials.
Both the XPS and EDAX results indicate that surface contaminants composed of aluminum compounds are also introduced
during polishing.  SEM analysis of the surfaces was unable to
determine whether or not these contaminants are firmly embedded
in the carbon surface.  From our XPS data, we have found that
approximately 85-90% of the carbons do not appear to be chemical-
ly-bonded to oxygen.  The remaining 10 to 15% contain three major
types of oxygen functionality.  From the chemical shift data, we
estimate that 60% of the carbon-oxygen functional groups are
either-like in nature, 25% are quinone-like, and the remaining
15% are ester-like or carboxylic in nature.  These results were
very consistent, regardless of the type of carbon surface investigated.
Though the materials examined are expected to be representative of the materials delivered to medical device manufacturers
it is possible that other changes could occur.  Thus these results
should not be extrapolated to carbon-containing medical devices
until the actual carbon components have been surface analyzed.

## Acknowledgement

The authors wish to thank Mr. Gary Iwamoto for his assistance.  This work was supported in part by NIH Grants HL16921 and
18519.

## Abstract

Low temperature isotropic (LTI) pyrolytic carbon has been
studied by X-ray photoelectron spectroscopy, scanning electron
microscopy, and energy dispersive X-ray analysis.  Both silicon-
alloyed and unalloyed carbons were studied, in both as-deposited
and polished (finished) forms.  The polished materials contain
significant amounts of surface oxygen.  Approximately 1 in 10 of
the carbon atoms in the surface volume analyzed by XPS are

oxidized. About 60% of the oxygen-containing functional groups are ether-like, 25% are quinone-like, and the remaining 15% are ester- or carboxyl-like. Polishing also resulted in small amounts of aluminum on the surface as well as several other impurities. The surface properties of LTI pyrolytic carbon are important in view of its success as a material for medical implant purposes. The literature on the surface properties of carbon is reviewed.

## Literature Cited

1. Bruck, S.D., Rabin, S. and Ferguson, R.J., Biomat., Med. Dev. Art. Org., 1973, 1, 191.

2. Nyilas, E., Morton, W.A., Federman, D.M., Chiu, T.H. and Cumming, R.D., Trans. Amer. Soc. Artif. Int. Organs., 1975, 21, 55.

3. Andrade, J.D., Med. Inst., 1973, 7, 110.

4. Kim, S.W., personal communication.

5. Chiu, T.H., Nyilas, E. and Federman, D.M., Trans. Amer. Soc., Artif. Int. Organs., 1976, 22, 498.

6. Fenstermaker, C.A., Grant, W.H., Morrissey, B.W., Smith, L.E. and Stromberg, R.R., Nat. Bur. Stds. Int. Report, NBSIR 74-470 (3/22/74).

7. Stromberg, R.R., Morrissey, B.W., Smith, L.E., Grant, W.H. and Fenstermaker, C.A., Nat. Bur. Stds., Int. Report, NBSIR 75-667 (1/15/75).

8. Kim, S.W., Lee, R.G., Oster, H., Coleman, D., Andrade, J.D., Lentz, D.J. and Olsen, D., Trans. Amer. Soc. Artif. Int. Organs., 1974, 29, 449.

9. Bokros, J.C. in "Chemistry and Physics of Carbon," Ed. P.L. Walker, Jr., 1969, 5, 9.

10. Biscoe, J. and Warren, B.E., J. Appl. Phys., 1942, 13, 364.

11. Hoseman, R., Polymer, 1962, 3, 349.

12. Kaae, J.L., Carbon, 1975, 13, 55.

13. Smisek, M. in "Active Carbon, M. Smisek and S. Cerny, Eds. (Elsevier Publishing Company, Amsterdam, 1970).

14.  Puri, B.R., in "Chemistry and Physics of Carbon," Ed.  P.L. Walker, Jr., 1970, 6, 191.

15.  Mattson, J.S. and Mark, H.B., Jr., "Activated Carbon-Surface Chemistry and Adsorption from Solution," (Marcel Dekker, New York, 1971).

16.  Snoeyink, V.L. and Weber, W.J., Jr., Prog. Surf. Memb.  Sci., 1972, 5, 63.

17.  Mattson, J.S., Lee, L., Mark, H.B., Jr., Weber, W., Jr., J. Colloid Interface Sci., 1970, 33, 284.

18.  Donnet, J.B., Carbon, 1968, 6, 161.

19.  Blurton, K.T., Electrochimica Acta., 1973, 18, 869.

20.  Elliott, C.M. and Murray, R.W., Anal. Chem., 1976, 48, 1247.

21.  Epstein, B.D., Dalle-Molle, E., and Mattson, J.S., Carbon, 1971, 9, 609.

22.  Thomson, R.H., "Naturally Occurring Quinones," (Butterworths, London, 1957).

23.  Garten, V.A. and Weiss, D.E., Aust. Chem. Soc. J., 1955, 8, 68.

24.  Mattson, J.S., Mark, H.B., Jr., Malbin, M.D., Weber, W.J., Jr., and Crittenden, J.C., J. Colloid Interface Sci., 1969, 31, 116.

25.  Akins, R.J. and Bokros, J.C., Carbon, 1974, 12, 439.

26.  Scofield, J.H., J. Elec. Spec. and Related Phen., 1976, 8, 129.

27.  Marsh, H., Foord, A.D., Mattson, J.S., Thomas, J.M., and Evans, E.J., J. Colloid Interface Sci., 1974, 49, 368.

28.  Evans, S. and Thomas, J.M., Proc. Royal Soc. London A., 1977, 353, 103.

29.  Clark, D.T., "Molecular Spectroscopy," A.R. West, Ed., Heydon and Sons, New York, 1977.

30.  Clark, D.T., "Polymer Surfaces," D.T. Clark and W.J. Feast, Eds., John Wiley and Sons, New York, 1978.

RECEIVED March 25, 1981.

# Oxidation of Polystyrene and Pyrolytic Carbon Surfaces by Radiofrequency Glow Discharge

G. K. IWAMOTO, R. N. KING, and J. D. ANDRADE

Surface Analysis Laboratory, College of Engineering, University of Utah, Salt Lake City, UT 84112

Plasma treatment is widely used commercially for polymer surface modification. Plasma discharge treatments are used to improve adhesiveness and printing properties, to improve cell adhesion to tissue culture substrates (1) and to etch or clean the surfaces of materials (removal of photoresist materials on semiconductors, for example (2). The surface characterization of plasma-modified surfaces is important in order to provide greater insight into how the properties are changed.

Plasma treatment involves the production of chemically active species and ultra-violet radiation. Conventional methods of surface modification are often limited by the temperature needed for surface treatment, the leaching or toxicity of chemical agents used, and the spectral and geometric limitation of UV treatments. Plasma treatment also provides a means of selectively modifying the surface while the bulk properties remain generally unaffected (1).

In this study polystyrene and pyrolytic carbon (3) were used to investigate the nature of plasma surface modification. Polystyrene is widely used as a material for bacteriological cell culture and, in a surface-treated, oxidized form, is widely used as a solid substrate for in vitro cell culture. It is generally assumed that commercial polystyrene cell culture substrates are surface-treated by a corona or radio frequency glow discharge (RFGD) process. Although these materials are extensively used, no general surface characterization is available.

Radio frequency glow discharge (RFGD) plasmas were used in this study. Glow discharge plasmas are characterized by average electron energies of 1 to 10 eV and electron densities of $10^9$ to $10^{12}$ cm$^{-3}$. Glow discharges, also called cold plasmas, are characterized by a lack of thermal equilibrium between electron temperature (Te) and gas temperature (Tg). Typical ratios are on the order of Te/Tg = 10 to $10^2$. Thus, the Tg of a glow discharge remains near ambient temperatures while the electrons are suffi-

0097-6156/81/0162-0405$05.00/0
© 1981 American Chemical Society

ciently energetic to rupture bonds. This makes this type of
plasma quite useful in applications involving thermally sensitive
materials (4).

Surface modification by a plasma usually results in changes
in surface wettability, molecular weight, and other chemical
changes. Molecular weight changes occur from chain scission and
crosslinking. Chemical changes occur from the addition or ab-
straction of groups on the surface, which in turn influence the
wettability of the surface. Almost all changes produced by plasma
modification are confined to the top 1 to 10 μ m of the surface
(1).

The reactions which occur are controlled by the pressure of
gas, electric field strength, reaction chamber dimensions and the
gas flow rate. The electric field strength determines the amount
of energy imparted to the electrons. The gas pressure and tube
dimensions affect the degree of ionization, atomic lifetimes,
mean free path and gas temperatures. The gas flow will affect the
rate that new reaction material can reach the solid surface (5).

X-ray photoelectron spectroscopy (XPS), scanning electron
microscopy (SEM), and air/octane underwater contact angles were
used to characterize the surfaces. XPS can provide both the
atomic composition and chemical bonding information from approximately the top 70Å or less of the sample surfaces (6). Additional
information may be gained from the XPS spectrum by observing the
presence of satellite lines. The ejection of a core level
electron from an atom changes the shielding of the nuclear charge
and is felt by the outer shell electrons. This perturbation in
the potential of the valence electrons is of sufficient energy
that an electron can be excited to a higher energy level (shake-
up) or be ejected (shake-off). For the C-1s line, satellite
structure is seen up to ~12 eV above the major peaks; any other
features would be lost in the inelastic tail which occurs ~15-20
eV above the major photoionization peak. Clark's studies on
polymer systems have shown that polymers must have an unsaturated
backbone or unsaturated pendant groups to have observable satellite 6.6 eV above the C-1s line (7). The satellite is attributed
to $\pi \rightarrow \pi^*$ transitions. Studies of alkane-styrene copolymers show
that the intensity of the satellite peak is related to the number
of styrene groups in the chain (8), and to the substituents on
the pendant phenyl group (8).

Contact angle measurements provide information on the wettability of the sample, the surface energetics of the solid, and
the interfacial properties of the solid-liquid interface. The
samples were immersed in water and captive air and octane bubbles
were determined by measuring the bubble dimensions. By measurement of both air and octane contact angles the surface free energy
($\gamma$) of the solid-vapor ($\gamma_{sv}$) interface may be calculated by use of
Young's equation and the harmonic mean hypothesis for separation
of the dispersive and polar components of the work of adhesion.
This method for determination of surface and interfacial proper-

ties has been discussed in detail (9, 10).  Because the measurement is made underwater, it is basically a receding angle measurement in the case of the air/water/solid measurement, thus $\gamma_{sv}$
values obtained are larger than those commonly reported, which
are generally advancing angle measurements (see (10) for a complete discussion).

Scanning electron microscopy was used to detect changes in
surface topography due to the plasma treatment.  Preferential
etching of the material will change the surface topography.  The
formation of volatile, low molecular weight species under the
surface of the material can produce bubbles (1).

Measurement of substrate surface charge was not performed in
this study.  A change in surface charge might be expected due to
the plasma treatment either by ion implantation or by formation of
ionizable functionalities on the substrate surface.

Change in contact angles as a function of storage time was
also not studied.  Studies on polystyrene indicate that the
contact angle does change with time after plasma treatment (11).

Experimental

Samples of polystyrene were cut from Petri dishes (Falcon
1008, Falcon Plastics, Oxnard, California).  Unalloyed low temperature isotropic (LTI) carbon samples were obtained from the
General Atomic Company (Pyrolite   - registered trademark of
General Atomic Company, now Carbo-Medics, Inc.).

The polystyrene samples were used as received.  Examination
by XPS showed only carbon on the surface (XPS does not detect
hydrogen).  The pyrolytic carbon samples were prepared by a
"steady state" fluidized bed process (12).  The carbon samples
were polished by the manufacturer using $\gamma$-alumina.  Before use in
this study the pyrolytic carbon samples were ultrasonically
cleaned in reagent grade methanol for five minutes.  Examination
by XPS of both the as received and ultrasonic methanol cleaned
samples showed removal of small amounts of chlorine, magnesium,
silicon, and sulfur by the cleaning procedure.

The samples were oxidized using a commercial plasma discharge unit (Plasmod   registered trademark of Tegal Corporation, Richmond, California), which operates at 13.56 MHz and has a
variable power output from 0 to 100 watts.  A variable leak valve
(Granville-Phillips Company, Boulder, Colorado), a three-way
valve and other modifications were added to provide a better
vacuum, to control the gas flow rate and to control the gas
pressure.

The samples were inserted into the RFGD unit in air and
placed on the bottom-center region of the reaction chamber.  In
all experiments the top surface was the analysis surface.  The
reaction chamber was evacuated to $10^{-2}$ torr pressure and then
backfilled with the reaction gas to above 2000mmHg pressure and
then re-evacuated.  This process was repeated three times.  The

gases used in this study were helium and oxygen. The helium was
liquid nitrogen cold trapped during backfill to remove condens-
able impurities. After the third backfill and pumpdown to $10^{-2}$
torr pressure, the gas flow rate was adjusted to correspond to 0.4
torr pressure, which was maintained during discharge. 0.4 torr
pressure was determined to give minimal deposition of silicon on
the sample surface. The silicon sputtering originated from the
Pyrex walls of the reaction chamber during plasma treatment.
Lower pressures gave appreciable amounts of silicon deposition on
the sample as determined by XPS examination. The samples were
exposed to the helium gas for various amounts of time at 50 watts
of power output. The samples were then exposed to a helium gas
purge for various amounts of time at above 4 torr pressure and
then exposed to oxygen at above 4 torr pressure for five minutes.
The samples were stored in Petri dishes in air until surface
characterizations were performed. It was found if the samples
were exposed to air instead of oxygen after the inert gas purge
that both oxygen and nitrogen functionalities were observed by
XPS examination.

Oxygen plasmas were used on the pyrolytic carbon. The
procedure was the same, except that the samples were exposed to an
oxygen purge for 5 minutes after the discharge. XPS spectra were
obtained with a Hewlett-Packard 5950 B instrument utilizing monochromatic Al $K\alpha_{1,2}$ radiation at 1487 eV. The samples were mounted
in air, inserted into the spectrometer, and analyzed at ambient
temperatures in a $10^{-9}$ torr vacuum. Power at the X-ray source was
800 watts. Instrument resolution was nominally 0.8 eV or less as
measured by the full width at half maximum of the C-1s line from
spectroscopic grade graphite. An electron flood gun operating at
0.3 mA and 5.0 eV supplied a flux of low energy electrons to
minimize charging artifacts in the resulting spectra.

Wide scans (0 to 1000 eV) were performed for surface elemental analyses. The wide scans were carefully inspected for
trace element contamination. Detailed 20 eV scans of the C-1s
(275 to 295 eV), O-1s (520 -540 eV) and Al-2s (105 to 125 eV)
regions for the pyrolytic carbon and of the C-1s and O-1s for the
polystyrene were run to determine both elemental stoichiometry
and chemical shifts. Standards were available to give accurate
chemical shift data for various carbon-oxygen functional groups.
These included poly(ethylene terephthalate), poly(ethylene oxide)
and anthraquinone (17). The latter was run at $-50°C$ in order to
minimize volatility under our high vacuum conditions. Table I
summarizes these results. All spectra were charge - referenced to
a C-1s line for an alkyl-like carbon at 284.0 eV. The Scofield
theoretical XPS photoelectric cross sections (13) were used for
elemental quantitation.

Scanning electron micrographs were obtained on a Cambride
Mark II Stereoscan SEM. The samples were mounted on the specimen
mounts with double-sided tape. Silver paint along the edge of the
sample provided electrical contact between the specimen mount and

Table I

Chemical shifts in the C-1s line for poly (ethylene terephthalate), polyethylene oxide, and anthraquinone. (See also Reference 17).

| Material | Observed Binding Energy (eV)* | $\Delta$BE** | Functional Group |
|---|---|---|---|
| | 284.0 | 0.0 | alkyl and/or aromatic |
| Polyethylene terephthalate | 285.6 | +1.6 | ether |
| | 288.0 | +4.0 | ester |
| Polyethylene oxide | 285.8 | +1.8 | ether |
| Anthraquinone | 284.0 | -- | alkyl or aromatic |
| | 286.5 | +2.5 | quinone |

\* Charge-referenced to C-1s at 284.0 eV
\*\* Binding Energy -284.0 eV = $\Delta$BE

the sample. The samples were coated with carbon and gold to
reduce charging. The analyzing vacuum was about $10^{-4}$ torr.

Contact angles were determined by immersing the sample in
doubly-distilled water and measuring the height and diameter of
both air and octane bubbles in water (9). The bubbles were
introduced on the sample surface using a microliter syringe. The
height and diameter were measured by use of micrometers which
manipulated a stage holding the sample immersed in water. The
bubbles were observed through a microscope using a 20X long
working distance objective and a 15X eye-piece equipped with a
crosshair reticle. The bubble was manipulated across the crosshair and the dimensions were read directly from the micrometers.
The sample box was back illuminated by a variable light source.
The bubble volume was approximately 0.1 to 0.2 μl and the bubbles
were applied to the surface by forming a bubble at the tip of a
micro syringe and then snapping the tip to allow the bubble to
float up to the water-sample interface. The bubble volume was
minimized in order to avoid buoyancy effects. The octane was
99.99% pure n-octane (Aldrich Chemicals –Gold Label Octane).
Temperature of the immersion bath was 26°C. Figure 1 schematically illustrates the geometry of the contact angle measurement.
The contact angles were calculated using the equation

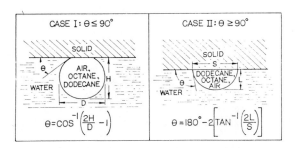

*Figure 1.   Schematic of both cases for contact angle calculation*

$$\phi = 180 - 2\tan^{-1}(\frac{2E}{S}) \text{ for } \phi \geq 90^{\circ} \text{ and } \phi = \cos^{-1}(\frac{2H}{D} -1) \text{ for } \phi \leq 90^{\circ}.$$

This technique measures the fully hydrated solid-water interface. In essence the air/water/solid angle is similar to a receding water contact angle in the conventional contact angle geometry. By probing the fully hydrated solid/water interface, the polar components of the solid surface are more optimally evaluated. As it is the solid/water interface which is of primary interest for biological interactions, we feel this method is more appropriate for surface characterization for our purposes. The use of air and octane angles also allows one to deduce the fully hydrated solid surface free energy and the solid/water interfacial free energy. Assumptions also allow the fully hydrated solid surface free energy to be decomposed into its apolar and polar components (10).

## Results and Discussion

Polystrene. Table 2 presents the carbon:oxygen ratios as determined by XPS. The as received and methanol-cleaned materials were essentially unoxidized, with carbon:oxygen ratios of the order of 100 to 1 or greater. The material which was helium plasma treated was extensively oxidized with a C:O ratio of approximately 3.5:1. The air and octane angles at the solid/water interface are also presented for those two cases. The air angle, which can be interpreted as a conventional water receding contact angle, is approximately 85$^{\circ}$ and decreases to approximately 14$^{\circ}$ on plasma treatment, indicating a substantial increase in surface wettability. The XPS spectra of the as received polystyrene show the presence of the C-1s aromatic satellite at 290.70 eV, 6.7 eV from the main carbon 1s line with approximately the correct

Table II

Carbon: Oxygen (C/O) Ratios, as determined by X-ray photoelectron spectroscopy and contact angle data for polystyrene.

| Sample | C/O Ratios | Contact Angle (10) | |
| --- | --- | --- | --- |
| | | Air | Octane |
| As received | 100/1 | 83±4$^{\circ}$ | 135±5$^{\circ}$ |
| Methanol Cleaned (Ultrasonic) | 98/1 | 82±4$^{\circ}$ | 135±5$^{\circ}$ |
| Helium Plasma | 3.5/1 | 14±4$^{\circ}$ | 14±4$^{\circ}$ |

intensity ratio (8). The satellite peak is reduced to background
level for the helium plasma treated, oxidized material, suggesting a considerable decrease in aromaticity of the polystyrene in
the surface volume examined by XPS (see Figure 2). Results from
helium plasma treated polystyrene are essentially identical to
those found with "tissue culture" polystyrene produced by Falcon
Plastics.

Scanning electron mcirographs of the sample showed no gross
etching of the surface. Polystyrene has no oxygen in the polymer
which can form atomic oxygen in the plasma, therefore etch rates
due to oxidative degradation are expected to be low. Plasma
exposure times were not long enough to create low molecular weight
volatile species under the surface. These can diffuse to the
surface and form bubbles, as is often seen in polyethylene (1).

Hansen (14) reports a decrease in molecular weight in a
helium plasma and Westerdahl (15) reports a change in contact
angle to higher wettability for both helium and oxygen plasmas.

Pyrolytic Carbon. Polished LTI carbon is composed of a
crystalline graphitic-like microstructure, combined with amorphous material (16). The polished samples have been shown to be
oxidized with a C:O ratio of about 10:1, containing three major
types of carbon-oxygen functionalities: quinone-like, ether-
like, and ester-like (17).

Electrochemical studies of carbon samples have shown that
both quinone-like and ester-like groups are present (18).

In this study various discharge times in oxygen and helium
gas plasmas were used. The carbon/oxygen ratio varied from 7 to 1
in the as received material to 1.2 to 1 for the treated samples
(see Table 3).

A rise in the amount of aluminum, as can be seen in Table 3,
is also noticed after plasma treatment. The increase in aluminum
at the surface may be due to preferential etching of surface
carbon, uncovering $Al_2O_3$, which is thought to be embedded in the
polishing process. M. Millard has seen this same type of
phenomena when plasma etching cells, i.e., the organic portion of
the cells is etched away concentrating inorganic trace elements
on the surface of the sample (19). The aluminum peak was
strongest after oxygen plasma treatment, probably due to higher
etching rates.

The increase in the aluminum peak is accompanied by the
growth of lower binding energy oxygen and carbon peaks (see Figure
3). All three peaks were observed to move up in binding energy
when the flood gun was turned off. Shifts in binding energy due
to the flood gun occur when the sample is non-conducting. The
flood gun is used to provide a source of low energy electrons to
the sample to counter positive charging of a sample due to the
electrons being ejected. The excess supply of electrons provided
by the flood gun charges the sample negatively and lowers the
apparent binding energies of the elements. For conducting

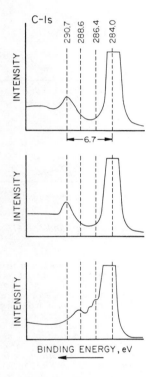

*Figure 2.   XPS spectra of the C-1s region of PS.*

*Flood gun conditions, 0.3 ma, 5 eV. The C-1s alkyl line was approximately 279.0 eV; the spectra above are charge-referenced to 284.0 eV for the alkyl carbon line. Spectrum A is the as-received material; note the presence of the aromatic satellite at 6.7 eV from the main carbon line at 284.0. Note also the absence of any carbon–oxygen functionalities as evidenced by the lack of structure between 284 and 290: (a) as-received material; (b) methanol-cleaned; (c) oxygen plasma-treated material. Note the decrease in the satellite line at 297 eV; it has disappeared to nearly background level. Also note the presence now of two carbon–oxygen functionalities as evidenced by apparent peaks at about 288.6 and 286.4 eV, characteristic of ester or carboxylic acid, and ether or hydroxyl carbon, respectively.*

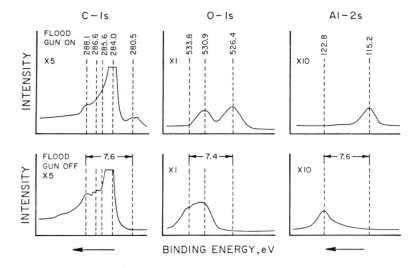

*Figure 3.* XPS spectra of the (a) C-1s region; (b) O-1s region, and (c) Al-2s regions of the oxygen plasma–treated pyrolytic carbon.

The upper spectra are with the electron flood gun on for charge compensation, the bottom spectra are with the electron flood gun off. Note in the upper spectrum that the main carbon peak appears at 284.0 eV, as expected for a conducting material such as pyrolytic carbon. Note also the presence of the weak apparent doublet in the vicinity of 280.5 eV. The Al-2s top far right appears at about 115.2 eV or charge shifted down stream from its apparent normal position. Comparing these upper spectra with the flood gun off spectra and looking at the relative peak positions, one can deduce (see *text*) that there is an insulating component in the surface region of the material that charge shifts to higher binding energies in the absence of the flood gun and is pushed to lower binding energies in the presence of the flood gun. Note also that there are a number of major lines that are not affected by the flood gun conditions. These, of course, are those intrinsic to the conductive pyrolytic carbon structure. The insulating material which is influenced significantly by the flood gun conditions is attributed to $Al_2O_3$ particles embedded in the carbon during the polishing process.

Table III

Carbon: Oxygen (C/O) ratio and atomic percentages as determined by X-ray photoelectron spectroscopy for pyrolytic carbon.

| Treatment Type | C/O Ratios | Atomic Percent | | | |
|---|---|---|---|---|---|
| | | C | O | Al | Trace* |
| As Received | 8 | 85.5 | 10.4 | 1.4 | Cl, Mg, |
| Methanol Cleaned (ultrasonic) | 6 | 85 | 13.4 | 1.6 | -- |
| Helium Plasma Treatment | 2-4 | 60-70 | 34-27 | 5-3 | -- |
| Oxygen Plasma Treatment | 1.2-1.4 | 56 | 36.5 | 7.5 | -- |

* Trace less than 1 atomic percent.

samples the flood gun will not affect the binding energy since the sample is in electrical contact with the grounded sample probe. Grunthaner has reported the effect of a flood gun on non-conducting samples and has used it to investigate the chemical composition of non-conducting oxides formed on metal catalyst systems (20).

In the pyrolytic carbon, aluminum, oxygen, and carbon peaks were observed to move in binding energy as a function of the flood gun conditions. Also a carbon peak and an oxygen peak remained unaffected by the flood gun (see Figure 3). The majority of the carbon and oxygen are unaffected by the flood gun and all the aluminum moves with the flood gun. From these results it is concluded that the major portion of the sample is conductive pyrolytic carbon; islands of non-conducting $Al_2O_3$, with some carbon and oxygen associated with it, are being exposed on the surface due to plasma etching.

SEM of the samples shows a change in the surface roughness after glow discharge treatment. The change is due to etching of the surface by the plasma treatment and supports the noticed increase in amount of aluminum detected by XPS. Energy dispersive analysis of X-rays of the sample could not distinguish between aluminum on the surface and aluminum embedded below the surface

due to higher analysis depths. The surface appeared rougher after
etching. The change in topography is probably due to preferential
etching of the amorphous portion of the pyrolytic carbon.

As the samples were highly oxidized from the beginning, due
to the polishing process (17), no real change in contact angle was
observed.

## Summary

Pyrolytic carbon and polystyrene surfaces were studied by X-
ray photoelectron spectroscopy (XPS), contact angles, scanning
electron microscopy (SEM), and energy dispersive analysis of X-
rays (EDAX). The materials were radiofrequency glow discharged
(RFGD) in helium and oxygen plasmas. RFGD of the polished carbons
increased the degree of oxidation and the Al content. The
increased Al content is interpreted as due to exposure of $Al_2O_3$
particles embedded in the polishing process. This is confirmed by
flood gun-charging results.

RFGD polystyrene was oxidized, wettable, and of a decreased
aromatic character as determined by analysis of XPS C-1s satellite spectra. The RFGD oxidation process etches the surface as
witnessed by scanning electron micrographs.

These data are of interest in understanding the behavior of
polished carbon and oxidized polystyrene in biomedical applications, including surgical implants and solid substrates for in
vitro cell cultures.

## Acknowledgements

Portions of this work were supported by NIH Grant #HL16921-
04 and the University of Utah Faculty Research Committee.

## Abstract

Surface characterization of RF plasma modified polymers is
necessary in order to understand and improve certain properties
including cell adhesion to tissue culture substrates. Radio
frequency glow discharge (RFGD) plasmas were used to modify the
surfaces of polystyrene and pyrolytic carbon. Surface characterization by X-ray photoelectron spectroscopy (XPS), scanning electron microscopy (SEM), and air and octane contact angles were
performed on the as received and plasma modified samples. Plasma
modification of polystyrene produced a number of carbon-oxygen
functional groups and decreased both air and octane contact
angles. Plasma modification of pyrolytic carbon showed an increase in aluminum and oxygen on the surface, probably due to
preferential etching of surface organics, exposing the inorganic
component of the sample. The aluminum and oxygen are probably
from $\gamma$-alumina, used in the polishing of pyrolytic carbon. SEM
also showed a change in topography indicating preferential etching.

## Literature Cited

1.  Hudis, M., in "Techniques and Applications of Plasma Chemistry," (J.R. Hollahan and A.T. Bell, Eds.), Chap. 3. John Wiley and Sons, New York, 1974.

2.  Kirk, Ralph W., in "Techniques and Applications of Plasma Chemistry," (J.R. Hollahan and A.T. Bell, Eds.), Chap. 9. John Wiley and Sons, New York, 1974.

3.  Bokros, J.D., Carbon, 1977, 15, 355.

4.  Bell, A.T., in "Techniques and Applications of Plasma Chemistry," (J.R. Hollahan and A.T. Bell, Eds.), Chap. 1. John Wiley and Sons, New York, 1974.

5.  Hollahan, J.R., J. Chem. Ed., 1966, 43, A401.

6.  Hall, S.M., Andrade, J.D., Ma, S.M., King, R.N., J. Electron Spectrosc., 1979, 17, 181-189.

7.  Clark, D.T., and Dilks, A., J. Poly. Sci., 1976, 14, 533.

8.  Clark, D.T., Adams, D.B., Dilks, A., Peeling, J., and Thomas, H.R., J. Elect. Spect., 1976, 8, 51.

9.  Andrade, J.D., King, R.N., Gregonis, D.E., Coleman, E.L., J. Poly. Sci. Symp C., 1978, 66, 313.

10. Andrade, J.D., Ma, S.M., King, R.N., Gregonis, D.E., J. Coll. Interface Sci., 1979, 72, 488.

11. Triolo, P., Thesis, Department of Bioengineering, University of Utah, June, 1980.

12. Akins, R.J. and Bokros, J.C., Carbon, 1979, 12, 439.

13. Scofield, J.H., J. Electron Spect., 1976, 8, 129.

14. Hansen, R.H., Pascale, J.V., DeBenedictus, T., and Rentzepis, P.M., J. Poly. Sci., 1965, A3, 2205.

15. Westerdahl, C.A.L., Hall, J.R., Schramm, E.C., and Levi, D.W., J. Colloid Interface Sci., 1974, 47, 610.

16. Biscoe, J., and Warren, B.E., J. Appl. Phys., 1942, 13, 364.

17. King, R.N., Andrade, J.D., Haubold, A.D., and Shim, H.,

"Surface Analysis of Silicon-Alloyed and Unalloyed LTI Pyrolytic Carbon," in this volume.

18.  Evans, R. and Kuwana, T., Anal. Chem., 1977, 49, 1632.

19.  Millard, M.M., "Surface Characterization of Biological Mate-
     rials  by  X-ray  Photoelectron  Spectroscopy,"  in  D.M.
     Hercules,  et  al.,  Eds.,  Cont. Topics in Anal, and Clin.
     Chem., 1978, 3, 1.

20.  Grunthaner, F., Ph.D. Dissertation, California Institute of
     Technology, 1974.

RECEIVED March 23, 1981.

# ESCA Studies of Polyimide and Modified Polyimide Surfaces

H. J. LEARY, JR. and D. S. CAMPBELL

IBM Corporation, General Technology Division, Essex Junction, VT 05452

The organic dielectrics known as polyimides have been studied extensively by a variety of bulk characterizational techniques as a perusal of the literature will illustrate. Little has been published on their surface properties. X-ray photoelectron spectroscopy (ESCA) has been extremely useful for polymer characterization ($\underline{1}$, $\underline{2}$, $\underline{3}$). In a previous paper ($\underline{4}$), we have reported the ESCA spectra of structurally different polyimides derived from both commercially available polyamic acid resins (DuPont's PI5878, PI2525, PI2550), and from laboratory synthesized polyamic acid resins.

In the present paper two types of results are reported. First, data are presented on the surface properties of cured polyimides derived from the Skybond 703 commercial resin manufactured by the Monsanto Co., and from the NR-055X resin manufactured by DuPont. The fluorine containing polyimide derived from the NR-055X resin is an ideal system for ESCA investigations because fluorine, being the most electronegative of the elements, induces the largest chemical shifts in the C 1s levels, and a large amount of background data is available in the literature ($\underline{5}$, $\underline{6}$, $\underline{7}$). The second type of information presented is on PI5878 polyamic acid/polyimide modifications produced by a variety of treatments (e.g., etching, plasma exposures).

The principal points to which this work has been focused are as follows:

1.  Are there any consistent surface features detectable by ESCA that are characteristic of the polyimides (cf., Skybond 703, NR-055X, PI5878, PI2550)?
2.  What types of information are extractable by ESCA on the changes in cured PI5878 produced by short (i.e., 2 minutes) exposures to $O_2$, and $O_2/CF_4$ plasma environments?
3.  Is reaction specificity (i.e., attack at the acid carbonyl functionality only) indicated in the interaction of KOH with the polyamic acid of PI5878?

0097-6156/81/0162-0419$05.00/0

4.  Will imidization occur if KOH treated polyamic acid films of PI5878 are subsequently subjected to standard cure-cycles?
5.  Are highly cured PI5878 films really stable (i.e., do they show invarient surface features) in high humidity, and aqueous environments?

It will become clear that most of our discussion (and conclusions) will center around detailed scrutiny of the C 1s core level spectra, binding energies (BE) and chemical shifts, even though other core level data are also presented. This is done for several reasons: (a) an extensive C 1s literature for polymeric systems exist for comparative purposes, (b) the key informational content for polymers is most frequently contained in the C 1s spectra, and (c) a good and workable theoretical model has evolved (8, 9, 10, 11) for determining C 1s charge distributions from experimentally determined binding energies as a result of the finding that the magnitude of the relaxation energy accompanying photoionization follows simple trends with binding energy for closely related chemical systems (12).

## Experimental

The methods used for collecting and analyzing the experimental data that were collected on the Hewlett-Packard 5950B electron spectrometer have previously been documented (4). Quoted binding energy values fall in a range of $\pm$ 0.2 eV based on evaluations of detailed curve analysis results. Previous work in our laboratory has shown that the binding energy values calculated for deconvolved components can show a small dependency on the signal/background ratio.

NR055X and 703 polyamic resin material were separately coated onto cleaned silicon substrates so as to produce a resin layer $\sim 2 \mu$ thick. Each sample type was subsequently heated on a hot plate for 10 minutes at $85^\circ$C to remove solvent from the films to produce the condensed phase amic acid. Specimens were sequentially step cured as follows: 10 minutes at $200^\circ$C; 30 minutes at $300^\circ$C on a hot plate having a lid in a $N_2$ flow; and 30 minutes at $400^\circ$C in a closed tube furnace in a flowing stream of $N_2$ (in separate experiments, a flowing stream of forming gas, i.e., 80% $N_2$-20% $H_2$, was used for the $300^\circ$C and $400^\circ$C cures). ESCA measurements were made after each curing step in order to evaluate differences caused by increasing the temperature of cure.

Plasma treatments of cured PI5878 samples were carried out in the LFE 1000 Plasma Apparatus operating at 500 watts. Gas flows were adjusted so as to obtain a pressure of 1.0 torr for the $O_2$-plasma experiment, 0.5 torr for the $CF_4$ (92%)-$O_2$ (8%)-plasma experiment, and 1.0 torr for the $CF_4$ (84%)-$O_2$ (16%)-plasma experiment. An exposure time of 2 minutes was used in each experiment.

For evaluations of the effect of KOH on PI5878 polyamic acid films, the as-applied resin film ($\sim 2 \mu$) after the $120^\circ$C solvent removal step (4) was exposed to KOH (0.23M-0.75M) for 5 seconds. ESCA measurements were made on the KOH treated surfaces. These data were compared with those obtained on non-KOH treated resin surfaces that had undergone the identical $120°C$ solvent removal step. Spectral data were also collected on several of the KOH-treated polyamic acid surfaces after *in situ*, and after air step cures to $300-325°C$. These data could then be compared to those obtained earlier (4) so that the effect of KOH on the imidization (or curing) could be assessed.

The approach used to characterize the humidity effects on cured PI5878 films was as follows: A film cured to $\sim 300°C$, affixed to a silicon substrate, was placed in a humidity chamber maintained at $85°C$ and 80% relative humidity for 280 hours. ESCA measurements were made on separate halves of the same wafer, before and after the 85/80 treatment. In addition, cured PI5878 films on silicon substrates were exposed to boiling water for 30 minutes and ESCA analyzed. These samples were then reanalyzed after a subsequent *in situ* ($10^{-5}$ torr) heating for 30 minutes at $280-300°C$.

## Results and Discussion

Cured Polyimide Film Surfaces - NR055X. To put the NR-055X and Skybond 703 results in perspective, we refer to Figure 1 taken from our earlier work (4) on PI5878 (spectra 1a, 1b) and on PI2525 (spectra 1c, 1d). Though structurally dissimilar both polyamic acid films (1a, 1c) show acid carbonyls at 289.4 eV, amide carbonyls at 288.4 eV and sizable amounts of partially oxidized carbon species (binding energy 286.0 eV, C 1s chemical shift $\sim 1.0$ eV). On curing, both polyimides showed imide carbonyl features (binding energy 288.9 eV, chemical shift $\sim 3.9$ eV) and a significant increase in the amount of partially oxidized carbon species ($\sim 42\%$ of the C 1s signal intensity). For the reasons cited in Reference 4, we concluded imide carbonyl deficiency and non-stoichiometric chemical structures for the cured film surfaces. Though isoimide structures have been proposed to exist in n-imide systems (13), we cannot say that this structure does exist because there is no way to differentiate its existence from polyimide molecular weight changes in the surface layer. Molecular weight changes could occur as a result of: (a) scissioning in the surface layer - a polymer dissociation process, (b) cross-linking, i.e., the production of a three-dimensional structure as a result of random covalent bond formation between adjacent polymer chains, or (c) branching in the surface layer, i.e., the creation of side chains which are attached to the main polymer molecule. All we can confidently say now about the partially oxidized carbon environments is that the $\sim 1.0$ eV chemical shift is consistent with C-OH, C-N-CO, C-O-C bond types, and that its magnitude is inconsistent with any reasonable assumptions we could make about the charge density on carbon atoms by reference to the chemical

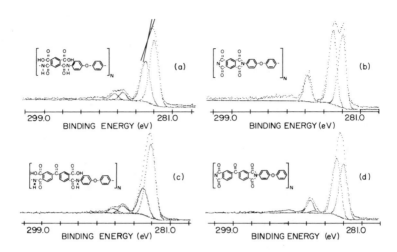

*Figure 1.   The C-1s spectra for two polyimides: (a) PI 5878 after 120°C solvent removal drying cycle; (b) PI 5878 after step cures 325° in N₂; (c) PI 2525 after 95°C solvent removal drying cycle; (d) PI 2525 after step cures to 325°C in N₂.*

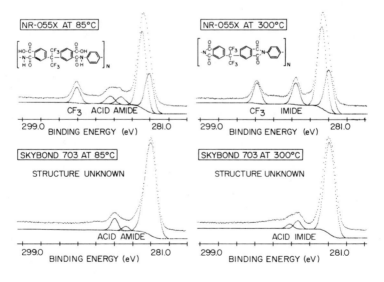

*Figure 2.   The C-1s spectra for NR-055X and Skybond 703*

Figure 2 shows C 1s spectra for NR-055X polyamic acid/polyimide, and for Skybond 703 polyamic acid/polyimide. The fluorinated polyamic acid showed $\underline{C}F_3$, acid carbonyl and amide carbonyl lines at binding energies of 293.3 eV, 289.4 eV, and 288.1 eV, respectively. The chemical shift and BE values are in excellent agreement with literature data for $\underline{C}F_3$ carbons. The most intense C 1s line is asymmetric on the low binding energy side of the peak maxima. The curve fit results indicated 284.9 eV - assigned to the ring carbons, and a peak at 285.8 eV. The chemical shift of ∿0.90 eV is consistent with that observed earlier (4) for PI5878. Thus, we have no reason to attribute the 285.8 eV peak to any type of carbon-fluorine bonding.

Only one carbonyl component appears in the NR-055X spectrum after the 300°C cure treatment. Its 3.8 eV chemical shift is consistent with imide. The $\underline{C}F_3$ binding energy and chemical shift values were identical to those observed for the polyamic acid. The most intense C 1s component peak characteristics showed little change with temperature spanning 85°-400°C.

For the cured polyimide, the $\underline{C}F_3/\underline{F}$ ratio was calculated using the integrated peak area results. A value of 0.33 was determined experimentally (vs. the value of 0.33 predicted). Extensions of this approach occurred by calculation of $=\underline{C}=O/\underline{N}$ and $=\underline{C}=O/\underline{F}$ ratios. Values of 1.08 and 0.32, respectively, were determined experimentally (vs. 2.0 and 0.67 predicted on the basis of structural formula considerations). These results indicated either carbonyl deficiency, an excess of nitrogen or an excess of fluorine in the surface region.

In an attempt to resolve this, we recognized that the $\underline{C}F_3/\underline{F}$ ratio was stoichiometric with the structure of NR-055X polyimide. Thus the corrected signal ratio reflects the ratio of the actual number of fluorine atoms to $\underline{C}F_3$ atoms. This would mean that ∿8% of the total carbon signal should be due to $\underline{C}F_3$ carbons, and implies that ∿16% of the carbon signal should be due to carbonyl groups (i.e., a 1:2 relation for $\underline{C}F_3$:imide carbonyl). Figure 2 shows that this 1:2 relation did not exist. The result was confirmed by calculations from the experimental data:

| NR-055X SAMPLE CURE | $CF_3/C_{TOTAL}$ (%) | $=C=O/C_{TOTAL}$ (%) |
|---|---|---|
| 300°C in N$_2$ | 12.25 ± 2.63 | 11.75 ± 1.50 |
| 400°C in Forming Gas | 10.50 ± 0.58 | 10.75 ± 0.50 |
| 400°C in N$_2$ | 10.75 ± 0.50 | 10.00 ± 0.00 |

which indicate enrichment in $\underline{C}F_3$ groups in the surface region, and a marked deficiency in carbonyl. The latter conclusion is consistent with previous results (4).

   Cured Polyimide Film Surfaces - Skybond 703. As shown in
Figure 2, the most intense line in the C 1s spectrum appeared at
284.9 eV and was nearly symmetrical in line shape for the polyamic
acid. For the 703 polyimide, no asymmetry was detectable. Two
bands are indicated by the asymmetrical character of the carbonyl
region of the polyamic acid:  one at BE 289.0 eV, and one at BE
288.1 eV, the area intensity ratio was 2.84:1.0. The bands are
assigned to "acid" and amide carbonyls, respectively though the
former could conceivably be due to ester or ester/acid linkages.
On curing of this resin to 300°C, the asymmetrical character of
the carbonyl region changed. A carbonyl component at 289.4 eV due
to pure acid is observed, in addition to a feature at 288.5 eV
(imide). The band area ratio was 0.50:1.0.
   Several important factors are inherent in these data.
First, the C 1s chemical shifts for imide and acid groups were
determined to be ∿3.64 eV and ∿4.2 eV, respectively.  These
results are in very close agreement with those determined for
PI5878/PI2550 systems (4). They are consistent with our model
compound results (14). Second, the low energy carbonyl peak of
the polyamic acids was assigned by inference on the basis of both
expected inductive effects and knowledge of resin formulation,
e.g., laboratory prepared PI5878 or PI2550, shows acid and imide
carbonyl features in its C 1s spectra. Its unique character is
also illustrated by its N 1s spectra (Figure 3) that shows ∿67%
imide when compared to ∿96% imide for a cured PI5878 sample
surface. The Skybond results would be consistent with, but do not
prove, a copolymer resin formulation having two different amines
that imidize at different rates.
   For all cured, non-modified polyimide films studied, the C
1s spectra showed extremely weak satellite features centered
about 292.8 ± 0.3 eV. To date, we have not studied these shake up
structures extensively. We have noted, however, that they frequently do not exist in the C 1s spectra of the polyamic acid
films. These low intensity features appear resolvable into two
Gaussian components of unequal intensity. Their chemical shifts
are consistent with assignments given to π* ← π transitions e.g.,
D.T. Clark et al, J. Electron. Spectrosc. Relat. Phenom., 8, 51
(1976). These observations suggest that the polyamic acid film
surfaces may not have appreciable unsaturated carbon environments.

   Polyamic Acid-KOH-Polyimide Interactions. Table I shows C
1s and K 2p data that demonstrates the effects of KOH and KOH-
acetic acid (0.5% by volume) treatments on PI5878 polyamic acid
surfaces. For the non-treated surface, C 1s chemical shifts of
4.23 eV, 3.34 eV and 1.02 eV reflect acid, amide and "partially
oxidized" bonding types. Treatment with KOH produced two
effects: (a) incorporation of potassium into the carbon matrix,
and (b) changes in both carbonyl binding energies. This indicates
that chemical reaction at both acid and amide occurred and that no

reaction specificity exists.   Since dilute KOH solution etch removes polyamic acid films, our data suggests the etching process to be associated with intramolecular chain scission. We have noted in our experiments that demineralized water rinses for 2 minutes did not alter the potassium level.  However, a 5 second acetic acid rinse did lower the potassium level. As indicated in Table I, a 5 minute acetic acid rinse removed detectable potassium from the surface, and reverted the carbonyl binding energies to their original values.

Table II shows comparisons of data for several cured PI5878 samples.   For the KOH treated sample that had no acetic acid rinse, bulk potassium diffused to the surface on direct heat curing to 280°C.  Though the data given reflect in situ results, the same phenomenon occurred in air or nitrogen ambients with virtually no change in the carbonyl binding energies for the sample.   By contrast, the non-KOH treated sample and the treated surface that had a 5 minute acetic acid rinse showed identical C 1s features – imide and "partially oxidized" carbon functionalities.   The results demonstrate that potassium incorporated in the polyamic acid matrix prevents the imidization process.

Boiling Water and Temperature and Humidity Effects on Polyimides.   Figure 4 shows the changes in the C 1s spectrum of cured PI5878 polyimide after exposure for 280 hours in a temperature-humidity chamber (85°C/80%), and after 30 minutes in boiling water.   Three points are worth noting in comparisons with the spectrum of the "control" sample.  First, the "partially oxidized" carbon component as well as the imide carbonyl intensities were lower after the T&H exposure.   Second, after the boiling water treatment, the "partially oxidized" carbon component had virtually disappeared (the main signal does, however, show asymmetrical character).   In addition, the half-width of the carbonyl band increased and showed shoulders (∿289.5 eV, ∿288.2 eV) of weak intensities on each side of the centroid (288.8 eV).   Third, in situ vacuum heating of the water-treated sample produced C 1s data that was identical to that of the "control" sample.

The observations made by analysis of the C 1s data show that the cured surfaces are not invarient to T&H or to boiling water exposures.   The data suggests that de-imidization, though not extensive, may be occurring after 15 minutes in boiling water. Very close scrutiny of the compositional (Table III) and binding energy (Table IV) results indicate a very complex mechanism – not within the objectives of our present studies – is operative in the "weathering" of cured polyimide films.  The results from our work do indicate that the surface is involved in this environmental-type of phenomena since nitrogen depletion and other compositional changes were detectable.   It was of interest to know if electrical property characteristics were affected.  Results (15, 16) from cursory experiments of the dissipation factor showed higher values for cured films that had been water soaked.

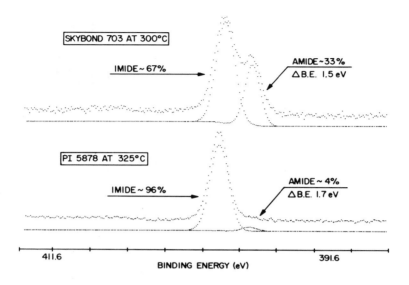

*Figure 3.   The N-1s spectra for cured Skybond 703 and PI5878 polyimides*

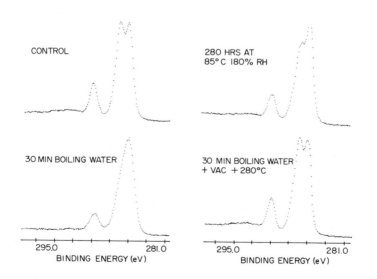

*Figure 4.   The C-1s spectra for PI5878 after T&H and boiling H₂O exposures*

**Table I.   The C-1$s$ and K-2$p$ Binding Energy Data for Polyamic Acid (PI5878)KOH–HAc Surfaces**

| SAMPLE | PROCESS | C 1s | | | | K 2P | $K \times 10^{22}$ |
| | | B.E. (eV) | $\triangle$B.E. (eV) | FWHM (eV) | % | B.E. (eV) | (A/cc) |
|---|---|---|---|---|---|---|---|
| 41–36 | PI + 120°C | 285.0 | | 1.50 | 73.00 | | |
| | | 286.0 | 1.0 | 1.10 | 18.69 | | |
| | | 288.3 | 3.3 | 0.90 | 4.29 | | |
| | | 289.2 | 4.2 | 0.90 | 4.02 | | |
| 31–17 | PI + 0.75M KOH[+] | 285.0 | | 1.50 | 80.02 | 293.2 * | 3.3 |
| | | 286.2 | 1.2 | 1.10 | 10.77 | 296.0 * | |
| | | 288.1 | 3.1 | 0.90 | 5.48 | | |
| | | 288.8 | 3.8 | 0.90 | 3.73 | | |
| 38–49 | PI + 5 MIN[+] | 285.0 | | 1.50 | 80.42 | K NOT EVIDENT | |
| | HAc RINSE | 286.1 | 1.1 | 1.10 | 11.94 | | |
| | (DIL. SOLN) | 288.4 | 3.4 | 0.90 | 3.36 | | |
| | | 289.3 | 4.3 | 0.90 | 3.97 | | |

* INTENSITY RATIO WAS 1:2.19
[+] DEMINERALIZED $H_2O$ RINSE DID NOT REMOVE K
[‡] 5 SEC. HAc RINSE DID NOT REMOVE K, IT LOWERED CONC. TO 1.1 x $10^{22}$ A/cc

**Table II.   The C-1$s$ and K-2$p$ Binding Energy Data for Cured Polyimide (PI5878)KOH–HAc Surfaces**

| SAMPLE | PROCESS | C 1s | | | | K 2P | $K \times 10^{22}$ |
| | | B.E. (eV) | $\triangle$B.E. (eV) | FWHM (eV) | % | B.E. (eV) | (A/cc) |
|---|---|---|---|---|---|---|---|
| 41–36 | PI + 120°C + 280°C | 285.0 | | 1.25 | 52.93 | | |
| | | 286.1 | 1.1 | 1.00 | 35.26 | | |
| | | 288.9 | 3.9 | 0.90 | 11.81 | | |
| 31–17 | PI + 120°C + 0.75M | 285.0 | | 1.40 | 74.05 | 293.1 * | 10.3 |
| | KOH + 280°C | 286.4 | 1.4 | 1.10 | 11.02 | 295.8 * | |
| | | 287.9 | 2.9 | 0.90 | 7.12 | | |
| | | 288.5 | 3.5 | 0.90 | 7.81 | | |
| 38–49 | PI + 120°C + 0.75M | 285.0 | | 1.25 | 52.66 | K NOT CLEARLY EVIDENT | |
| | KOH + HAc RINSE + | 286.1 | 1.1 | 1.00 | 35.29 | | |
| | 280°C | 288.9 | 3.9 | 0.90 | 12.04 | | |

* INTENSITY RATIO WAS 1:2.16

**Table III.  Binding Energy Results: PI 5878 (Cured) Surface Changes in H$_2$O Environments**

| TREATMENT | C 1s | | | O 1s | | N 1s | |
|---|---|---|---|---|---|---|---|
| | B.E. (eV) | ΔC 1s | % | B.E. (eV) | % | B.E. (eV) | % |
| 280°C – AS RECEIVED | 285.0 | | 47.46 | 532.2 | 74.50 | 400.7 | 100 |
| (CONTROL) | 286.1 | 1.1 | 40.12 | 533.5 | 25.50 | | |
| | 288.9 | 3.9 | 12.43 | | | | |
| 280°C + 280 HRS | 285.0 | | 50.09 | 532.2 | 80.27 | 398.9 | 100 |
| AT 85°C/80% RH | 286.1 | 1.1 | 39.13 | 533.6 | 19.73 | | |
| | 289.0 | 4.0 | 10.79 | | | | |
| 280°C + 30 MIN. | 285.0 | | 65.46 | 532.2 | 76.68 | 399.5 | 9.38 |
| IN BOILING H$_2$O | 286.1 | 1.1 | 24.58 | 533.6 | 23.32 | 400.6 | 90.62 |
| | (288.3) | (3.3) | | | | | |
| | 288.8 | 3.8 | 9.96 | | | | |
| | (289.5) | (4.5) | | | | | |
| ABOVE + SUBSEQUENT | 285.0 | | 47.74 | 532.3 | 77.79 | 400.4 | 26.64 |
| IN–SITU VAC. HEAT | 286.1 | 1.1 | 39.98 | 533.6 | 22.21 | 400.8 | 73.36 |
| TO 280°C | 288.9 | 3.9 | 12.28 | | | | |

**Table IV.  PI5878 (Cured) Surface Changes in H$_2$O Environments: Relative Composition**

| TREATMENT | C 1S | O 1S | N 1S |
|---|---|---|---|
| 280°C – AS RECEIVED (CONTROL) | 11.36 ± 0.77 | 2.57 ± 0.20 | 1.00 |
| 280°C + 280 HRS AT 85°C/80% RH | 14.22 ± 0.16 | 3.20 ± 0.11 | 1.00 (−16.1%) |
| 280°C + 30 MIN IN BOILING H$_2$O | 14.40 ± 1.12 (−6.5%) | 3.07 ± 0.24 (−11.8%) | 1.00 (−26.2%) |
| ABOVE + SUBSEQUENT IN–SITU VAC. HEAT TO 280° | 10.93 ± 0.29 | 2.49 ± 0.04 | 1.00 |
| EXPECTED VALUE (EMPIRICAL FORMULA) | 11.0 | 2.5 | 1.00 |

THE UNCERTAINTY IN THE UNIT SCAN NORMALIZATIONS OF THE 12 INTEGRATED PEAK INTENSITIES AVERAGED TO 4.08 ± 1.91%.

VALUES IN PARENTHESES ARE REAL DIFFERENCES IN LINE INTENSITY VALUES RELATIVE TO THE LINE INTENSITY FOR THE AS–RECEIVED SAMPLE.

As indicated earlier, the in situ vacuum heating of the boiling water soaked sample produced data identical to that measured for the "control" surface that did show imide bond formation. This result suggests to us that the surface of a cured film is a degraded polyimide that likely has an average molecular weight significantly smaller than that expected for the bulk material. Our viewpoint centers around the following concepts. Presumably a long-chain linear polymer, e.g., a polyimide, has an average molecular weight in the tens of thousands range. Though imide bonds are formed in the surface region, the likelihood of the surface duplicating the molecular weight of the bulk seems remote. The molecular weight range should be lower in the surface region for any one of the reasons cited in an earlier paragraph.

Recuring of a water-soaked polyimide films is thought to produce a bulk polymer of lower molecular weight (17). The fact that recuring of the water-soaked film produced a surface nearly identical to that of the "control," and since boiling water treatment showed no dramatic change in the imide chemical (C 1s) shift (imides are known to resist hydrolysis), we feel confident in believing that the "control" surface is a low molecular weight polyimide of degraded character. The indication is that water and T&H is somehow attacking non-polyimide material incorporated in the polymer matrix.

<u>Modifications of Polyimides by Plasma Exposures</u>.   Table V contains a summary of binding energy results obtained on cured PI5878 surfaces after 2 minute exposures to three different plasma gases. C 1s spectra for the polyimide surface after $O_2$-plasma ($O_2$ ashing treatment), and after $O_2$ (8%)/$CF_4$ (92%)-plasma (discharge) exposures are shown in Figures 5 and 6, respectively. As the results show, extensive modifications occurred in the chemical make-up of the carbon matrix. Acid or ester (BE $\sim$ 289.1) groups and two additional carbon-oxygen bond types were formed as a result of the $O_2$-plasma treatment. Changes in the O 1s and N 1s binding energies also occurred.

$CF_3$ (293.5 eV), $CF_2$ (291.6 eV), and CF (289.8 eV) bonds in large quantities existed after exposures of the cured PI5878 film to $O_2/CF_4$-plasma environments. Lesser amounts of $-CF-CF$ (288.0 eV), $-C-CF_2$ (286.4 eV) and virtually no $\underline{C}H$ bonding is also illustrated. Note that $O_2/CF_4$-plasma treated surfaces were identical even though the oxygen content of the gas increased from 8% to 16% (Table V). For both the $O_2/CF_4$- and the $O_2$-plasma treated surfaces, the resulting N 1s signal intensities were quite intense.

The data indicate that extensive oxidation occurred due to the $O_2$-plasma treatment and oxidation/fluorination occurs for the $O_2/CF_4$-plasma treatment of cured PI5878 films. No overlayer formation is indicated as evidenced by the strong N 1s signals that resulted after plasma treatments. Thus these treatments seem to involve double bond additions of oxygen and fluorine to the polymeric benzene rings.

## Table V.  Binding Energy Results: Polyimide (PI5878)/Polyimide —Plasma-Exposed Surfaces

| TREATMENT | C 1s | | | O 1s | | N 1s | |
|---|---|---|---|---|---|---|---|
| | B.E. (eV) | $\Delta$C 1s | % | B.E. (eV) | % | B.E. (eV) | % |
| 325°C | 285.0 | | 43.66 | 532.3 | 74.58 | 399.3 | 3.82 |
| | 286.1 | 1.1 | 44.18 | 533.5 | 25.42 | 400.8 | 96.18 |
| | 288.9 | 3.9 | 12.16 | | | | |
| 325°C + $O_2$ PLASMA | 285.0 | | 24.39 | 532.6 | 61.82 | 402.4 | 100.00 |
| | 286.1 | 1.1 | 29.57 | 534.0 | 38.18 | | |
| | 287.5 | 2.5 | 17.32 | | | | |
| | 289.1 | 4.1 | 28.71 | | | | |
| 325°C + $O_2/CF_4$ PLASMA (8%/92%)[+] | 285.1 | | 1.37 | 533.7 | 16.54 | 402.0 | 100.00 |
| | 286.5 | 1.5 | 2.36 | 535.4 | 83.46 | | |
| | 288.0 | 2.9 | 3.85 | | | | |
| | 289.8 | 4.7 | 23.81 | | | | |
| | 291.7 | 6.6 | 55.01 | | | | |
| | 293.5 | 8.4 | 13.59 | | | | |
| 325°C + $O_2/CF_4$ PLASMA (16%/84%)[‡] | 285.0 | | 1.74 | 533.6 | 17.48 | 402.0 | 100.00 |
| | 286.4 | 1.4 | 4.66 | 535.4 | 82.52 | | |
| | 288.0 | 3.0 | 4.72 | | | | |
| | 289.7 | 4.8 | 22.96 | | | | |
| | 291.7 | 6.7 | 51.47 | | | | |
| | 293.6 | 8.5 | 14.46 | | | | |

[+] SINGLE F 1s PEAK AT B.E. = 688.36 eV
[‡] SINGLE F 1s PEAK AT B.E. = 688.40 eV

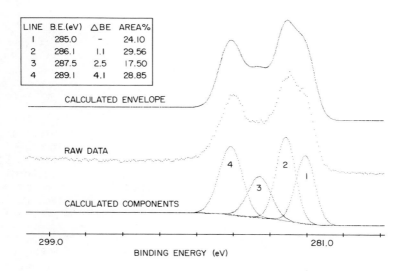

| LINE | B.E.(eV) | △BE | AREA% |
|------|----------|-----|-------|
| 1 | 285.0 | – | 24.10 |
| 2 | 286.1 | 1.1 | 29.56 |
| 3 | 287.5 | 2.5 | 17.50 |
| 4 | 289.1 | 4.1 | 28.85 |

CALCULATED ENVELOPE

RAW DATA

CALCULATED COMPONENTS

299.0

281.0

BINDING ENERGY (eV)

*Figure 5. The C-1s spectrum of PI5878 after O₂-plasma exposure*

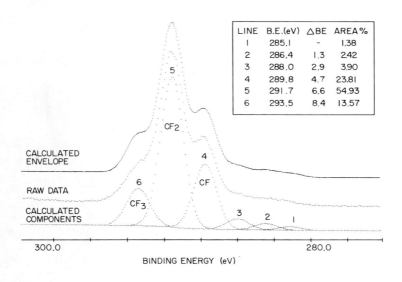

| LINE | B.E.(eV) | △BE | AREA% |
|------|----------|-----|-------|
| 1 | 285.1 | – | 1.38 |
| 2 | 286.4 | 1.3 | 2.42 |
| 3 | 288.0 | 2.9 | 3.90 |
| 4 | 289.8 | 4.7 | 23.81 |
| 5 | 291.7 | 6.6 | 54.93 |
| 6 | 293.5 | 8.4 | 13.57 |

CALCULATED ENVELOPE

RAW DATA

CALCULATED COMPONENTS

300.0

280.0

BINDING ENERGY (eV)

*Figure 6. The C-1s spectrum of PI5878 after O₂/CF₄-plasma exposure*

## Acknowledgements

Thanks are due to F. Soychak, W. Motsiff and J. Schiller for providing the samples. We deeply appreciated the support given to our work by B. Bertelsen and C. Wyand.

## Abstract

The surface chemical structure of several thin polyimide films formed by curing of polyamic acid resins was studied using X-ray photoelectron spectroscopy (ESCA or XPS). The surface modifications of one of the polymer systems after exposure to KOH, after exposure to temperature and humidity, after exposure to boiling water, and after exposure to $O_2$ and $O_2/CF_4$ plasmas were also evaluated. The results showed imide bond formation for all cured polyimide systems. It was found that: (a) K on the surface of the polyamic acid alters the "normal" imidization process, (b) cured polyimide surfaces are not invarient after T&H and boiling water exposures, and (c) extensive modifications of cured polyimide surfaces occur after exposures to plasma environments. Very complex surfaces for these polymer films were illustrated by the C 1s, O 1s, N 1s and F 1s line characteristics.

## Literature Cited

1.    Clark, D.T. and Thomas, H.R., J. Polym. Sci. Polym. Chem., 1978, 16, 791.

2.    Kardos, J.L. and Fountain, R., J. Polym. Sci. Polym. Lett., 1974, 12, 161.

3.    Millard, M.M., Windle, J.J. and Pavlath, A.F., J. Appl. Polym. Sci., 1973, 17, 2501.

4.    Leary, H.J., Jr. and Campbell, D.S., Surface and Interface Analysis, 1979, 1, 75.

5.    O'Kane, D.F. and Rice, D.W., J. Macromol. Sci. Chem., 1976, A10, 567.

6.    Clark, D.T., "Advances in Polymer Friction and Wear," Ed. L.H. Lee, Plenum Press, New York, Vol. 5A, 1975.

7.    Clark, D.T. and Shuttleworth, D., J. Polym. Sci., Polym. Chem. Ed., 1978, 16, 1093.

8.    Clark, D.T., "Electron Emission Spectroscopy,", Ed. W. Dekeyser and D. Reidel, D. Reidel Publ. Co., Dordrecht, 1973.

9.  Koopmans, T.A., <u>Physika</u>, 1934, 1, 104.

10. Clark, D.T., "Handbook of Electron Spectroscopy," Ed.  D. Briggs, Heydess and Sons, London, 1977.

11. Clark, D.T., Kilcast, D., Feast, W.J., and Musgrave, W.K.R., <u>J. Polym. Sci. Polym. Chem. Ed.</u>, 1972, 10, 1637.

12. Cromarty, B.J., Ph.D. Thesis, University of Durham, Durham, England, 1978.

13. Gay, F.P. and Berr, C.E., <u>J. Polym. Sci. Part A-1</u>, 1968, 6, 1935.

14. Leary, H.J., Jr. and Campbell, D.S., <u>J. Electron. Spectrosc. Relat. Phenom.</u>, submitted, 1980.

15. Gregoritsch, A.J., unpublished results.

16. Rothman, L., unpublished results.

17. Dine-Hart, R.A., Parker, D.B.V. and Wright, W.W., <u>Br. Polym. J.</u>, 1971, 3, 222-236.

RECEIVED February 18, 1981.

# INDEX

# INDEX

*Jacket design by Carol Conway.*
*Production by Susan Moses and Cynthia E. Hale*

*Elements typeset by Service Composition Co., Baltimore, MD.*
*The book was printed and bound by Maple Press Co., York, PA.*